DEVELOPMENT OF NOVEL ANTIMICROBIAL AGENTS:
EMERGING STRATEGIES

Edited by

Karl Lohner
Graz, Austria

Copyright © 2001
Horizon Scientific Press
P.O. Box 1
Wymondham
Norfolk NR18 0EH
England

www.horizonpress.com

Distributed exclusively in the United States, its dependent territories, Canada, Mexico, Central and South America, and the Caribbean by Springer-Verlag New York Inc, 175 Fifth Avenue, New York, USA, by arrangement with BIOS Scientific Publishers Ltd, 9 Newtec Place, Magdalen Road, Oxford OX4 1RE, UK.

Distributed exclusively in the rest of the world by BIOS Scientific Publishers Ltd, 9 Newtec Place, Magdalen Road, Oxford OX4 1RE, UK.

British Library Cataloguing-in-Publication Data

A catalogue record for this book is available from the British Library

ISBN: 1-898486-23-9

Printed and bound in Great Britain

Dedicated to my family for all their encouragement and support:

To Uschi, my wife,
to Thomas and Andreas, my sons,
to my parents and brothers,

and to the late Christopher J.Pratt.

The Front Cover

This pen-and-ink illustration is an artist's representation (Mr. Thomas Lohner) of a bacterial colony in cells. The scheme of a DNA (at the bottom) and of an antimicrobial peptide (at the right hand corner) represent major novel strategies to fight antibiotic resistant bacteria, microbial genomics and peptide antibiotics based on host defense peptides.

Contents

Books of Related Interest

For further information on these books contact:

Horizon Scientific Press   Tel: +44(0)1953-601106
P.O. Box 1, Wymondham   Fax: +44(0)1953-603068
Norfolk   Email: mail@horizonpress.com
NR18 0EH England   Internet: www.horizonpress.com

Our Web site has details of all our books including full chapter abstracts, book reviews, and ordering information:

www.horizonpress.com

Contributors

Arden Aspedon,
Division of Applied Health,
Southwestern Oklahoma State Univ.,
100 Campus Drive,
Weatherford,
OK 73096, USA

Thomas G. Blanchard,
Department of Pediatrics,
Case Western Reserve University,
School of Medicine,
11100 Euclid Ave.,
Cleveland,
OH 44106, USA

Sylvie E. Blondelle,
Torrey Pines Institute Molecular Studies,
3550 General Atomics Ct,
San Diego,
CA92121, USA

Trinad Chakraborty,
Institute of Medical Microbiology,
University Giessen,
D-35392 Giessen,
Germany

Christina M. Collis,
CSIRO Molecular Science,
Sydney Laboratory,
North Ryde,
NSW 1670,
Australia

Steven J. Czinn,
Department of Pediatrics,
Case Western Reserve University,
School Medicine,
11100 Euclid Ave.,
Cleveland,
OH 44106, USA

Tomas Ganz,
Department of Medicine,
10833 Le Conte Avenue,
UCLA School of Medicine,
Los Angeles,
CA 90095, USA

Eduardo A. Groisman,
Howard Hughes Medical Institute,
Washington Univ. School Medicine,
660 S. Euclid Ave.,
St. Louis,
MO 63110, USA

Ruth M. Hall,
CSIRO Molecular Science,
Sydney Laboratory,
North Ryde,
NSW 1670,
Australia

Scott J. Hultgren,
Washington Univ. School Medicine,
660 S. Euclid Ave.,
Dept. Molecular Microbiology,
St. Louis,
MO 63110-1093, USA

Raymond J. Jackson,
The Immunobiology Vaccine Center,
Dept. Microbiology,
University Alabama at Birmingham,
845 19th Street South,
Birmingham,
AL 35294-2170, USA

Leonard Katz,
Kosan Biosciences Inc.,
3832 Bay Center Place,
Hayward,
CA 94545, USA

Alan M. Krensky,
Div. Immunol. Transplantation Biol.,
Dept. of Pediatrics,
Stanford University,
CA 94305-5208, USA

Solomon Langermann,
MedImmune Inc.,
Dept. Immunol. Mol. Genetics,
35 West Watkins Mill Road, Gaithersburg,
MD 20878, USA

Robert I. Lehrer,
Department of Medicine,
10833 Le Conte Avenue,
UCLA School of Medicine,
Los Angeles,
CA 90095, USA

Karl Lohner,
Inst. Biophy. Röntgenstrukturforschung,
Österreichische Akademie de Wissenschaften,
Schmiedlstrasse 6,
A-8042 Graz,
Austria

Yi-An Lu,
Vanderbilt University,
Dept. Microbiol. Immunology,
1161 21st Avenue South,
Nashville,
TN 37232-2363, USA

Katsumi Matsuzaki,
Graduate School of Biostudies,
Sakyo-ku,
Kyoto 606-8501,
Japan

Robert McDaniel,
Kosan Biosciences Inc.,
3832 Bay Center Place,
Hayward,
CA 94545, USA

Jerry R. McGhee,
The Immunobiology Vaccine Center,
Dept. Microbiol.,
University Alabama at Birmingham,
845 19th Street South,
Birmingham,
AL 35294-2170, USA

Robert L. Modlin,
Div. of Dermatology,
Dept. Microbiol. Immun. Mol. Biol. Inst.,
UCLA School Medicine,
10833 Le Conte Ave.,
Los Angeles,
CA 90095, USA

Matthew A. Mulvey,
Washington Univ. School Medicine,
Dept. Molecular Microbiology
660 S. Euclid Ave.,
St. Louis,
MO 63110-1093, USA

Mari Ohmura,
The Immunobiology Vaccine Center,
Dept. Microbiol.,
University Alabama at Birmingham,
845 19th Street South,
Birmingham,
AL 35294-2170, USA

Ziv Oren,
Dept. Biological Chemistry,
Weizmann Inst. Science,
Rehovot,
76100 Israel

Ulrike Pag,
Inst. Med. Mikrobiol. Immunol.,
University of Bonn,
Sigmund-Freud-Str. 25,
D-53105 Bonn,
Germany

David S. Pisetsky,
Durham VA Medical Center,
508 Fulton St.,
Durham,
NC 27705, USA

Hans-Georg Sahl,
Inst. Med. Mikrobiol. Immunol.,
Univ. Bonn,
Sigmund-Freud-Str. 25,
D-53105 Bonn,
Germany

Yechiel Shai,
Dept. Biological Chemistry,
Weizmann Inst. Science,
Rehovot,
76100 Israel

Erich Staudegger,
Inst. Biophy. Röntgenstrukturforschung,
Österreichische Akademie Wissenschaften,
Schmiedlstrasse 6,
A-8042 Graz,
Austria

Steffen Stenger,
Inst. Klin. Mikrobiol., Immunol. and Hygiene,
Universitat Erlangen-Nürnberg,
D-91504 Erlangen,
Germany

Yoshifumi Takeda,
National Institute of Infectious Disease,
1-23-1 Toyama,
Shinjuhu-ku,
Tokyo 162-8640,
Japan

James P. Tam,
Vanderbilt University,
Dept. Microbiol. Immunol.,
1161 21st Avenue South,
Nashville,
TN 37232-2363, USA

Qitao Yu,
Vanderbilt University,
Dept. Microbiol. Immunol.,
1161 21st Avenue South,
Nashville,
TN 37232-2363, USA
Siegfried Weiss,
Molecular Immunology,
GBF-National Research Center Biotechnol.,
Mascheroder Weg 1,
D-38124 Braunschweig,
Germany

Jin-Long Yang,
Vanderbilt University,
Dept. Microbiol. Immunol.,
1161 21st Avenue South,
Nashville,
TN 37232-2363, USA

Michael Zasloff,
Magainin Pharmaceuticals Inc.,
5110 Plymouth Meeting,
PA 19462, USA

Preface

After their discovery, antibiotics rapidly spread throughout the whole world and it was thought that the problem of treating infectious diseases had been overcome. However, the fight against infectious bacteria still continues today, because we are facing a world-wide rapid increase in pathogenic bacteria which show multi-resistance to currently available antibiotics. Therefore, the development of novel antimicrobial agents has become an urgent matter so as to combat the spread of such pathogens. The relevance and importance of this serious global health problem, partly caused by our current lifestyle, is best reflected by the numerous warnings of the World Health Organization and, recently, by the more frequently occurring reports in the daily press. This is further emphasized by the strong current interest in this field by scientists of quite diverse disciplines, covering a wide range from biophysics and biochemistry, molecular biology and genetic engineering, microbiology, medicine to pharmacology.

It was a great challenge to edit a book that will be useful and of interest for such a wide readership, ranging from the scientists performing the basic research to the physicians being confronted with antibiotic resistance in their daily work and awaiting new powerful drugs in order to save thousands of lives. Moreover, both the rapid progress and expansion of this research area made it a difficult task to cover the whole field completely. Therefore, this book attempts to provide the reader with an overview on emerging strategies for the development of novel antimicrobial agents, with contributions from scientists from academia and industry. However, it was not always that easy to keep the balance between all these different needs and indeed, some readers will find more detail given in certain chapters. A consequence of this wide spectrum of approaches is the fact that some scientists may not be aware of other approaches and hopefully this book will also help to fill this gap.

The book is organized into three sections, subdivided into various chapters. It was my intention to interrelate the chapters and, therefore, there is some overlap, although each chapter can be read independently. Section 1 is devoted to the problem of antibiotic resistance, Section 2 to the field of microbial genomics and finding new targets and novel vaccines and Section 3 focuses on the development of novel peptide antibiotics based on host defense peptides which evolved in nature as part of the innate immune system. I hope that the book will provide a coherent picture of emerging strategies for novel antimicrobial compounds.

Of course, many people are involved before such a project is complete. Some are directly related to the work, others create the right environment to do it in. All these people I would like to thank by naming a few. First of all, I want to thank my family, who have been very encouraging during the long and sometimes seemingly endless process of putting this book together. They have always been a great support to me and showed a lot of patience and understanding, when I spent my free time working on the book. Special gratitude goes to Christopher J. Pratt, who was not only a dear friend, but carefully read the manuscripts with a lot of enthusiasm, checking the language and giving me his opinion on the texts from the point of view of an outsider to the field. Unfortunately, he deceased far too prematurely and was no longer able to see the final product, which he was so much looking forward to. I also wish to thank my collaborators, who contributed in one form or another to this book, in particular Erich Staudegger, who assisted in the final proofreading and in making the index, and Petra Still for the invaluable secretarial assistance provided throughout the whole project. Moreover, I would like to thank all authors for their excellent contributions to this book and for their patience in finalizing this project, which has taken longer than initially anticipated.

Finally, I want to express my sincere thanks to Annette Griffin from Horizon Scientific Press for her confidence in my work, as well as for her continuous support, and to the publishers, who gave me the opportunity to edit a book on such an important global issue. Not only will the efforts of researchers and pharmaceutical companies be required to ensure that we are not on the threshold of the post-antibiotic era as we enter the new millennium, but rather we should look to the combined efforts of us all in avoiding an excessive and inappropriate use of antibiotics in human and animal health care, in practicing good hospital as well as common hygiene, in distributing educational material and in improving the public health infrastructure etc. If this book can contribute towards improving the awareness of the global problem of antibiotic resistance and stimulate the development of novel antimicrobial agents, it will have achieved its goal.

Karl Lohner
Graz, Austria

From: *Development of Novel Antimicrobial Agents: Emerging Strategies*
ISBN 1-898486-23-9 © 2001 Horizon Scientific Press, Wymondham, UK.

1

Are We on the Threshold of the Post-Antibiotic Era?

Karl Lohner and Erich Staudegger

Abstract

At the end of the 20th century we face a world-wide rapid increase in pathogenic bacteria which are multi-resistant to antibiotics. This alarming situation has its origin in the excessive and often inappropriate use of antibiotics in human and animal health care for the treatment and prevention of bacterial infections. Since the development of the first commercially available antibiotic penicillin in the 1940s, the high expectations by man in the healing power of these "wonder drugs" has not been fulfilled, as resistance – which is not a new problem – is a vital part of the survival strategy of bacteria. Some stategies for the containment of antimicrobial resistance are discussed, of which the continuous research and development of new classes of antibiotics with novel mechanisms of action, the topic of this book, and the awareness of a more sophisticated and prudent use of antibiotics are the most important ones.

Status Quo of Antibiotic Resistance at the End of the 20th Century

"Infectious diseases are the leading cause of death world-wide and the third leading cause of death in the United States." This alarming statement by James M. Hughes, M.D., Director of the National Center for Infectious Diseases & Centers for Disease Control and Prevention, Department of Health and Human Services, in front of the Committee on Health, Education, Labor, and Pension, of the U.S. Senate, in February 25, 1999, is in sharp contradiction of the optimistic statement of the Surgeon General of the U.S.A. made in 1979: "It´s time to close the book on infectious diseases." The latter statement reflects the general misjudgment also regarding the control of bacterial infections by the wonder drug antibiotics in the late seventies. Nowadays, the fight against infectious bacteria continues and antibiotic resistance will be one of the major public health threats in the 21st century.

At the end of the 19th century in the year 1897 Ernest Duchesne (Figure 1), a French military doctor, showed with his remarkable thesis "Contribution to the study of vital competition in microorganisms – Antagonism between the moulds and the microbes." the antibacterial activity of mildews (*Penicillium glaucum*), which he tested on laboratory animals (1). Guinea-pigs were intravenously infected with pathogenic strains of *Coli* or *Typhus* bacteria and died within 24 hours, whereas these microbes did not show any lethal effect in the presence of *P. glaucum*. Duchesne concluded that certain moulds are able to attenuate in very notable proportions the virulence of these bacterial cultures: "*que certaines moisissures (Peni-*

cillium glaucum), inoculées à un animal en même temps que des cultures très virulentes de quelques microbes pathogènes (B. coli et B. thyposus d'Ebert), sont capable d'atténuer dans de très notables proportions la virulence de ces cultures bactériennes". He further stated very modestly that he hoped that his contribution would be useful and applicable to prophylactic hygiene and therapeutics. Apparently, Duchesne and the scientific community were not aware of the whole importance of his discovery. Tragically, Duchesne died very early of tuberculosis and his important findings fell into oblivion and remained unknown to the public, which could have already saved thousands of lives, in particular during Word War I. It took a further thirty years until the understanding of the range of antibacterial compounds (from fungi) for medication was initiated by the discovery of penicillin by Alexander Fleming in 1928 and further developed by the work of Cecil Paine, Howard Florey and Ernst Chain, who strongly contributed to the commercial availability of antibiotics in the 1940s. Since then, 9 major chemical classes of antibiotics have been developed (2) and approximately 160 antibiotics of natural, semi-synthetic or synthetic origin have been used for medical treatment (3). These compounds have been the cornerstone of the therapy of bacterial infections. Most interestingly, several pioneers, who – among many others – contributed essentially to antibiotic research, were awarded the Nobel Prize: Robert Koch, Alexander Fleming, Gerhard J.P. Domagk, Howard W. Florey, Ernst B. Chain and Selman A. Waksman.

The remarkable healing power of these "wonder drugs" lead to widespread and often inappropriate use in human and animal health care. Thus, the "Pax Antibiotica" has not been long lived (4) and we are facing a rapid increase in bacteria which are multi-resistant to conventional antibiotics. Lately, reports from the World Health Organization, statements of national governments and health organizations as well as reports in the Daily Press (e.g. 4,5) have contributed to an increased awareness of the public to this serious global health problem. This affects virtually all the pathogens, which were previously considered to be readily treatable. In fact, the threat of the growing resistance of bacteria causing common infectious diseases to frequently used antibiotics pressed WHO forward with its program for the global monitoring of antibiotic resistance. Thus the increasing number of antibiotic-resistant bacterial strains has been one of the dominant matters for the WHO in the last few years, which had already published a warning to this respect on the occasion of the Word Health Day, April 7, 1997. It should be mentioned that all WHO documents, press releases, fact sheets and features can be easily obtained from the WHO home page (see also Appendix).

Drug resistant bacterial strains have a deadly impact on the fight against infectious diseases, which are the world's biggest killer of children and young adults (63% deaths of young

Figure 1. In 1974 Monaco published a stamp commemorating the 100th birthday of Ernest Duchesne (1874-1912), who first discovered antibacterial activity of mildews (*Penicillium glaucum*).

children up to an age of 4 years and 48% main cause of premature deaths up to an age of 44 years (6)). Despite the enormous advances in health care in the last half-century, estimations for 1998 showed that infectious diseases still account for about 25% of deaths world-wide. In low-income countries this percentage can range up to 45%. More than 13 million people died from these conditions in 1998. Of about 52 million causes of death in 1995 more than 17 million were due to infectious and parasitic diseases (7). The real hot zone encompasses a handful of familiar infectious diseases such as pneumonia, cholera, diarrhea, tuberculosis, and malaria. The biggest killers in 1995 (7) and 1996 (8), respectively, were pneumonia (4.4 and 3.9 million people), diarrhoeal diseases including cholera (3.1 and 2.5 million) and tuberculosis (3.1 and 3 million). While the formers mostly affect children, latter affects mainly adults. Diarrhoeal diseases and tuberculosis are also leading the list of the most common infections. The World Health Report 1996 (7) concludes:

"Disastrously, this is happening at a time when too few drugs are being developed to replace those that have lost their effectiveness. In the race of supremacy, microbes are sprinting ahead."

The dramatic situation is further emphasized by reports of the kind that e.g. in the United States antibiotic resistant bacteria are responsible for up to 70% of hospital-acquired infections (9). By 1995 *Enterococcus* strains being resistant to many antibiotics have evaded every hospital in New York (10). The real chill arises from the fact that the bacteria are resistant to vancomycin, a drug of last resort, and that it will pass its gene to even more dangerous pathogens like *Pneumococcus*, the leading bacterial cause of death in the U.S., and *Staphylococcus aureus*, the most common cause for nosocomial wound infections world-wide. In fact, in 1997 infections caused by methicillin-resistant *S. aureus* clinical strains with reduced susceptibility to vancomycin were reported for the U.S.A. (11) and for Japan (12), although resistance in the latter strain does not seem to be a result of the transfer of enterococcal vancomycin resistance genes (13). Nevertheless, these strains represent a potentially serious threat to public health. Meanwhile they have been reported from Japan (multiple strains), the U.S.A. (four strains), Western Europe as well as from Hong Kong and Korea suggesting that this phenomenon will continue to occur world-wide (14). Nosocomial spread of *S. aureus* with reduced vancomycin susceptibility seems to have occurred so far only in Japan (14). There, heterogeneously vancomycin-resistant *S. aureus* was found throughout the hospitals. It has been proposed that the frequent therapeutic failure of methicillin-resistant *S. aureus* infection with vancomycin is at least partially due to the presence of this strain, which is a preliminary stage that allows development into vancomycin-resistant *S. aureus* upon exposure to this glycopeptide antibiotic (15). In this respect it is also worthwhile to note that in Japan about 60% of *Staphylococcus* strains are multidrug resistant (6).

Furthermore, the following trends in the antimicrobial susceptibility of bacterial respiratory tract pathogens were derived by the Alexander Project during the period 1992-1996 (16,17). In this multicenter surveillance study isolates of *Streptococcus pneumoniae* were collected from geographically separate centers in various states of the U.S.A. and countries of the European Union. In 1996, the project was extended to centers in Mexico, Brazil, Saudi Arabia, South Africa, Hong Kong and other European countries not included previously. Centers with a high prevalence of penicillin resistance were found in Hong Kong (resistant 50%, intermediate 9.1%), Mexico (resistant 15.7%, intermediate 31.4%) and combined resistance rates in excess of 40% were reported for Toulouse (France) and Barcelona (Spain). Furthermore, it was shown that penicillin resistance has evolved in the U.S.A. increasing from 5.6% in 1992 to 16.4% in 1996. The SENTRY antimicrobial surveillance program for

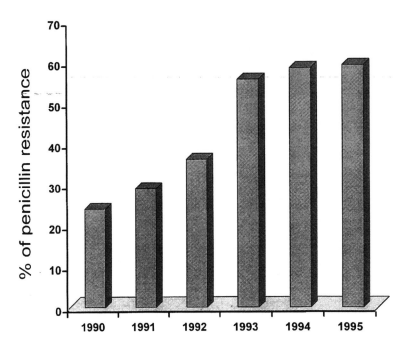

Figure 2. Susceptibility study to penicillin on 1176 isolates from the Chiba Children's Hospital, Japan (data taken from 19).

pneumococcal isolates, performed in North-America in 1997, reported similar resistance data (27.8% of strains with intermediate and 16.0 % with high-level resistance to penicillin) for the United States (18). However, it should be noted that among the centers in the U.S.A. and Canada which contributed at least 19 isolates, the combined rate varied between 24% and 67.8%! During the early 90s a strong increase in penicillin-resistant strains of *S. pneumoniae* was also found in Japan, e.g. as demonstrated by a representative study on 1176 clinical isolates from the pediatric clinics of Chiba Children's Hospital between 1990 and 1995 (Figure 2, (19)). Moreover, all these studies indicated that resistance to non-beta lactam antibiotics and cephalosporins was considerably higher in penicillin-resistant strains than with penicillin-susceptible isolates. Similar findings were reported from surveillance studies in Spain (20) and Singapore (21) as summarized in Table 1. A selection of the most serious antibiotic-resistant bacteria, given in Table 2, further stresses that urgent action is needed to combat the spread of antibiotic resistant bacterial strains. The question remains: How can this be achieved? Some steps regarding the containment of antibiotic resistance will be discussed below.

Socio-Economic Impact of Antimicrobial Resistance on Society

Antimicrobial resistance has a strong socio-economic impact. Resistance costs money but most importantly human lives (22). Despite the extraordinary advances in the 20th century, a significant component of the burden of illness globally still remains attributable to infectious diseases. Hundreds of millions of people are disabled by infectious diseases. These conditions are primarily concentrated in the poorest countries afflicting populations that are trapped in a vicious circle of poverty and ill-health (many people are sick because they are poor, and poor because they are sick) often caused by bad housing and environmental conditions in poor rural areas (6). In many developing countries, the availability and use of antibiotics are hardly controlled, which results in a high rate of resistance, particularly to the older and relatively inexpensive (generic) antibiotics, or the distribution is completely impaired by conflicts. The high cost of the few remaining second-line antibiotics makes them an unrealistic choice where they are needed most.

However, it has to be emphasized that antimicrobial resistance is a global problem, affecting also developed countries. Not only an increase in morbidity and mortality is observed but also considerable economic effects. In the U.S.A., antibiotic-resistant bacteria generate costs of a minimum of $4 billion to $5 billion yearly and these costs are likely to be much higher in developing countries (23). Prolonged hospital stays, greater direct and indirect costs, longer periods during which individuals are infectious, and greater opportunities for the spread of pathogens to other individuals. When infections become resistant to antibiotics administered, treatment has to be switched to intravenous or intramuscular administration of second-line antibiotics (e.g. glycopeptide vancomycin) which are more costly, have added indirect costs (needles, syringes, etc.) and are often associated with a higher risk of superinfection (e.g. catheter-associated sepsis) and toxic side-effects.

Emergence and Origins of Antibiotic Resistance

Antimicrobial resistance is neither a new, nor a surprising problem. It occurred contemporary with the first antibiotics on the market, but has become more serious in the last decades. Resistance has been detected among all types of microorganisms – bacteria, fungi, parasites – and is a vital part of their self-defense and survival strategy. Already in 1940, before the introduction of penicillin into clinical practice, Abraham and Chain (24) identified a bacterial enzyme, that hydrolizes the β-lactam ring of the antibiotic, and they noted that this enzyme might interfere with penicillin therapy. All bacteria possess an inherent flexibility that enables them, sooner or later, to evolve genes that render them resistant to any antimicrobial agent (see also Hall and Collis this monography). The alarming number of antibiotic resistant bacterial strains has several origins. Many of them are related to our current lifestyle (Figure 3) and it is likely that these factors will still contribute to further emergence of infectious diseases in the 21st century (25).

Table 1. Overall combined resistance rates (%) of *Streptococcus pneumoniae* to selected antibiotics.

Country	South-West Germany	Singapore	Spain	USA	Canada
period	1992-1994	1995	1996-97	02-06/1997	02-06/1997
numbers of isolates	174	144	1113	845	202
reference	52	21	20	18	18
penicillin	6.3[1]	25	37	43.8	30.2
macrolides	3.5	26.4[2]	33	13.0	6.2
tetracycline	-	52.1	-	10.2	10.9
chloramphenicol	-	20.1	-	3.9	4.0
trimethoprim-sulfamethoxazole	-	33.3	-	19.8	15.8
cefuroxime	-	-	46	19.5	12.9
ceftriaxone	2.3	14.0	8	-	-
cefotaxime	-	-	13	4.0	3.0

[1] a similar low value (4.9 %) was reported from an Austrian surveillance study on 1385 streptococcal isolates during 1995 (53)
[2] data for erythromycin

Table 2. Overview of most serious antibiotic-resistant bacteria (data taken from Davies (45) and studies quoted in this chapter).

Organism(s)	Resistance to	Diseases
Bacteroides spp.	Clindamycin, penicillin	Anaerobic infections, septicaemia
Enterobacteriaceae	Aminoglycosides, β-lactams, chloramphenicol, trimethoprim, vancomycin	Bacteriaemia, pneumonia, urinary tract and surgical wound infections
Enterococcus spp.	Aminoglycosides, erythromycin, penicillin, tetracycline, vancomycin	Catheter infections, blood poisoning
Haemophilus influenza	Ampicillin, β-lactams, chloramphenicol, tetracycline, trimethoprim, sulfonamide	Pneumonia, sinusitis, epiglottitis, meningitis, ear infections
Mycobacterium spp.	Aminoglycosides, ethambutol, isoniazid, pyrazinamide, rifampin	Tuberculosis
Neisseria gonorrhoeae	β-Lactams, penicillin, tetracycline, spectinomycin	Gonorrhoea
Shigella dysenteriae	Ampicillin, chloramphenicol, tetracycline, trimethoprim, sulfonamide	Severe diarrhoea
Pseudomonas aeruginosa	Aminoglycosides, β-lactams, chloramphenicol, ciprofloxacin, tetracycline, sulfonamides	Bacteriaemia, pneumonia, urinary tract infections
Salmonella enterica serotype typhimurium	Ampicillin, chloramphenicol, streptomycin, sulfonamides, tetracycline	Salmonellosis
Staphylococcus aureus	β-Lactams, chloramphenicol, ciprofloxacin, clindamycin, erythromycin, rifampin, tetracycline, trimethoprim, vancomycin	Bacteriaemia, pneumonia, surgical wound infections
Streptococcus pneumoniae	Chloramphenicol, cefuroxime, erythromycin, penicillin, trimethoprim, tetracycline	Meningitis, pneumonia

A predominant factor is overmedication and inappropriate use of antibiotics in humans, animals and agriculture. It is estimated by the Centers for Disease Control (CDC) and Prevention in Atlanta, U.S.A., that about 50% of all prescribed antibiotics are unnecessary. Antibiotic prescribing is usually empiric, without time-consuming laboratory confirmation of infection. In fact, the contemporary medical practice of grossly overprescribing antibiotics against diseases e.g. upper respiratory viral infections, where they are useless, is an important factor in the emergence of multidrug resistant strains. Easy availability of antibiotics in some countries, erratic antibiotic supply, counterfeit or low-quality antibiotics combined with the poor patient compliance, as they often fail to complete antibiotic therapy, also contribute to the recent situation. Furthermore, a large amount of the total production of antimicrobial agents is used in farm animals, in fish farming and in agriculture, mainly for growth promotion. This has been paralleled by an increase in resistance in zoonotic bacteria, i.e. those that can spread from animals to cause infections in humans. In addition, the increase in mass population movements, the massive growth of international travel, and transportation of animals and animal products may carry diseases into areas where they have never been seen

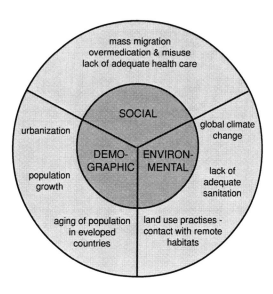

Figure 3. Major factors that contribute to the emergence of infectious diseases (25).

before and concomitantly further enhance the world-wild spread of antibiotic resistant bacteria (6). Also environmental changes, e.g. deforestation or climate changes, has brought man into closer contact with animals and insects that harbour (unknown) diseases. It is interesting to note that surveys of many ecosystems indicate that so far, only 1% of microorganisms have been cultivated (26, 27)!

Until recently, physicians and pharmaceutical companies were confident that another antibiotic in stock would work, if bacteria developed resistance to a particular antibiotic. Unfortunately, the antibiotic arsenal is now close to empty, partly due to the fact that in the mid 80s, many drug companies decided to shift resources from antibiotic research to the development of drugs for chronic conditions such as neurological and immunological diseases (28). Other companies hesitated designing completely new chemical classes of antibacterial drugs and rather preferred to make chemical modifications on existing antibiotic classes which had been already approved by the Food and Drug Administration. Very recently, this trend has begun to reverse itself, as can also be deduced from various contributions in this book. Hence, several novel antimicrobial agents are in phase I-III clinical trials, but their usefulness is still awaited.

Preventing Emerging Infectious Diseases: Strategies for the 21st Century

Preventing infections in the first place may be the best defense against bacterial infections. An improvement of the current situation could be already achieved by minor efforts, such as the rational use of antibiotics in human and animal medicine or good hospital as well as common hygiene practices, e.g. kitchen: cooking, washing fruit and vegetables thoroughly, washing hands etc. A first step in this direction is to produce awareness of a more prudent use and proper administration of antibiotics by physicians and patients to achieve maximum therapeutic effect with minimum selective pressure for resistance. Respective activities have to come from the pharmaceutical industry, daily press, governmental and health organizations. For example, the Centers for Disease Control and Prevention in Atlants, U.S.A., has started a campaign to inform about the proper use of antibiotics by distributing videos, pamphlets and other educational material (29). Global efforts, however, have to be devoted to the improvement of public health infrastructure, especially in developing countries, in order to raise the level of immunity in populations by specific immunization programmes and to improve both their nutritional status and clean water supply. This will certainly require strong international collaboration, because this means combatting poverty, but it will be necessary to reduce the global selective pressure of antibiotics on bacteria and to prevent the spread of infectious diseases. The achievement of these goals is an educational process, that will, of course, take time. Finally, we have to keep in mind that bacteria are a natural part of life and that bacteria often protect us via the skin, mucous membranes or intestinal tract from diseases, because they compete with, and thus limit the proliferation of pathogenic bacteria ("antibiosis", terminus from Pasteur). Therefore, "a guiding principle of antibiotic use for any purpose must be the protection of the susceptible commensal bacteria in our environment. They are the chief allies in reversing the crisis of multidrug resistance. (30)"

Prudent use of Antibiotics in Livestock

A very important aspect in the containment of antimicrobial resistance is to change the treatment strategy for non-human medicine. As mentioned above, antimicrobial agents are also widely used in farm animals, in fish farming and in agriculture. There are no exact figures available, but CDC estimates that over 40% of the total production of antibiotics in the U.S.A. is used in livestock, whereby more than 80% of this amount is used in subtherapeutic doses as growth promoters for animals (30, 31)! Most importantly, six classes of antibiotics out of nineteen, which are approved for use as growth promoters, are important in human medicine. There is increasing evidence also in Europe, especially the United Kingdom, that most antimicrobial resistance among *Salmonella* strains, and thus most human *Salmonella* infections, is the result of excessive antimicrobial use in food-producing animals and there exists little correlation between treating *Salmonella* infections in humans and antibiotics! In the United States the prevalence of *S. typhimurium enterica serotype typhimurium* DT104, being resistant to ampicillin, chloramphenicol, streptomycin, sulfonamides and tetracycline, increased from 0.6% to 34% between 1979-1980 to 1996 (32). Already at a very early stage of this development, Levy (30) urged the reevaluation of the subtherapeutic use of valued human antibiotics for growth promotion, but little has changed so far. Furthermore, fluoroquinolone-resistant *S. typhimurium* DT104 (33) and *Campylobacter* (34) have appeared after the introduction of the use of quinolone in animals which was the topic of the 1998 special WHO meeting in Geneva on the "Use of Quinolones in Food Animals and the Potential Impact on Human Health". Moreover, it is suggested that vancomycin-resistant enterococci (VRE) exist in animal feces and human foods of animal origin. That a nosocomial pathogen like VRE may be transmitted from animal to human hosts via the food-chain illustrates the impact of using antibiotics on food-producing animals. At a WHO meeting on the "Medical Impact of the Use of Antimicrobial Drugs in Food Animals" in Berlin, 1997, health officials concluded that the practice of giving antimicrobial agents to food-producing animals for growth promotions should be terminated and that "programs of prudent antimicriobial use in food animals should be developed to reduce the risks of selection and dissemination of antimicrobial resistance" (35). In 1998 the European Union made a first step in this direction and banned among its member states the further use of avoparcin, a closely related glycopeptide to vancomycin, which was used for years in many European countries as a growth promotor in food-producing animals, although the overall resistance rate among Gram-positive bacterial pathogens against vancomycin in nine European countries in 1995 was less than 0.5%, with geographical diversity (36). While WHO, EU and CDC favor the immediate phase-out of antibiotics as growth promoters that are used in human medicine, the U.S. Department of Agriculture, as well as the Animal Health Institute, claim for more research before reducing the use of antibiotics in livestock (31).

Finally, it should be mentioned that the emergence of antibiotic resistant bacterial strains has also been found in plants. For example, *E. coli* O157:H7 infections, usually associated with animal products, were recently epidemiologically linked to unpasteurized apple juice. Standard procedures at a state-of-the-art plant were inadequate to eliminate contamination with this bacterial strain, which was transferred to the fruit by infected deers that frequented the orchards (37). This led to widespread changes in the fresh juice industry and represents only one example of the economic burden caused by resistant strains.

Antibiotic Rotation

An interesting alternative strategy for the containment of antibiotic resistance was brought up again in a recent letter to Science in response to researchers hope that nisin, an antimicrobial peptide, and related compounds might trumpet the end to the problem of bacterial resistance to antibiotics. However, Roccanova and Rappa III (38) argued that the hope of finding such compounds violates the principle of natural selection. Therefore, they suggest international cooperation to rotate antibiotic use for treatment of disease under the guidance of the World Health Organization or the United Nations. It could be expected that, when a particular antibiotic agent is removed from the microbial environment, the selective pressure for maintaining resistance to the respective antibiotic will also be removed. This would allow us, after a time without selection pressure, to bring the same compound onto the market again. These authors further state that such continual rotation of antibiotics used to combat a disease organism could guarantee that the arsenal of treatment would always be effective.

The following studies in two European countries show that bacterial resistance may indeed be reversed. In the 1980s Hungary was highly dependent on penicillin, as it was cheap and because patients demanded it (39). Hungarians used even more penicillin per capita than e.g. people from Spain, a country that is infamous for its high antibiotic consumption. This also resulted in a very high level of penicillin resistant pneumococcal infections (50%). However, Hungarian physicians became aware of this resistance problem and switched to other antibiotics. Surprisingly, in 1992 the level had fallen down to 34%, which is still a high percentage of resistance, but definitely reflects an improved situation. Another example is Finland. There, nation-wide reductions in the use of macrolide antibiotics in the treatment of respiratory and skin infections resulted in a significant decline in the frequency of erythromycin resistance among group A streptococci isolates from 16.5% in 1992 to 8.6% in 1996 (40). These observations suggest that the precarious situation of bacterial resistance can improve again after reducing the antibiotic doses or changing the medication.

However, in a recent article, evidence is presented that bacteria that have become resistant to overused antibiotics will not evolve backward once the drug is removed (41). For example, multidrug-resistant *E. coli* persisted among chickens even after the removal of the antibiotics selecting for resistance (42). Furthermore, streptomycin resistant *E. coli* evolved a second site mutation to compensate for the Darwinian fitness loss of the pathogen in the absence of the drug. This may account for the persistence of streptomycin resistance in populations maintained for more than 10,000 generations in the absence of the antibiotic and shows that these genetic adaptions to the cost of resistance can virtually preclude resistant

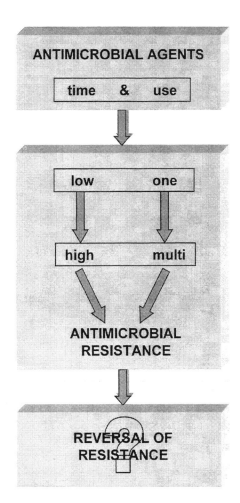

Figure 4. Scheme of underlying principles of value in understanding antimicrobial resistance deduced from Levy (30).

lineages from reverting to sensitivity (43). Therefore, these authors ask if a more prudent use of antimicrobials will lead to a decline in the incidence of antibiotic resistant microorganisms. In contrast, Levy (30) argues that resistance to antibiotics will emerge given sufficient time and drug use. Furthermore, it will evolve from low, through intermediate to high levels, frequently accompanied by occurence of multidrug resistance. Thus, increasing minimal inhibitory concentrations have to be considered as a marker for future resistance. Finally, resistance will decline only slowly, if at all. These principles may be of value in understanding the problem and approaching a solution (Figure 4).

International Surveillance

The containment of antimicrobial resistance requires world-wide surveillance. The establishment of national and global surveillance networks is a prerequisite for rapid information sharing, collecting of resistance data and linking them to treatment outcome. Additionally, monitoring of antibiotic usage must be a part of surveillance. It can be expected that modern communication technology will facilitate the achievement of this goal. Currently, WHO is working with its partners to achieve international consensus for standards for surveillance of antimicrobial resistance, to strengthen national surveillance systems and to create a repository of information about resistance in key pathogens (44). A selection of important organizations and networks, dealing with antimicrobial resistance, is listed in the Appendix.

Detection of antimicrobial resistant bacterial strains has to be one of the main focuses of an international surveillance system. In this regard Davies (27) pointed out that the application of the molecular version of Koch's rules to identify the actual pathgogens in an increasing number of situations will be a future challenge as the available technology is still a long way from providing routine analysis, except for well-characterized disease agents. Moreover, there exists considerable diversity in the laboratory methodologies used for susceptibility testing. For example, the disc-diffusion method is very popular in Europe. Due to variations in the media used, the size of the inocula, and the antibiotic content of the discs, the comparability of results from different studies is questionable. For dilution tests, testing methodologies tend to be less varied, but in some countries automation is more common than in others. But even with a convergence in methodologies, there still remains differences in interpretation, difficulty in performing accurate data analyses, insufficient government support for laboratory networks, insufficient numbers of trained professionals to perform surveillance work, and limited awareness by clinicians of the problem of antimicrobial resistance (6,16).

As there still exists no global standard method neither for performing the tests, nor for interpreting the results regarding pathogenic bacterial strains, uncertainty about the correlation and confidence between *in vitro* tests and *in vivo* results in some infections is inevitable (22). This has significant implications for the comparison of resistance data across the world. Therefore, studies have been performed towards the development of harmonized, simplified and rapid diagnostic tests which has to be a main goal for global antimicrobial resistance surveillance. So far, some promising results have been obtained. For example, comparison of resistance data of *E. coli* isolates tested at the Laboratory of Enteric Pathogens, London, and at Regional Health Authorities did show a good confidence interval, although the proportions of resistant isolates for each antibiotic tested were in poorer agreement than the trends (46). From this study it was concluded that routine susceptibility data for ampicillin, ciproflaxin, gentamicin and trimethoprim appear sound for *E. coli* and might be suitable for correlation with other data, e.g. for prescribing. Similar observations were described in a study by 18 national reference laboratories within the EU-funded Enter-net group (47). Forthy eight strains of Salmonella were tested for resistance to 11 antimicrobials using disc diffusion, agar breakpoint or full minimum inhibitory concentration. Again a high degree of concordance for the detection of resistance to most drugs was found. Substantial nonconcordance was reported for decreased sensitivity to ciprofloxacin and also in the detection of sensitivity to streptomycin. However, for some pathogens e.g. vancomycin intermediate-resistant *S. aureus*, only a small number of acceptable techniques for screening are available which influence the ability of laboratories to perform surveillance for such organisms (9). In summary, these results emphasize that caution has to be involved in the establishment and interpretation of harmonized methods used for sensitivity testing.

Research and Development of Novel Antibiotics

In addition to the actions for containment of antimicrobial resistance described above, it will be essential to encourage industry and academic groups to discover and develop novel antibiotic drugs and vaccines with novel mechanisms of action, as well as identifying the emerging molecular mechanisms of drug resistance and epidemiological risk factors, and to explore alternative therapies which boost host immunity to infection. Advances in molecular biology, biotechnology and combinatorial chemistry open up new avenues for these challenges. Many of these aspects will be addressed in the following chapters of this book, though it cannot cover the whole area of research devoted to antimicrobial resistance and agents (see also preface). Companies are also using novel approaches to search for new structures and novel antimicrobial substances in nature, e.g. in soil or plants. An interesting example for the latter is a recent report on hyperforin from *Hypericum perforatum*, which is a folk remedy commonly used for the treatment of skin injuries, burns, neuralgia, and depression (48). This compound showed not only antiviral but also antibacterial activity against multiresistant *Staphylococcus aureus* and Gram-positive bacteria. Nature has many undiscovered resources and it is up to us to find novel natural antibiotics, to elucidate their mechanism(s) of action and also to use this knowledge for the development of novel antimicrobial agents.

Concluding Remarks

"Recent increases of antimicrobial resistance are cause for serious concerns but not pessimism." This realistic statement by James M. Hughes rather characterizes the situation of bacterial resistance against antibiotics on the threshold of the 21st century. It is evident now that bacterial pathogens quickly react with their genetic machinery and adapt to environmental changes. Also it is inevitable and only a matter of time that resistance will emerge to new derivatives of existing classes of antibiotics, e.g. the glycopeptide vancomycin (49, 50) and to new antimicrobial agents. Therefore, urgent steps have to be taken for the containment of antimicrobial resistance by minimizing the environmental impact of antibiotics, which will reduce selection and transfer of resistance traits among environmental bacteria, thereby lowering the probability of the occurrence of antibiotic resistant bacteria, by reducing the inappropriate use of antibiotics in human health care as well as in livestock and by devoting global efforts to the improvement of the world-wide health infrastructure. However, in addition it will be essential to search for novel mechanisms of antimicrobial activity and hence for "antibiotics that bacteria have never seen before (51)" in order to prevent that resistant bacteria will take us back to the pre-antibiotic era.

References

1. Duchesne, E. 1897. Thesis. Contribution à l'étude de la concurrence vitale chez les microorganismes – Antagonisme entre les moisissures et les microbes. Faculté de Médecine et de Pharmacie de Lyon. France.
2. Robbins-Roth, C. 1995. Just when you thought it was safe to go back in the shower. Bioventure View. 10(5): 1-3.
3. Service, R.F. 1995. Antibiotics that resist resistance. Science 270: 724-727.
4. Wade, N. 1995. Pax antibiotica. The New York Times Magazine, Nov. 15.
5. Fisher, L.M. 1996. Biotech counterattack on resistant bacteria. The New York Times, Business Day. April 26.
6. WHO Report on Infectious Diseases. 1999. Removing obstacles to healthy development. Geneva, World Health Organization.
7. The World Health Report. 1996. Fighting diseases, fostering development. Geneva, World Health Organization.
8. The World Health Report. 1997. Conquering suffering, enriching humanity. Geneva, World Health Organization.

9. Stephenson, J. 1998. Emerging infections on center stage at First Major International Meeting. JAMA. 279: 1055-1056.
10. Saltus, R. 1995. Antibiotics: Overused & misunderstood. American Health. October: 50-55.
11. Smith, T.L., Pearson, M.L., Wilcox, K.R., Cruz, C., Lancaster, M.V., Robinson-Dunn, B., Tenover, F.C., Zervos, M.J., Band, J.D., White, E. and Jarwis, W.R. 1999. Emergence of Vancomycin Resistance in *Staphylococcus aureus*. Glycopeptide-Intermediate *Staphylococcus aureus* Working Group. N. Eng. J. Med. 340(7): 493-501.
12. Hiramatsu, K., Hanaki, H., Ino, T., Yabuta, K., Oguri, T. and Tenover, F.C. 1997. Methicillin-resistant *Staphylococcus aureus* clinical strain with reduced vancomycin susceptibility. J. Antimicrob. Chemother. 40 (1): 135-136.
13. Hiramatsu, K. 1998. The emergence of *Staphylococcus aureus* with reduced susceptibility to vancomycin in Japan. Am. J. Med. 104(5A): 7S-10S.
14. Tenover, F.C. 1999. Implications of vancomycin-resistant *Staphylococcus aureus*. J. Hosp. Infect. 43: S3-S7.
15. Hiramatsu, K., Aritaka, N., Hanaki, H., Kawasaki, S., Hosoda, Y., Hori, S. Fukuchi, Y., and Kobayashi, I. 1997. Dissemination in Japanese hospitals of strains of *Staphylococcus aureus* heterogeneously resistant to vancomycin. Lancet. 350(9092):1670-1673.
16. Felmingham, D. and Washington, J. 1999. Trends in the antimicrobial susceptibility of bacterial respiratory tract pathogens - findings of the Alexander Project 1992-1996. J. Chemother. 11:5-21.
17. Goldstein, F.W. and Acar, J.F. 1996. Antimicrobial resistance among lower respiratory tract isolates of *Streptococcus pneumoniae*: results of a 1992-93 Western Europe and USA collaborative surveillance study. The Alexander Project Collaborative Group. J. Antimicrob. Chemother. 38: 71-84.
18. Doern, G.V., Pfaller, M.A., Kugler, K., Freeman, J. and Jones, R.N. 1998. Prevalence of antimicrobial resistance among respiratory tract isolates of *Streptococcus pneumoniae* in North America: 1997 results from the SENTRY antimicrobial surveillance program. Clin. Infect. Dis. 27(4): 764-770.
19. Nakamura., A. 1997. Prevalence of antimicrobial resistance among clinical isolates of *Streptococcus pneumoniae* in a children's hospital. Kansenshogaku Zasshi. 71(5):421-429.
20. Baquero, F. Garcia-Rodriguez, J.A., Garcia de Lomas, J. and Aguilar, L. 1999. Antimicrobial resistance of 1,113 *Streptococcus pneumoniae* isolates from patients with respiratory tract infections in Spain: results of a 1-year (1996-1997) multicenter surveillance study. The Spanish Surveillance Group for Respiratory Pathogens. Antimicrob. Agents Chemother. 43(2):357-359.
21. Koh, T.H. and Rin, R.V. 1997. Increasing antimicrobial resistance in clinical isolates of *Streptococcus pneumoniae*. Ann. Acad. Med. Singapore. 26(5):604-608.
22. Williams, R.J. and Heymann, D.L. 1998. Containment of antibiotic resistance. Science. 279: 1153-1154.
23. Harrison, P.F. and Lederberg, J. 1998. Antimicrobial resistance: issues and options. Institute of Medicine. Washington D.C., National Academic Press.
24. Abraham, E.P. and Chain, E. 1940. An enzyme from bacteria able to destroy Penicillin. Nature. 146: 837.
25. Binder, S., Levitt, A.M., Sacks, J.J. and Hughes, J.M. 1999. Emerging infectious diseases: public health issues for the 21st century. Science. 284: 1311-1313.
26. Relman, D.A. 1999. The search for unrecognized pathogens. Science. 284:1308-1310.
27. Davies, J. 1999. Millennium bugs. Trends Biochem. Sci. 24(12): M2-M5.

28. Kelley, K.J. 1996. Using host defenses to fight infectious diseases. Nature Biotech. 14: 587-590.
29. Nemecek, S. 1997. Beating bacteria. Sci. Am. 276(2): 38-39.
30. Levy, S.B. 1998. Multidrug resistance – A sign of our time. N. Engl. J. Med. 338: 1376-1378.
31. Hileman, B. 1999. Livestock antibiotic debate heats up. Chem. Eng. News. Oct.25: 32-35.
32. Glynn, M.K., Bopp, C., Dewitt, W., Dabney, P., Mokhtar, M. and Angulo, F.J. 1998. Emergence of multidrug-resistant *Salmonella enterica serotype typhimurium* DT104 infections in the United States. N. Engl. J. Med. 338(19):1333-1338.
33. Threlfall, E.J., Hampton, M.D., Schofield, S.L., Ward, L.R., Frost, J.A. and Rowe, B. 1996. Epidemiological application of differentiating multiresistant *Salmonella typhimurium* DT 104 by plasmid profile. Commun. Dis. Rep. CDR Rev. 6: R155-R169.
34. Endtz, H.P., Ruijs, G.J., van Klingeren, B., Jansen, W.H., van der Reyden, T. and Mouton, R.P. 1991. Quinolone resistance in campylobacter isolated from man and poultry following the introduction of fluoroquinolones in veterinary medicine. J. Antimicrob. Chemother. 27(2): 199-208.
35. The medical impact of antimicrobial use in food animals. Report of a WHO Meeting, Berlin, 1997. Document No.: WHO/EMC/ZOO/97.4.
36. Gruneberg, R.N. and Hryniewicz, W. 1998. Clinical relevance of a European collaborative study on comparative susceptibility of Gram-positive clinical isolates to teicoplanin and vancomycin. Int. J. Antimicrob. Agents. 10: 271-277.
37. Cody, S.H., Glynn, M.K., Farrar, J.A., Cairns, K.L., Griffin, P.M., Kabayashi, J., Fyfe, M., Hoffman, R., King, A.S., Swaminathan, B., Bryant, R.G. and Vugia, D.J. 1999. An outbreak of *Escherichia coli* O157:H7 infection from unpasteurized commercial apple juice. Ann. Intern. Med. 130(3): 202-209.
38. Roccanova, L. and Rappa III, P. 2000. Antibiotic rotation. Science. 287: 803.
39. Novak, R. 1994. Hungary sees an improvement in penicillin resistance. Science. 264: 364.
40. Seppala, H., Klaukka, T., Vuopio-Varkila, J., Muotiala, A., Helenius, H., Lager, K. and Huovinen, P. 1997. The effect of changes in the consumption of macrolide antibiotics on erythromycin resistance in group A streptococci in Finland. Finnish Study Group for Antimicrobial Resistance. N. Engl. J. Med. 337 (7): 441-446.
41. Morell V. 1997. Antibiotic resistance: road of no return. Science 278:575-576.
42. Levy, S.B. 1992. The antibiotic paradox: how miracle drugs are destroying the miracle. New York. Plenum.
43. Schrag, S.J., Perrot, V. and Levin, B.R. 1997. Adaptation to the fitness costs of antibiotic resistance in *Escherichia coli*. Proc. R. Soc. Lond. B. Biol. Sci. 264(1386):1287-1291.
44. Williams, R.J. and Ryan M.J. 1998. Surveillance of antimicrobial resistance – an international perspective. B.M.J. 317: 651.
45. Davies, J. 1996. Bacteria on the rampage. Nature. 383: 219-220.
46. Livermore, D.M., Threlfall, E.J., Reacher, M.H., Johnson, A.P., James, D., Cheasty, T., Shah, A., Warburton, F., Swan, A.V., Skinner, J., Graham, A. and Speller, D.C. 2000. Are routine sensitivity test data suitable for the surveillance of resistance? Resistance rates amongst *Escherichia coli* from blood and CSF from 1991-1997, as assessed by routine and centralized testing. J. Antimicrob. Chemother. 45(2): 205-211.
47. Threlfall, E.J., Fisher, I.S., Ward, L.R., Tschape, H. and Gerner-Smidt, P. 1999. Harmonization of antibiotic susceptibility testing for Salmonella: results of a study by 18 national reference laboratories within the European Union-funded Enter-net group. Microb. Drug Resist. 5(3): 195-200.

48. Schempp, C.M., Pelz, K., Wittmer, A., Schopf, E. and Simon, J.C. 1999. Antibacterial activity of hyperforin from St John's wort, against multiresistant *Staphylococcus aureus* and Gram-positive bacteria. Lancet. 353(9170): 2129.

49. Ge, M., Chen, Z., Onishi, H.R., Kohler, J., Silver, L.L., Kerns, R., Fukuzawa, S., Thompson, C. and Kahne D. 1999. Vancomycin derivatives that inhibit peptidoglycan biosynthesis without binding D-Ala-D-Ala. Science. 284: 507-511.

50. Walsh, C. 1999. Deconstructing Vancomycin. Science. 284: 442-443.

51. Rouhi M. 1995. Steps urged to combat drug-resistant strains. Chem. Engineering News. May 22: 7-8.

52. Abb, J., Breuninger, H. and Kommerell, M. 1994. Prevalence of antimicrobial resistance of *Streptococcus pneumoniae* in Southwest Germany as determinded by the E test. Eur. J. Epidemiol. 10(5):621-623.

53. Georgopoulos, A., Buxbaum, A., Straschil, U. and Graninger, W. 1998. Austrian national survey of prevalence of antimicrobial resistance among clinical isolates of *Streptococcus pneumoniae* 1994-96. Scand. J. Infect. Dis. 30(4): 345-349.

From: *Development of Novel Antimicrobial Agents: Emerging Strategies*
ISBN 1-898486-23-9 © 2001 Horizon Scientific Press, Wymondham, UK.

2

Origins and Evolution of Antibiotic and Multiple Antibiotic Resistance in Bacteria

Ruth M. Hall and Christina M. Collis

Abstract

The selective pressure of antibiotic use has brought to prominance strains of many important pathogens that are resistant to more than one antibiotic. Acquisition of antibiotic resistance genes is the predominant factor in the emergence, evolution and spread of multiply antibiotic resistant bacteria. Two processes are important. Horizontal gene transfer enables genes to move from one bacterium to another, and translocation enables them to move from a location on one DNA molecule to another. The accumulation of genes conferring resistance to different antibiotics on the same horizontal gene transfer vehicle leads to the ability to simultaneously transfer multiple antibiotic resistance determinants to further bacterial strains, species or genera. Since genes for toxins and other pathogenicity and virulence determinants, genes conferring resistance to heavy metals and many others use the same highways and byways, the emergence of antibiotic resistance cannot be viewed in isolation from many other selective forces.

Introduction

Antibiotic resistance arises in two fundamentally distinct ways. The first involves alteration of relevant genes in the bacterial chromosome either by mutation or incorporation of small fragments of foreign DNA, and the second involves the acquisition of new genes that confer resistance but were not formerly part of the genetic complement of the bacterium. The relative importance of these two mechanisms differs for each organism; for example, resistance in Mycobacteria arises predominantly by mutation, whereas for the Enterobacteriaceae and Staphylococci acquisition of resistance genes is most common in the clinical setting. The relative importance of these two mechanisms also depends on the antibiotic; for example, though resistance of Enterobacteriaceae to most antibiotics commonly used in human therapy is largely due to acquired genes, resistance to quinolones and fluoroquinolones arises predominantly by mutation.

Though resistance to many, if not most, antibiotics can arise by alteration of chromosomal genes, the mutant bacteria can generally pass on the resistance trait only to their progeny. In contrast, acquired antibiotic resistance genes can often be passed on not only to progeny but also to other bacteria of the same or different species (Figure 1A). This phenomenon is known as horizontal gene transfer and makes use of various mechanisms for the movement of DNA from one bacterium to another. The impact of horizontal gene transfer is greatly enhanced by a number of processes that allow discrete units of DNA known as "mobile elements" to translocate from one location in a DNA molecule to another location in the same replicon or in any other replicon present in the same cell (Figure 1B). Translocation permits resistance genes that are within a mobile element to hop onto DNA molecules such as plasmids and conjugative transposons that are able to move into new bacteria, and hence spread broadly amongst quite unrelated bacterial species. Various translocation processes also provide the mechanistic framework to create genomes that contain several genes that each confer resistance to a different antibiotic. These two gene mobility phenomena, horizontal gene transfer and translocation, dramatically influence the dynamics of the emergence and spread of antibiotic resistant and multiply antibiotic resistant bacteria.

As a consequence of all of these factors, the details of the resistance problem are enormously complex. Each important human pathogen and each antibiotic or antibiotic family has its own specific story. Here, only the most important common principles of the emergence and spread of resistance will be addressed and selected detailed examples, mainly from Enterobacteriaceae, are used to illustrate each point. The themes are the same, though the details are different, for other important pathogens such as the Staphylococci (1, 2) and Clostridia (3). A large number of reviews that cover specific aspects in more detail are available in the published literature and some of these are cited.

The Power of Selection

Because of infidelity in DNA replication and repair, mutations arise spontaneously and bacterial populations contain many mutants, some of which are more resistant to particular antibiotics. Bacterial populations can also contain one or more organisms that carry acquired antibiotic resistance determinants. When such bacterial populations are exposed to the appropriate level of that antibiotic, the mutants or the cells that contain resistance genes sur-

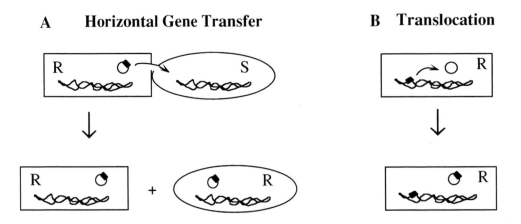

Figure 1. Gene movement. A. Horizontal gene transfer, in which DNA, here a replicon such as a plasmid or conjugative transposon represented by the circle, is transferred to another bacterial cell. B. Translocation, in which a gene contained within a mobile element (filled box) moves from one replicon to another within the same cell.

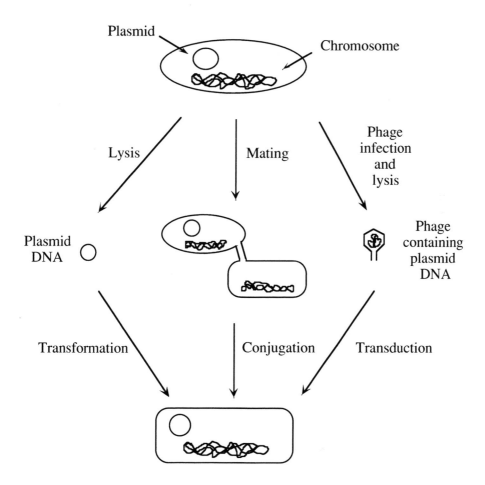

Figure 2. Horizontal gene transfer. Three ways in which DNA can be transferred from one cell to another using a plasmid (circle) are shown as an example. Transformation: free plasmid DNA generated by cell lysis is released and is taken up by another cell. Conjugation: conjugative plasmids and conjugative transposons have the ability to promote cell-to-cell contact and direct their own self-transfer to another cell. Both the donor and recipient cell then contain the plasmid. Transduction: the plasmid is packaged into a virus particle during bacteriophage infection and injected into a new host.

vive, while the susceptible cells either die (bacteriocidal) or cease to divide (bacteriostatic). In both cases, the resistant bacteria can continue to divide so long as nutrients are available. This leads to amplification of the resistant sub-population and thus, the final population after antibiotic exposure contains a greater proportion of resistant bacteria (4).

It is generally not appreciated that overall the selective pressure of antibiotic use is far more important in determining the prevalence of resistant bacteria than the frequency with which any of the events that lead to the emergence of resistant strains of bacteria either by mutation or by gene acquisition occur. In the case of acquired resistance genes, selection will also favour any other bacterium that has gained the resistance gene from the original host. And, as nothing succeeds like success, the processes and vehicles that move acquired resistance genes are likely to be examples of the most successful strategies for sharing genes amongst bacteria. They have come to our attention as a consequence of the selective pressure of over 50 years of antibiotic use.

Horizontal Gene Transfer

Cells can acquire DNA from exogenous sources in three ways (Figure 2). Firstly, naked DNA (released by cell lysis) can, under certain conditions, be taken up by bacterial cells in a process known as transformation. Secondly, when a bacterium is infected by certain bacteriophage, fragments of DNA other than the phage genome can be packaged into a proportion of the newly formed virus particles which are released when the cell is lysed. These phages can inject the packaged DNA into a new host and this process is known as transduction. New genes can also be incorporated into a bacteriophage genome and stable resistance can result if that genome is maintained in a bacterial cell. Thirdly, direct transfer of DNA from one bacterium to another can occur via a process known as conjugation. The machinery for bringing the two cells into direct contact and creating a mating pair and the functions required for conjugative transfer of the DNA from the donor to the recipient cell are encoded by certain plasmids and also by conjugative transposons.

The fate of incoming DNA is determined by a number of factors. When the DNA has homology with a region of DNA in the genome (chromosome + extrachromosomal elements) of the recipient, homologous recombination can lead to stable incorporation of the new DNA into that genome. A translocatable element (transposon, gene cassette or conjugative transposon) can also jump into the genome of the recipient, and a plasmid can establish itself independently in the recipient.

The relative importance of the three gene transfer processes in the emergence of antibiotic resistant organisms varies from species to species. Some species such as Neisseria and Streptococci are naturally transformable, and the evolution of resistance to β-lactam antibiotics occurs by the creation of mosaic genes for penicillin-binding proteins (5). However, overall the most important vehicles in the emergence and spread of antibiotic resistant bacteria appear to be plasmids and conjugative transposons. Though plasmids can move into new cells by transformation, transduction and conjugation, conjugation is the dominant process that drives the spread of conjugative transposons.

Conjugation and Mobilization

Some plasmids, known as conjugative or self-transmissible plasmids, have the ability to direct their own transfer from one cell to another. The first identified and the best studied of these is the *E. coli* F-factor (6). Conjugative plasmids encode a set of functions that construct a sex pilus. This pilus can connect with a second cell and create a mating pair. Conjugative plasmids also encode mobilization functions (Mob), and an origin of DNA transfer *(oriT)*. The plasmid is nicked at *oriT* and a copy of the plasmid is then transferred through the pilus from the donor to the recipient cell. The net result is that both the donor and the recipient now contain the plasmid (Figure1A) and both can transfer it again to further recipients. Conjugative transposons also encode conjugation (mating pore formation and DNA transfer) functions that facilitate their transfer to new hosts (7). Indeed, the conjugative transposons differ from conjugative plasmids mainly in that they are unable to replicate and hence cannot be stably maintained in a free form in the cell. To avoid confusion with transposons (see below) the term "conjugon" is hereafter used to replace "conjugative transposon".

A further group of plasmids do not encode transfer machinery, but are able to transfer to another cell when mating has already occurred due to the presence of a conjugation-proficient element in either the donor or the recipient. These plasmids generally encode mobilization (Mob) functions and contain an origin of transfer *(oriT)* and are designated mobilizable plasmids. Likewise, relatives of conjugons that do not themselves include genes for conjugative transfer but are able to utilize the transfer apparatus of conjugons have been identified (8). Finally, plasmids that are neither conjugative nor mobilizable can nonetheless be co-transferred to a new host if they are first integrated into a mobilizable element. Formation of cointegrates can occur in many ways, with homologous recombination and transposition being best known. After entry into the new host, the cointegrate can be resolved to re-form two separate plasmids.

Translocation

The term translocation is used here to describe all processes that lead to the relocation of genes to a new genetic (DNA) context (9). From the perspective of resistance gene spread, the most important form of translocation occurs when the new location is in a different DNA molecule e.g. chromosome to plasmid, plasmid to plasmid, plasmid to chromosome or in-coming DNA fragment to either chromosome or plasmid. Two quite distinct mechanisms are known to be important in translocation, namely transposition and site-specific recombination and these are described below. However, other processes that achieve the same end are also known but less well studied, and it is likely that more remain to be discovered.

Transposition

Insertion sequences (IS) and transposons are discrete genetic elements that encode proteins that are able to recognize the defined ends of the element, cut it out and splice it into a new location (9, 10). The IS are the simplest of these elements. While the delineation between IS and transposon is not currently clearly defined, the simple functional definition used here is that an IS carries only genes for one or more proteins that catalyze or regulate translocation of the IS (transposition functions) and the terminal inverted repeat sequences that are recognized by the transposition machinery. A transposon also contains these functional elements but in addition carries further genes e.g. antibiotic resistance genes. Also in the interests of clarity, the well known term "transposon" is used only for elements that move by a specific type of mechanism, namely transposition, and thus excludes the conjugons. Transposition generates a short direct duplication of the new target site as a consequence of the fact that cleavage of this site by the transposase occurs at slightly different positions on the two DNA strands. The size of this target duplication is a characteristic property of any individual IS or transposon. A detailed description of a variety of transpositional mechanisms can be found in a recent review (10).

Site-specific Recombination

Two important groups of mobile DNA elements that use site-specific recombination to move from one location to another are the conjugons (11-13), which are prevalent in Gram-positive bacteria, and gene cassettes that are an important source of resistance genes found in many Gram-negative organisms (14-16). Certain plasmids and bacteriophage (e.g. λ) can also integrate into another replicon using this mechanism (17). Site-specific recombination differs from transposition in a number of important ways. Firstly, these elements integrate preferentially at one or a limited number of target sites and both the target sites and a site in the mobile element include features that are recognized by the enzyme, a site-specific recombinase belonging to the bacterial (or λ) integrase superfamily, that catalyzes the reaction. Secondly, the integration (and excision) reactions are conservative in that no new bases are generated (i.e. there is no target site duplication), nor are bases lost.

Resistance due to Acquisition of Antibiotic Resistance Genes

Antibiotic resistance due to the acquisition of genes not normally present on the chromosome of the relevant species was first encountered as *Staphylococcus aureus* became resistant to penicillin not long after its introduction for therapy. This resistance is due to the *blaZ* gene that encodes a β-lactamase and is found on a plasmid (18). Multiple antibiotic resistance was first noted in Japan in the late 1950s when *Shigella flexneri* that were resistant to more than one antibiotic became commonplace. The resistance determinants could be transferred simultaneously to other bacteria and these studies led to the identification of resistance transfer factors or R-factors (19, 20). R-factors are now known to be a specific sub-group of the independently-replicating, extra-chromosomal genomes collectively known as plasmids; that is R-factors are plasmids that carry antibiotic resistance genes. More recently, related elements, the conjugons, have been identified and these are most prevalent in Gram-positive species.

Studies of bacteria isolated in the pre-antibiotic era have revealed that plasmids are present (21) but only rarely have these plasmids been found to contain antibiotic resistance genes (22). This finding is explained in large part by the existence of mobile elements (transposons and gene cassettes) that can relocate onto plasmids. Mobile DNA elements are now known to come in many forms, but the first to be identified and studied were transposons. Indeed, plasmids that contain one or more antibiotic resistance genes commonly contain one transposon or more (see below). More recently, the role of integrons in the acquisition of the many antibiotic resistance genes found in Enterobacteriaceae and Pseudomonads that are part of small mobile elements known as gene cassettes has also been uncovered.

Plasmids

Plasmids are important vehicles for the transfer of novel genes from one cell to another (horizontal gene transfer). They are discrete small genomes that are physically distinct from the bacterial chromosome because they are able to replicate as independent entities (replicons). Plasmids are found in many bacterial species, both Gram-positive and Gram-negative. Together they form a large pool of genetic information that is part of a "floating genome" that now appears to be shared by all bacteria. Plasmids come in a large range of sizes and the functions they carry also vary. The unifying feature of this group is the ability to replicate in the host cell. To be stably maintained, plasmids must also be reliably passed on to progeny cells, and a number of different strategies to achieve this are known. Conjugation and mobilization functions are optional extras, though plasmids that carry them have a distinct advantage in spreading genes to new hosts.

Plasmids can be characterized using a number of other features, and one that has been important historically in the classification of plasmids is incompatibility. Two plasmids of the same incompatibility group cannot be stably maintained in the same cell, whereas plasmids from different incompatibility groups can. Stable co-existence provides an important opportunity for reassortment of plasmid genomes and the acquisition of new genes, but even transient co-existence can achieve the same end. A further important feature is host range. Some plasmids can replicate and be stably maintained in only a limited number of bacterial species, while others such as IncP and IncQ plasmids, the so-called promiscuous plasmids, can survive in many hosts. Indeed, the capacity of plasmids to cross species and genera boundaries, including from Gram-negative to Gram-positive bacteria and vice versa (23, 24), and to be maintained in their new host, permits them to disseminate genes broadly in the bacterial kingdom. Clearly, combining broad host range and either conjugation or mobilization func-

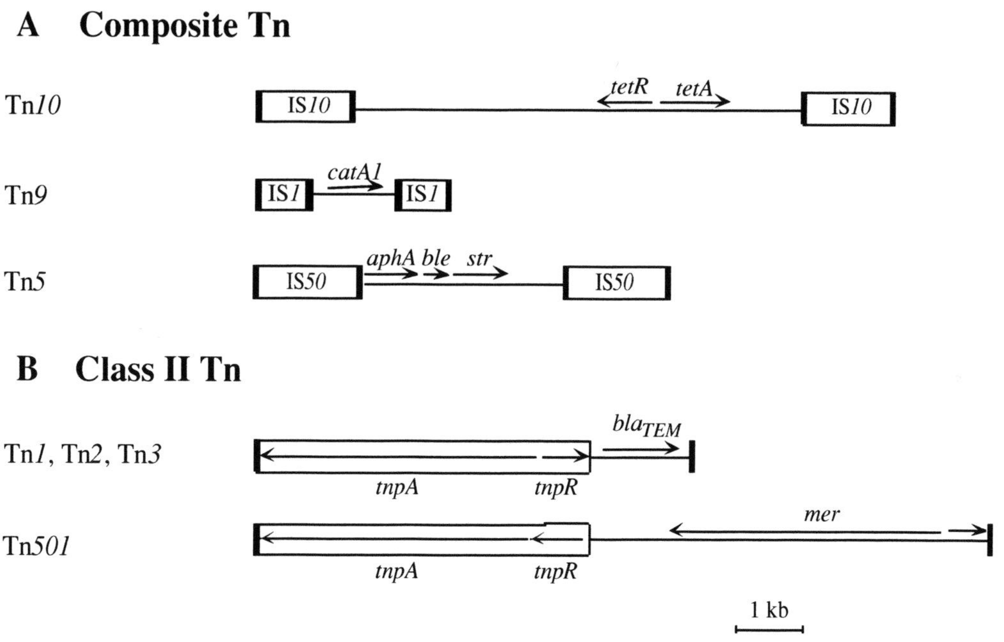

A   Composite Tn

Tn*10*

Tn9

Tn5

B   Class II Tn

Tn*1*, Tn2, Tn3

Tn*501*

1 kb

Figure 3. Transposons. A. Composite transposons consist of two IS elements flanking a DNA sequence which includes an antibiotic resistance gene. Transposition functions are encoded by the IS sequences. B. Class II Tn consist of a region containing genes essential for transposition (open box) grouped at one end and a region containing antibiotic resistance genes (line) flanked by short terminal inverted repeats (black boxes).

tions in the same plasmid creates a potent vehicle for gene spread, so it is hardly surprising that conjugative and mobilizable broad host-range plasmids are frequently found to carry the resistance determinants in antibiotic resistant bacteria.

Conjugons -Conjugative Transposons

Conjugons combine the features necessary for horizontal gene transfer and those necessary for translocation in the same element. They can translocate within the same cell to a new location in the chromosome of the host bacterium or to a plasmid, if one is present. They can also move into a new cell by conjugation and integrate into the chromosome of that cell. Thus conjugons also form part of the readily shared bacterial gene pool. Conjugons were identified only recently, initially as agents of transfer of tetracycline resistance (*tetM*) in Gram-positive species (7, 11-13). However, conjugons are also found in Bacteroides (8) and the STX element of *Vibrio cholerae* O139, that confers resistance to trimethoprim, sulphonamides and streptomycin, appears to be a conjugon (25).

Conjugons are generally large elements that include conjugation functions but, because they lack the ability to replicate independently, they can exist only transiently as separate DNA molecules. To overcome this constraint, they integrate into another replicon (bacterial chromosome or plasmid) and are normally found integrated in the bacterial chromosome. Conjugons also encode a site-specific recombinase that belongs to the bacterial integrase superfamily and contain a site adjacent to the integrase gene that is recognised by the integrase (12, 13). Conjugons readily cross species and genera boundaries, including from Gram-negative to Gram-positive bacteria and vice versa (7, 11, 24), and this extremely broad host range permits them to carry antibiotic resistance genes far and wide in the bacterial kingdom. Like plasmids, they can acquire new genes such as antibiotic resistance genes, and these travel as passive passengers when horizontal transfer or translocation occurs.

Transposons

Transposons are particularly important in the spread of antibiotic resistance genes. Two distinct groups of transposons are known (Figure 3). The composite transposons consist of two IS elements flanking a DNA segment that contains the antibiotic resistance genes. Well-known examples of this group (Figure 3A) are Tn*10*, which carries the tetB tetracycline resistance determinant and Tn9 which carries the *catA1* (*cat₁*) chloramphenicol resistance gene. The second group, which includes mainly members of the Tn3 family (26, 27) and also Tn*7* and its relatives (28), are bounded by short inverted repeats, and contain genes essential for transposition grouped at one end (Figure 3B). Tn3 includes the TEM β-lactamase (*bla*$_{TEM}$) gene and this transposon and its variants Tn*1*, Tn*2* etc are found in many different plasmids isolated from many different bacterial species (29). A further important example is Tn*1546* found in Enterococci which carries a set of genes that confer resistance to the glycopeptide antibiotics vancomycin, avoparcin and teichoplanin (30).

Gene cassettes and integrons

Gene cassettes are the smallest mobile elements known (Figure 4A). They consist of a single gene and a recombination site (14, 16, 31). Cassettes differ from most other mobile elements in that they do not encode the protein(s) required to catalyse their movement. The necessary enzyme, a member of the bacterial integrase superfamily, is encoded by a companion element known as an integron (32) and integrons provide the standard location for integrated gene cassettes (Figure 4B). Cassettes generally contain one antibiotic resistance gene and to

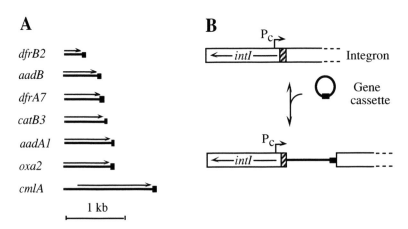

Figure 4. Gene cassettes and integrons. A. Some representative gene cassettes are shown in the linear form. Gene cassettes consist of a gene coding region (indicated by an arrow) and a 59-base element recombination site (depicted by a black box). The cassettes shown all contain antibiotic resistance genes: *dfrB2* and *dfrA7* (trimethoprim resistance), *aadB* and *aadA1* (aminoglycoside resistance), *oxa2* (β-lactam resistance) and *catB3* and *cmlA* (chloramphenicol resistance). B. Integration of a cassette into an integron. Circular gene cassettes may be incorporated into the *attI* receptor recombination site of the integron (shaded) by site-specific recombination catalysed by the *intI* integrase encoded by the integron. The cassette-encoded genes are transcribed from a promoter, P$_c$, in the integron.

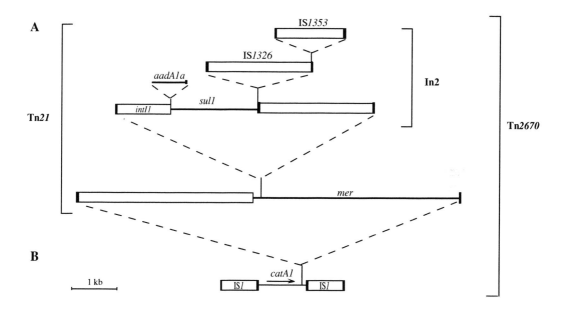

Figure 5. A complex multidrug resistance transposon from plasmid NR1. Tn*2670* includes a number of discrete transposons and IS that are able to move independently or together in different combinations. Features of IS and transposons are depicted as in Figure 3 and of the integron and *aadA1* gene cassette as in Figure 4.

date over 60 distinct antibiotic resistance genes that determine resistance to β-lactam antibiotics, trimethoprim, aminoglycosides, chloramphenicol, erythromycin, and rifampicin as well as genes that determine resistance to the quaternary ammonium compounds used as antiseptics and disinfectants have been found in this configuration (14, 33). However, as most cassettes do not contain promoters, the genes are expressed from a promoter in the integron (Figure 4B) (34). Gene cassettes can be excised from an integron (Figure 4B), but as they are unable to replicate can exist only transiently as independent covalently-closed circular molecules (35). However, they can be integrated either back into the source integron, into any other integron present in the same cell, or rarely into secondary sites (14-16).

Multiple Antibiotic Resistance due to Acquisition of Antibiotic Resistance Genes

When clinical strains that are resistant to more than one unrelated antibiotic are examined, it is common to find plasmids that carry more than one antibiotic resistance gene. In many cases this results from the fact that such plasmids carry more than one transposon. However, transposons that carry more than one antibiotic resistance gene are also well known. A simple example is the composite transposon, Tn*5* (Figure 3A) which carries three genes encoding resistance to the aminoglycosides kanamycin and neomycin (*aphA*), to cancer chemotherapeutic agents of the bleomycin group (*ble*) and to streptomycin (*str*).

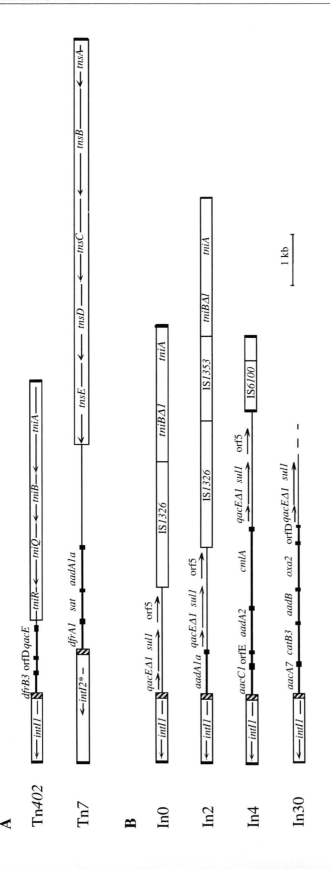

Figure 6. Integrons with multiple antibiotic resistance genes. A. Integrons which include a full set of transposition functions and are active transposons. Tn*402* (In28) is a class 1 integron and Tn*7* belongs to class 2. Cassettes are represented as in Figure 4 and both the integrase (*intl*) and the transposition genes (*tni* or *tns*) regions are also shown as open boxes. The *attl* site is hatched. Cassettes shown contain genes that determine resistance to trimethoprim (*dfrA1*, *dfrA7*), quaternary ammonium compounds (*qacE*), streptothricin (*sat*) and streptomycin and spectinomycin (*aadA*) and an open reading frame of unknown function (orfD). B. Class 1 integrons that contain the 3'-conserved segment. The *sul1* sulphonamide resistance determinant, the *qacEΔ1* gene and orf5 are in the 3'-CS. In0 which contains no cassettes is from plasmid pVS1; In2 is in Tn*21* and is from plasmid NR1 (see Figure 5); In4 is in Tn*1696* and is from R1033 and In30 is from pBWH301. These integrons are transposition-defective due to loss of part or all of the *tni* genes and contain various IS. IS and transposition genes are represented by open boxes, and the terminal 25 bp inverted repeat sequences that delineate the ends of the integron are shown as black bars. Further cassettes shown contain genes that determine resistance to gentamicin (*aacC1*), streptomycin and spectinomycin (*aadA2*), gentamicin, kanamycin and tobramycin (*aadB*), amikacin (*aacA7*), β-lactams (*oxa2*) and chloramphenicol (*cmlA* and *catB3*) and open reading frames of unknown function, orfE and orfD.

It is also not uncommon for transposons and IS to be found located within other transposons. One particularly complex example (Figure 5) is found in plasmid NR1 (R100) which was recovered from one of the multidrug resistant *Shigella flexneri* strains isolated in the late 1950s in Japan (36). This composite transposon includes several discrete IS or transposons that are able to move either independently or in consort in different combinations. The first in the stack is a transposon that is closely related to Tn*9* (see Figure 3A) and contains the chloramphenicol resistance determinant (*catA1*) flanked by two IS*1* elements. The well-studied transposon Tn*21* (37) is located within this composite transposon. Tn*21* includes determinants for resistance to streptomycin, spectinomycin (*aadA1*), sulphonamides (*sul1*) and mercuric ions (HgII) (*mer* genes), but the antibiotic resistance genes are contained in an internal passenger, the integron, In2. The backbone of Tn*21* is a mercury resistance transposon similar to Tn*501* (Figure 3B) and the integron is itself a discrete transposon-like element that is transposition-defective due to loss of some of the essential transposition genes (38). In2 contains three further discrete mobile elements, the *aadA1* gene cassette and two IS. Thus the resistance determinant region on NR1 is a complex collection of mobile elements, each of which can move in its own right, but can also move as part of all larger structures that include it. As the broader context of individual antibiotic resistance genes is being examined, it is becoming clear that this theme is repeated in many specific arrays of resistance genes.

Integrons are particularly important in the emergence of multiple antibiotic resistant strains (39) because more than one cassette can be incorporated into an integron. More than one class of integron has been identified (see 16) though the class 1 integrons are most prevalent in Enterobacteria and Pseudomonas. Active transposons that are members of both class 1 and class 2 are known (Figure 6A) and integrons thus combine the ability to translocate and the ability to incorporate new genes in gene cassettes. Tn*402*, the prototype class 1 integron that includes the transposition genes (40) contains cassettes encoding resistance to trimethoprim and to antiseptics and disinfectants. Tn*7* which is the prototype class 2 integron contains three cassettes that confer resistance to trimethoprim, streptothricin and streptomycin and spectinomycin.

However, the integrons most frequently isolated belong to a specific subgroup of class 1 integrons that contain a sulphonamide resistance determinant as part of the fixed backbone (Figure 6B) and the presence of even a single cassette encoding an antibiotic resistance determinant in this type of class 1 integron leads to the linkage of resistance to that antibiotic with sulphonamide resistance. Though known members of this group are so far all defective transposon derivatives, it is presumed that they can move if transposition proteins are supplied in trans and there is clear evidence that this has happened in the past (38).

The Antibiotic Resistance Genes Found in Multiple Hosts

Though the sequencing of antibiotic resistance genes began late and got off to a slow start, the sequences of an enormous number of antibiotic resistance genes are now available. This has permitted the use of hybridization, PCR and even direct sequencing to track the spread of individual genes in the real world situation. Many examples of the same gene found in a wide variety of bacteria have been reported (3, 11, 28, 41), but perhaps the most data have accumulated for the *sul1* gene found in the backbone of the predominant group of class 1 integrons. This gene, usually together with one or more cassette-encoded genes, has been found in many Gram-negative bacteria including species of Escherichia, Salmonella, Morganella, Klebsiella, Enterobacter, Proteus, Providencia, Acinetobacter and Vibrio, and also in Pseudomonas and Xanthomonas species. But occasional sitings have also been made elsewhere in

Mycobacterium smegmatis (42), *Corynebacterium glutamicum* (43) and *Enterococcus faecalis* (44). In all cases the sequence of *sul1* is identical, indicating that its interspecies spread occurred recently. This is undoubtedly only one example of many where a single gene has become broadly disseminated and it is likely that the large gene pool built up by the continuous use of this and other antibiotics in human medicine, animal husbandry, aquaculture and agriculture has facilitated this spread. Apart from providing a large reservoir of bacteria containing one or more resistance genes, antibiotic use also selects for and amplifies the bacterial products of these potentially rare events, allowing them to become established in the bacterial community.

Concluding Remarks

The combination of two forces, horizontal gene transfer (i.e. transfer of DNA from one bacterium to another) and translocation (i.e. movement of DNA from one discrete DNA molecule or replicon to another) has allowed genes that determine resistance to antibiotics to emerge and then spread into and between important human pathogens. As more and more sequences are determined, it is becoming clear that there are many different genes that confer resistance to each individual antibiotic or antibiotic family. Some of these have first been found in one or a few bacterial species, but with time and continued selective pressure they have spread to new bacterial species. Thus, ultimately all of these genes are likely to be available to all bacterial species and development of resistance in one particular pathogen or commensal organism cannot be viewed as without effect on other pathogens. The physical linkage of more than one resistance gene and consequent ability to cotransfer to a new host means that the use of one antibiotic cannot be viewed as without effect on the development of or spread of resistance to other unrelated antibiotics. As new antibiotics are developed, it seems reasonable to predict that genes that can confer resistance to them will emerge and spread in the same way.

References

1. Skurray, R.A. and Firth, N. 1997. In: Ciba Foundation Symposium 207. Antibiotic Resistance: Origins, Evolution, Selection and Spread. D.J. Chadwick and J. Goode, eds. John Wiley and Sons, Chichester. p.167-191.
2. Paulsen, I.T., Firth, N. and Skurray, R.A. 1996. Resistance to antimicrobial agents other than β-lactams. In: The Staphylococci in Human Disease. K.B. Crossley and G.L. Archer, eds. Churchill Livingstone, New York. p. 175-212.
3. Lyras, D. and Rood, J.I. 1997. Transposable genetic elements and antibiotic resistance determinants from *Clostridium perfringens* and *Clostridium difficile*. In: The Clostridia: Molecular Biology and Pathogenesis. J.I. Rood, B.A. McCane, J.Q. Songer and R.W. Titball, eds. Academic Press, Ltd., London. p. 73-92.
4. Baquero, F., Negri, M-C., Morosini, M-I. and Blázquez, J. 1998. Antibiotic-selective environments. Clinical. Infect. Diseases. 27 (Suppl. 1):S5-11.
5. Dowson, C.G., Coffey, T.J. and Spratt, B.G. 1994. Origin and molecular epidemiology of penicillin-binding-protein-mediated resistance to β-lactam antibiotics. Trends Microbiol. 2: 361-366.
6. Firth, N., Ippen-Ihler, K. and Skurray, R.A. 1996. Structure of the F-factor and mechanism of conjugation. In: *Escherichia coli* and *Salmonella*. F. C. Neidhart, ed. American Society for Microbiology, Washington, D.C. p. 2377-2401.

7. Clewell, D.B., Flannagan, S.E. and Jaworski, D.D. 1995. Unconstrained bacterial promoscuity: the Tn*916*-Tn*1545* family of conjugative transposons. Trends Microbiol. 229: 229-236.

8. Salyers, A.A., Shoemaker, N.B. and Li, L-Y. 1995. In the driver's seat: the *Bacteroides* conjugative transposons and the elements they mobilize. J. Bact. 177: 5727-5731.

9. Kleckner, N. 1977. Translocatable elements in procaryotes. Cell. 11:11-23.

10. Craig, N. 1996. Transposition. In: *Escherichia coli* and *Salmonella*. F. C. Neidhart, ed. American Society for Microbiology, Washington, D.C. p. 2339-2362.

11. Rice, L.B. 1998. Tn*916* family conjugative transposons and dissemination of antimicrobial resistance determinants. Antimicrob. Agents Chemother. 42: 1871-1877.

12. Scott, J.R. and Churchward, G.G. 1995. Conjugative Transposition. Annu. Rev. Microbiol. 49: 367-397.

13. Salyers, A.A., Shoemaker, N.B., Stevens, A.M. and Li, L-Y. 1995. Conjugative transposons: an unusual and diverse set of integrated gene transfer elements. Microb. Rev. 59: 579-590.

14. Recchia, G.D. and Hall, R.M. 1995. Gene cassettes: a new class of mobile element. Microbiol. 141:3015-3027.

15. Hall, R.M. and Collis, C.M. 1995. Mobile gene cassettes and integrons: capture and spread of genes by site-specific recombination. Mol. Microbiol. 15:593-600.

16. Hall, R.M. and Collis, C.M. 1998. Antibiotic resistance in Gram-negative bacteria: the role of gene cassettes and integrons. Drug Resist. Updates. 1: 109-119.

17. Nash, H. 1996. Site-specific recombination: integration, excision, resolution, and inversion of defined DNA segments. In: *Escherichia coli* and *Salmonella*. F. C. Neidhart, ed. American Society for Microbiology, Washington, D.C. p. 2363-2376.

18. Lyon, B.R. and Skurray, R. 1987. Antimicrobial resistance of *Staphylococcus aureus*: genetic basis. Microbiol. Rev. 51:88-134.

19. Watanabe, T. 1963. Infective heredity of multiple drug resistance bacteria. Bacteriol. Rev. 27: 87-115.

20. Falkow, S. 1975. Infectious multiple drug resistance. Pion Ltd, London, UK.

21. Datta, N. and Hughes, V.M. 1983. Plasmids of the same Inc groups in Enterobacteria before and after the medical use of antibiotics. Nature. 306: 616-617.

22. Smith, D.H. 1967. R factor infection of *Escherichia coli* lyophilized in 1946. J. Bact. 94: 2071-2072.

23. Mazodier, M. and Davies, J. 1991. Gene transfer between distantly related bacteria. Annu. Rev. Genet. 25: 147-171.

24. Courvalin, P. 1994. Transfer of antibiotic resistance genes between Gram-positive and Gram-negative bacteria. Antimicrob. Agents Chemother. 38: 1447-1451.

25. Waldor, M.K., Tschäpe, H. and Mekalanos, J. 1996. A new type of conjugative transposon encodes resistance to sulfamethoxazole, trimethoprim, and streptomycin in *Vibrio cholerae* O139. J. Bact. 178: 4157-4165.

26. Sherratt, D. 1989. Tn*3* and related transposable elements: site-specific recombination and transposition. In: Mobile DNA. D.E. Berg and M.M. Howe, eds. American Society for Microbiology, Washington, D.C. p. 163-184.

27. Grinsted, J., de la Cruz, F. and Schmitt, R. 1990. The Tn*21* subgroup of bacterial transposable elements. Plasmid. 24: 163-189.

28. Craig, N. 1996. Transposon Tn*7*. Curr. Top. Microbiol. Immunol. 204: 27-48.

29. Roberts, M.C. 1989. Gene transfer in the urogenital and respiratory tract. In: Gene Transfer in the Environment. S.B. Levy and R.V. Miller, eds. McGraw-Hill Publishing Company, New York. p. 347-375.

30. Arthur, M., Molinas, F., Depardieu, F. and Courvalin, P. 1993. Characterization of Tn*1546*, a Tn*3*-related transposon conferring glycopeptide resistance by synthesis of depsipeptide peptidoglycan precursors in *Enterococcus faecium* BM4147. J. Bact. 175: 117-127.

31. Recchia, G.D. and Hall, R.M. 1997. Origins of the mobile gene cassettes found in integrons. Trends Microbiol. 5:389-394.

32. Stokes, H.W. and Hall, R.M. 1989. A novel family of potentially mobile DNA elements encoding site-specific gene integration functions: integrons. Mol. Microbiol. 3: 1669-1683.

33. Partridge, S.R. and Hall. R.M., unpublished.

34. Collis, C.M. and Hall, R.M. 1995. Expression of antibiotic resistance genes in the integrated cassettes of integrons. Antimicrob. Agents Chemother. 39:155-162.

35. Collis, C.M. and Hall, R.M. 1992. Gene cassettes from the insert region of integrons are excised as covalently closed circles. Mol. Microbiol. 6: 2875-2885.

36. Womble, D.D. and Rownd, R.H. 1988. Genetic and physical map of plasmid NR1: comparison with other IncFII antibiotic resistance plasmids. Microbiol Rev. 52: 433-451.

37. Liebert, C.A., Hall, R.M. and Summers, A.O. 1999. Transposon Tn*21*, flagship of the floating genome. Microb. Mol. Biol. Rev. 63: 507-522.

38. Brown, H.J., Stokes, H.W. and Hall, R.M. 1996. The integrons In0, In2 and In5 are defective transposon derivatives. J. Bact. 178: 4429-4437.

39. Hall, R.M. 1997. Mobile gene cassettes and integrons: moving antibiotic resistance genes in Gram-negative bacteria. In: D.J. Chadwick and J. Goode, eds. Ciba Foundation Symposium 207. Antibiotic Resistance: Origins, Evolution, Selection and Spread. John Wiley and Sons, Chichester. p. 192-205.

40. Rådström, P., Sköld, O., Swedberg, G., Flensburg, J., Roy, P.H. and Sundström, L. 1994. Transposon Tn*5090* of plasmid R751, which carries an integron, is related to Tn*7*, Mu, and the retroelements. J. Bact. 176:3257-3268.

41. Roberts, M.C. 1997. Genetic mobility and distribution of tetracycline resistance determinants. In: D.J. Chadwick and J. Goode, eds. Ciba Foundation Symposium 207. Antibiotic Resistance: Origins, Evolution, Selection and Spread. John Wiley and Sons, Chichester. p. 206-222.

42. Martin, C., Timm, J., Rauzier, J., Gomez-Lus, R., Davies, J. and Gicquel, B. 1990. Transposition of an antibiotic resistance element in mycobacteria. Nature. 345:739-743.

43. Nesvera, J., Hochmannova, J. and Patek, M. 1998. An integron of class 1 is present on the plasmid pCG4 from Gram-positive bacterium *Corynebacterium glutamicum*. FEMS Microbiol. Lett. 169: 391-395.

44. Clark, N.C., Olsvik, O., Swenson, J.M., Spiegel, C.A. and Tenover, F.C. 1999. Detection of a streptomycin/spectinomycin adenylyltransferase gene (*aadA*) in *Enterococcus faecalis*. Antimicrob. Agents Chemother. 42: 157-160.

From: *Development of Novel Antimicrobial Agents: Emerging Strategies*
ISBN 1-898486-23-9 © 2001 Horizon Scientific Press, Wymondham, UK.

3

Antimicrobial Peptide Resistance Mechanisms in Bacteria

Arden Aspedon and Eduardo A. Groisman

Abstract

Small cationic peptides with broad-spectrum antimicrobial activity are produced by a wide variety of plants and animals and represent a nonspecific arm of the immune systems of these organisms. Antimicrobial peptides are believed to kill bacteria by damaging the cytoplasmic membrane by a mechanism that does not appear to involve interaction of the peptide with discrete protein targets. Bacteria employ several mechanisms to resist the toxic effects of these agents: exclusion of the peptide from its target (membrane), inactivation of the peptide, modification of the target, and the ability of the cell to overcome peptide-induced sublethal damage. The expression of resistance determinants is often under transcriptional control in a manner that presumably enhances the chances of survival of the bacterial cell in a given environment. The propensity of bacteria to acquire antibiotic-resistance genes suggests a potential for the evolution of peptide hyperresistant strains analogous to the multiantibiotic resistant forms present today.

Introduction

Antimicrobial peptides are now recognized as being widely distributed in nature having been found in numerous plant and animal species, including humans (1). The sheer ubiquity of these compounds coupled with their broad-spectrum of antimicrobial activity, presence in specific tissues and the induction of their expression subsequent to infection or tissue injury implies a role for these peptides in defense against microbes (2, 3). Indeed, antimicrobial peptides are thought to constitute a nonspecific arm of the immune system of animals and their activity is referred to as 'innate immunity' (2). In regard to host-pathogen interactions, one can envisage the requirement of a microbe to be adapted to resist killing by these host-derived peptides, at least in those situations where an interaction with the peptide cannot be evaded. The ecological significance of these peptides must not be limited to plant-microbe or animal-microbe interactions. Many bacteria are known to produce peptide antibiotics, which implies an importance of these compounds in microbe-microbe interactions as well (4). In this minireview, we will describe the mechanisms employed by bacteria to resist killing by antimicrobial peptides. Although intended to be broad in scope, most of what is known about resistance mechanisms have been described for human pathogens. However, by analogy to what is seen with resistance to the classical nonpolypeptide antibiotics (e.g., penicillins, tetracycline, etc.) those peptide resistance mechanisms seen in pathogens may reflect common themes employed by all bacteria that encounter these peptides.

Mode of Action of Antimicrobial Peptides

A competent description of the mechanisms employed by bacteria to resist killing by antimicrobial peptides necessitates that the molecular structure and mode of action of these compounds be described. To date, more than 130 different antimicrobial peptides have been characterized biochemically and have been shown to exhibit considerable variation both in amino acid sequence as well as in their secondary and tertiary structures (1). In spite of these recognized differences, some generalizations can be drawn in regard to conserved structural motifs and antimicrobial activity. Antimicrobial peptides are small—most range from 14 to 40 amino acids in length—polycationic amphipathic compounds that are believed to kill bacteria by damaging the cytoplasmic membrane and/or inhibiting membrane-associated functions like nutrient transport and energy transduction (1, 5-7). Most of these peptides exhibit broad-spectrum antimicrobial activity in that they kill both Gram-negative and Gram-positive bacteria, and some are even effective at killing fungi or inactivating enveloped viruses (1).

There are some notable exceptions to this broad-spectrum activity: the animal-derived peptides magainin 2, Bac5 and Bac7 are more effective at killing Gram-negative bacteria, whereas the insect-derived peptides mastoparan, royalisin, phormicin and insect defensin are more potent against Gram-positives (1, 2). In accordance with the role of antimicrobial peptides as an arm of innate immunity, the relative specificity in killing exhibited by these agents may reflect an evolved response to infections by specific types of bacterial parasites. A specificity of action is also seen with many bacteria-derived antimicrobial peptides, i.e., the bacteriocins. For example, nisin and subtilin—peptides produced by *Lactococcus lactis* and *Bacillus subtilis,* respectively—permeabilize the cytoplasmic membrane of Gram-positive bacteria in general, whereas other bacteriocins are active against specific genera or species of bacteria (4).

Up to a point, one can draw an analogy between the mode of action of the classical antibiotics and antimicrobial peptides because both classes of antimicrobials serve to kill or inhibit the growth of the bacterium via interference with some aspect of the microbe's physiology. Indeed, antimicrobial peptides found in animals and plants are sometimes referred to as peptide antibiotics (2, 8), even though these proteinaceous compounds are not produced by microbes and hence do not fit the classical definition of an antibiotic, i.e., a compound produced by a microbe that inhibits the growth of or kills another microbe. Furthermore, there are examples of "true" peptide antibiotics with a mode of action similar to that seen with antimicrobial peptides from plants and animals. For example, the polymyxins are cyclic amino acid-containing compounds produced by soil-borne bacteria that kill other bacteria via their membrane damaging effects (8). Where the analogy to antibiotics falters, however, is the very limited way in which antimicrobial peptides have been found to exert their antimicrobial activity.

Antibiotics in chemotherapeutic use today are usable because, for the most part, they inhibit enzymes in biosynthetic pathways that only exist in the bacterium and not the human or animal host. For instance, penicillin derivatives (e.g., methicillin, amoxicillin) bind to and inactivate enzymes involved in the biosynthesis of peptidoglycan, a compound unique to the bacterial cell wall; whereas fluoroquinolones (e.g., ciprofloxacin) bind specifically to bacte-

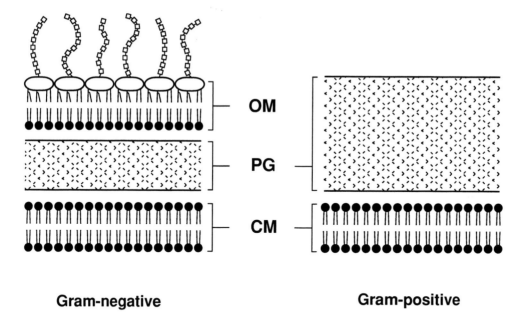

Gram-negative   **Gram-positive**

Figure 1. A cross-sectional view of the cell envelopes of Gram-negative and Gram-positive bacteria. CM, cytoplasmic membrane; PG, peptidoglycan cell wall; OM, outer membrane. In Gram-negative bacteria, the region between the cytoplasmic membrane and the outer membrane is called the periplasmic space. Cell envelope proteins have been omitted for clarity.

rial DNA gyrase thus inhibiting DNA replication (9). In contrast, no plant- or animal-derived antimicrobial peptide to date has been unequivocally shown to specifically inhibit a biosynthetic process within the bacterial cell. With very few exceptions, in those cases where the mode of peptide action has been studied, antimicrobial peptides have been shown to exert their activity through interaction with the cytoplasmic membrane where they may cause leakage of cytoplasmic material, inhibition of energy transduction, and/or cell lysis (1, 5, 7, 10, 11). In contrast to the membrane-as-target premise, there are some peptides that do not permeabilize the cytoplasmic membrane at their minimal inhibitory concentration and appear to exert their antibacterial activity through interaction with a target (e.g., nucleic acids) in the cytoplasmic compartment of the cell (12).

Although the analogy to mode of action has its limitations, there is some overlap in the mechanisms employed by bacteria to resist the toxic effects of classical antibiotics and antimicrobial peptides. The three basic resistance strategies are: (i) exclusion of the peptide from the target site; (ii) inactivation of the peptide; and (iii) chemical modification of the target that precludes a peptide-target interaction (13, 14). A fourth aspect of resistance that may not be readily identifiable, but one that subtends all other resistance mechanisms, is the cell's innate ability to repair or overcome the damage done by a sublethal concentration of peptide. For example, permeabilization of the cytoplasmic membrane can result in the loss of cytoplasmic K^+ and ATP, materials that the cell must recover from the extracellular medium or regenerate in order to remain viable (10, 15, 16). In addition, expression of peptide resistance determinants is often under transcriptional control (17, 18) and a cell's capacity to resist the toxic effects of these peptides may be largely determined by the extracellular environmental conditions governing the regulatory proteins that control expression of resistance loci.

Mechanisms of Resistance to Antimicrobial Peptides

Exclusion of the Peptide from the Target

The primary target of an antimicrobial peptide appears to be the cytoplasmic membrane. Studies with artificial planar membranes, liposomes, membrane vesicles, spheroplasts and whole cells have amply demonstrated the membrane-damaging effects of these peptides (1, 6, 7, 10, 15, 16). Although there is agreement on the ability of these peptides to disrupt membranes, a specific target within the cytoplasmic membrane (or the cytoplasm) has not been demonstrated. Indeed, both the D and L forms of four different peptides were equally effective at killing bacteria, which argues against these peptides interacting with stereo-specific targets such as proteins (2).

However, in contrast to the cytoplasmic membrane as the target of peptide activity, there have been reports of antimicrobial peptides exerting a lethal effect solely through an interaction with the cell surface outside the cytoplasmic membrane, e.g., by disrupting the outer membrane of Gram-negative bacteria (Figure 1). For instance, experiments with immobilized magainin 2 showed the peptide to have antibacterial activity even though attachment to an insoluble polymer prevented access of the peptide to the cytoplasmic membrane of the test bacterium (19). The immobilized magainin 2 was equally active against the Gram-negative bacterium *Escherichia coli* and the Gram-positive bacteria *Staphylococcus aureus* and *Bacillus subtilis*. Results from this study are equivocal, however, because it was not clear whether the apparent loss of viability, as measured by colony counting, was due to actual killing or merely the result of cells adhering to the polymer that gave artificially low colony counts on agar plates (19). Moreover, time course experiments with human defensin HNP-1 activity against the Gram-negative bacterium *E. coli* revealed that a loss of viability only occurred when the cytoplasmic membrane, and not the outer membrane, was permeabilized by the peptide (20). Furthermore, studies on the mode of action of magainin 2 and protamine showed that disruption of the outer membrane was not in itself sufficient to elicit cell killing (5, 21).

Given that the actual target of antimicrobial peptides is the cytoplasmic membrane, or even a cytoplasmic component, it would follow that the bacterial cell would adapt by evolving an outer cell envelope to be less permeable to these peptides. The apparent specificity of some antimicrobial peptides for killing Gram-negative or Gram-positive bacteria may only reflect differences in permeability (i.e., access to the cytoplasmic membrane) toward a given peptide rather than the presence of a specific target inherent to the cell type. For instance, the polymyxins are very effective at killing Gram-negative bacteria but have relatively little effect against Gram-positives (8). The fundamental structural differences between these two groups of bacteria are that Gram-negatives possess an outer membrane and a relatively thin peptidoglycan cell wall whereas Gram-positives lack an outer membrane and have a thick cell wall (Figure 1). The outer membrane, which is otherwise a very effective permeability barrier to many hydrophilic and hydrophobic compounds (22), is readily breached by polymyxin, and many other antimicrobial peptides (23). Once past the outer membrane, polymyxin may have unimpeded access to the cytoplasmic membrane, a freedom of access that may not exist in the Gram-positive bacterium because of resistance factors associated with the cell wall. Drawing once more on an analogy to antibiotic resistance, the relative ineffectiveness of the penicillins and other antibiotics against Gram-negative bacteria can be attributed to the impermeability of the outer membrane to these drugs rather than the absence of a specific target molecule in the Gram-negative cell (22). Despite gross differences in the cell envelope of Gram-negative and Gram-positive bacteria, both groups employ similar strategies in preventing access to the cytoplasmic membrane.

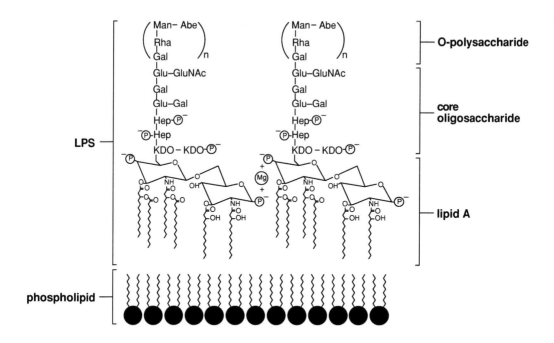

Figure 2. A cross-sectional view of the outer membrane of the Gram-negative bacterium *Salmonella typhimurium*. The outer leaflet of the outer membrane is composed almost exclusively of lipopolysaccharide (LPS), a structural arrangement that confers a permeability barrier property to the outer membrane. The lipid A component is an acylated disaccharide. The negatively charged phosphate groups are associated with the lipid A and the KDO and heptose moieties of the core oligosaccharide. Depending on growth conditions, the net negative charge of the LPS is reduced by adding 4-aminoarabinose or ethanolamine to phosphate groups of the lipid A or the core oligosaccharides, respectively. Adjoining LPS molecules are stabilized by the divalent cation magnesium. Outer membrane proteins have been omitted for clarity. KDO, 2-keto-3-deoxyoctonate; Hep, heptose; Glc, glucose; Gal, galactose; GlcNAc, N-acetylglucosamine; Rha, rhamnose; Man, mannose; Abe, abequose.

Antimicrobial peptides may differ in their amino acid sequence and three-dimensional structure but most are alike in that they carry a net positive charge (1), some exceptions being the bacteria-derived peptides gramicidin D and bacitracin, which are neutral (8). Gram-negative and Gram-positive bacteria differ in cell envelope architecture but they are alike in that their cell surface usually carries a net negative charge. Thus, the initial bacterium-peptide interaction is an electrostatic attraction between the positively-charged peptide and a negatively-charged cell surface (1). In Gram-positive bacteria, the acidic groups (e.g., phosphates) conferring this net negative charge reside in teichoic acids associated with the cell wall (24); whereas in Gram-negatives it is the lipopolysaccharide (LPS) in the outer membrane that bears the negative charge (Figure 2; 23). As one might expect, a resistance mechanism employed by bacteria is a reduction in the net negative charge of the cell surface. This reduction is effected by the substitution of negatively-charged groups in the cell envelope with neutral or positively-charged moieties.

The relatively high level of peptide resistance seen in *Staphylococcus spp.* can be abrogated by inactivation of genes in the *dltABCD* operon, which results in strains defective in the ability to esterify phosphate residues in teichoic acids with D-alanine (24). Compared to the wild-type, the *dlt* mutants bound more of the cationic antimicrobial peptide gallidermin and were 10- to 20-fold more susceptible to a variety of animal- and bacteria-derived antimicrobial peptides. This is the first demonstration of a direct connection between cell surface charge and peptide resistance in Gram-positive bacteria. Although one should exercise caution in drawing any generalizations in regard to this specific resistance mechanism, the presence of *dlt* homologs in other genera of Gram-positive bacteria (24) may reflect a mechanism of peptide resistance common to bacteria in a variety of habitats where they encounter antimicrobial peptides.

Gram-negative bacteria can also adjust their level of peptide resistance by modifying the net charge of their cell surface, i.e., the net negative charge residing in the LPS of the outer membrane. For instance, polymyxin resistance in *Proteus mirabilis* and *Salmonella typhimurium* is attributed to the substitution of phosphate groups in the LPS with 4-aminoarabinose and/or ethanolamine (23). Although reducing the net negative surface charge is recognized as important to peptide resistance, it is not the only mechanism available to the Gram-negative cell. The relative complexity of the outer membrane provides a basis for many different modifications that alter its permeability to antimicrobial peptides (see below). The outer membrane is a bilaminar structure composed largely of LPS and phospholipids (Figure 2; 25). Acidic phosphate groups are associated with the lipid A or inner core components of the LPS. Mg^{2+} serves to stabilize the LPS, and ultimately outer membrane integrity, by bridging adjacent LPS molecules through ionic association with phosphate residues on the lipid A core (23). Antimicrobial peptides disrupt outer membrane integrity by displacing Mg^{2+} from the LPS and inserting themselves into the bilayer effectively "loosening" the outer membrane such that it becomes permeable to the peptide or other compounds (e.g., detergents) that would normally be excluded from the cell interior (23). This mechanism by which antimicrobial peptides gain access to the cell interior is referred to as 'self-promoted uptake' (26).

Another mechanism employed by the cell to mitigate the self-promoted uptake of antimicrobial peptides is stabilization of the LPS in the bilayer such that the LPS is less prone to displacement by the peptide. Stabilization is achieved through the interaction of the polar O-antigen polysaccharide (21) and by the increased acylation of the lipid A component of the LPS, i.e., hexa-acyl to hepta-acyl lipid A conversion (Figure 3). Wild-type *Salmonella*

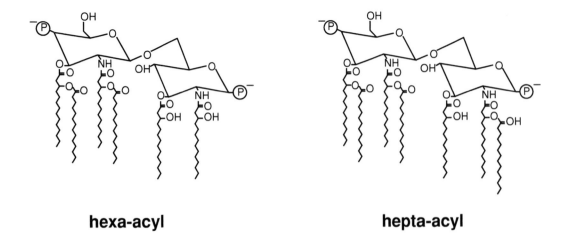

hexa-acyl   **hepta-acyl**

Figure 3. Structures of the hexa-acyl and hepta-acyl forms of the lipid A component of the *Salmonella typhimurium* outer membrane LPS. The figures are modeled on those presented in reference 27.

typhimurium that is capable of adding palmitate to the lipid A component is more resistant to a variety of antimicrobial peptides than is the *pagP* mutant which lacks the palmitoylation ability (27). The increased resistance attributed to the palmitoylated (hepta-acyl) form of lipid A is thought to occur by a decrease in fluidity of the outer membrane inhibiting insertion of the peptide (27). Modification of the acylated state of lipid A is not restricted to *Salmonella typhimurium* and may be a resistance mechanism common among Gram-negative bacteria as this phenomenon was also found to occur in *E. coli* and *Yersinia enterocolitica* (27).

The density of negative charges in the cell surface and the acylation of lipid A has been shown to be under regulatory control and can (presumably) be adjusted for optimal survival of the cell under a given set of environmental conditions. For example, growth under Mg^{2+}-limitation promotes the palmitoylation of lipid A in *Salmonella typhimurium* (27). And, Mg^{2+}-limitation or low pH results in the conjugation of 4-aminoarabinose and/or ethanolamine moieties to acidic phosphate groups in the LPS (28-30). These modifications effectively reduce the availability of peptide-binding sites and confer a level of peptide (i.e., polymyxin) resistance that may be orders of magnitude greater than that seen in cells that do not possess the 4-aminoarabinose or ethanolamine substitutions (18, 29).

The expression of genes encoding enzymes involved in those substitution reactions are under the control of the PhoP/PhoQ (Mg^{2+}-inducible) and PmrA/PmrB (pH-inducible) two-component regulatory systems (30, 31). These systems, and other two-component systems in general, are composed of an integral membrane sensor kinase that responds to an environmental signal or condition by phosphorylating/ dephosphorylating its cognate response regulator (32). In turn, the phosphorylated response regulator may effect transcriptional control by activating or repressing the expression of its target genes (30-32). In the PhoP/PhoQ regulatory system, the sensor kinase PhoQ responds to the extracellular Mg^{2+} concentration by phosphorylating PhoP when the Mg^{2+} concentration is low and dephosphorylating PhoP when the extracellular Mg^{2+} concentration is high (31). This regulatory system plays a key role in peptide resistance because *phoP* mutants of *S. typhimurium* are hypersusceptible to magainin 2, defensin NP-1, cecropin A and other antimicrobial peptides (33).

The importance of extracellular Mg^{2+} as an environmental signal controlling peptide resistance was demonstrated in *S. typhimurium*. When wild-type bacteria were grown with micromolar Mg^{2+} they were fully resistant to magainin 2, but bacteria grown under millimolar Mg^{2+} were nearly as susceptible as a *phoP* null mutant (31). When these results are considered with the observation that *phoP* mutants are also attenuated for virulence in mice, one might conclude that, (i) the cell compartment in which *S. typhimurium* resides during infection of the host must be a low Mg^{2+} environment, and (ii) these results strengthen the premise that antimicrobial peptides play an important role in immunity (2, 31, 33).

Two-component regulatory systems can interact in response to different environmental signals. For example, polymyxin resistance in *S. typhimurium* is regulated by pH and Mg^{2+} in a PmrA-dependent manner. PmrA-activated genes are transcriptionally induced by mild acidification (e.g., pH 5.8) and growth in micromolar concentrations of Mg^{2+}, whereas growth at pH 7.7 with millimolar concentrations of Mg^{2+} represses expression of PmrA-activated genes (30). However, transcription of PmrA-regulated genes occurs at pH 7.7 when cells are grown in micromolar concentrations of Mg^{2+} because of a PhoP-regulated activation of PmrA (30). The net effect of the interaction between the PhoP/PhoQ and PmrA/PmrB regulatory systems is that wild-type *S. typhimurium* is susceptible to polymyxin when grown under millimolar concentrations of Mg^{2+} at pH 7.7 but is resistant when grown under millimolar Mg^{2+} at pH 5.8 (18). Hence, through two-component regulatory systems a bacterium is able to sense its environment and regulate the expression of genes in a manner that would enhance its chances of survival in that environment.

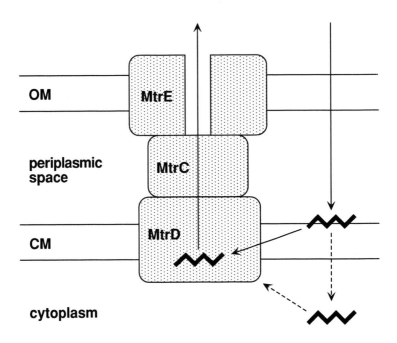

Figure 4. A representation of the MtrC-MtrD-MtrE efflux pump in the Gram-negative bacterium *Neisseria gonorrhoeae*. CM, cytoplasmic membrane; OM, outer membrane. The MtrC-MtrD-MtrE efflux pump can presumably translocate antimicrobial peptides out of the cytoplasmic membrane (and/or the cytoplasm). The peptidoglycan cell wall has been omitted for clarity. The figure is modeled on that presented in reference 22.

A decreased envelope permeability towards peptides is an obviously important mechanism of resistance but by itself may not be sufficient to ensure survival of the cell. Depending on the extracellular concentration some amount of peptide would be expected to reach the cell interior. The probable function of the permeability barrier is to slow the entry of the offending substance to a point that can be tolerated by intracellular resistance mechanisms. Thus, the physiological damage caused by a sublethal concentration of peptide would have to be repaired and the existing peptide molecules removed to prevent recurring injury. Effective removal of the peptide may involve degradation by proteases (see below) or translocation away from the target site by active efflux. Efflux pumps with broad substrate specificity are now recognized to play a role in elevating resistance to classical antibiotics and other toxic compounds including detergents and basic dyes (22). These energy-dependent pumps are multicomponent systems that remove toxins from the cytoplasmic membrane and/or cytoplasm and translocate them to the extracellular environment, e.g., beyond the outer membrane in the case of Gram-negatives (Figure 4). These pumps are widely distributed in nature and are found not only in Gram-negative and Gram-positive bacteria but in some fungi and animal cells as well (22). In spite of the prevalence of these systems in bacteria there is but one documented case showing the involvement of an efflux pump in peptide resistance.

In the Gram-negative human pathogen *Neisseria gonorrhoeae* the MtrC-MtrD-MtrE efflux pump confers resistance to protegrin-1, a cyclic peptide from porcine leukocytes, and to the human α-helical peptide LL-37; *mtr* mutants were about 10-fold more susceptible to these peptides than the isogenic wild-type strain (34). Moreover, mutations in the gene encoding MtrR, a transcriptional repressor that regulates expression of the *mtrCDE* operon, caused enhanced expression of the efflux system resulting in elevated resistance to protegrin-1. As with the modulation of surface charge, this is yet another example of the transcriptional control of genes encoding peptide resistance determinants and further illustrates the potential of bacteria to evolve hyperresistance to antimicrobial peptides in analogy to that seen in the appearance of multidrug resistant strains.

Efflux systems may not be the only mechanism of resistance involving peptide transport. Protamine resistance in *S. typhimurium* is mediated by SapABCDF, a transport system of the ATP-binding cassette (ABC) family that shows homology to oligopeptide transporters in Gram-negative and Gram-positive bacteria (35). The SapABCDF system may confer resistance to protamine by transporting the peptide into the cytoplasm—away from the putative membrane target—where it could either be degraded by proteases or initiate a regulatory cascade resulting in the activation of resistance determinants (35). A Sap-like system has also been implicated in peptide resistance in the plant pathogen *Erwinia chrysanthemi*. The *E. chrysanthemi* SapABCDF system is 71% identical to its *Salmonella* homolog and is required for wild-type levels of resistance to the plant-derived antimicrobial peptides alpha-thionin and snakin-1 (36). The level of identity between these two Sap systems may reflect an evolved specificity for the transport of the different kinds of antimicrobial peptides these bacteria may encounter in their respective ecological niches.

Inactivation of the Peptide

Among the repertoire of antibiotic resistance mechanisms employed by bacteria is the enzymatic inactivation of the antibiotic. The transfer and covalent attachment of a phospho group to aminoglycosides by the *aph* gene product and hydrolysis of penicillins by β-lactamases are prevalent examples of drug inactivation (14). Antimicrobial peptide inactivation by group transfer has not been demonstrated to exist in bacteria and may or may not be a realistic means of controlling peptide activity. Derivatization of an antibiotic inactivates the drug by blocking a stereospecific interaction between the antibiotic and its protein target (14). These derivatization reactions are usually carried out by enzymes in the cytoplasm of the cell. Since antimicrobial peptides seem to exert their activity against the cytoplasmic membrane, it is difficult to conceive that this type of chemical modification would be effective in neutralizing these agents. However, the recognition that some antimicrobial peptides may have targets in the cytoplasmic compartment invokes the possible existence of this type of resistance mechanism (12).

In contrast to these speculations, bacterial cells harbor numerous proteases with differing substrate specificity, and inactivation by proteolysis is a likely means of protection against these toxic peptides. One might expect the greatest degree of protection to be provided by proteases that are localized to the outer membrane, the periplasm, or the periplasmic side of the cytoplasmic membrane where they could hydrolyze peptides before they could do irreparable damage to the cytoplasmic membrane. Or, in a more preemptive fashion, proteases may be released into the extracellular milieu where they could inactivate peptides before any contact with the cell at all. For instance, culture supernatants of *Pseudomonas aeruginosa* PAO1 contain proteolytic activity against protamine, magainin 2 and LL-37 (A. Aspedon and E. A. Groisman, unpublished data), and extracellular proteases have been implicated in the protection of anaerobic oral bacteria against peptides carried into the mouth via saliva and migrating leukocytes (37).

While proteases associated with the outer membrane of Gram-negative bacteria may be an important resistance factor, their relative contribution to resistance has yet to be universally established. Indeed, the outer membrane protease OmpT of *Escherichia coli* is the only demonstrated example of this type of resistance mechanism; OmpT plays a direct role in

A   **B**

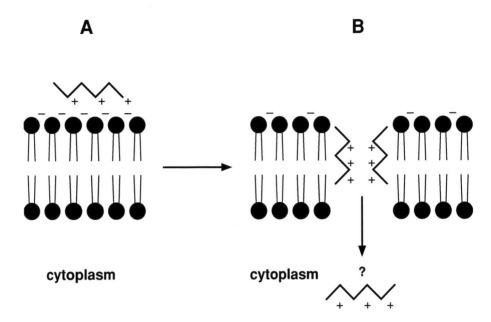

Figure 5. A representation of how cationic antimicrobial peptides may penetrate the cytoplasmic membrane of bacteria. Initially, there is an electrostatic attraction between the positively charged amino acids in the peptide and the negatively charged head groups of acidic membrane phospholipids. Subsequent to adsorption of peptide molecules to the membrane surface conformational changes in the peptide and/or electrophoretic forces imposed by the electrical membrane potential promotes the insertion of the peptide into the membrane. Some peptide molecules may enter the cytoplasm of the cell where they could interact with (bind to) negatively charged nucleic acids or other anionic compounds (12).

peptide resistance because wild-type cells were able to resume growth after treatment with the highly cationic peptide protamine whereas *ompT* mutants could not recover growth (38). It is instructive to note that optimal protamine resistance required both OmpT protease activity and the ability of cells to recover the cytoplasmic K^+ released into the extracellular medium upon exposure to the peptide (15, 38). This illustrates the point that peptide resistance involves many elements of cell physiology and indicates the necessity of adequate repair and recovery systems for optimal resistance.

In spite of the seemingly obvious role proteases might play in peptide resistance, there have been remarkably few studies implicating their activity. This may be due to the fact that antimicrobial peptides when present at a lethal concentration exert their killing effect too rapidly for proteases to have any obvious influence over cell survival. The relatively high concentrations of antimicrobial peptides used for *in vitro* studies may not be physiologically relevant in terms of what the bacterium would normally encounter in its natural setting. It may be that the true role of the protease is to protect the cell from chronically low, sublethal concentrations of the peptide, perhaps facilitating repair of incurred damage and vigilantly hydrolyzing peptides before damage ensues.

Modification of the Target

Some antimicrobial peptides are toxic to both bacterial and mammalian cells yet others are only active against bacterial cells (2, 8). This specificity for interaction with the cytoplasmic membrane of bacteria is thought to be largely attributable to the difference in lipid composition in the membranes of these two cell types (see also Lohner, this monograph). Eukaryotic cell membranes contain cholesterol and a large proportion of zwitterionic phospholipids (e.g., phosphatidylcholine) whereas bacterial membranes are rich in acidic phospholipids and do not contain cholesterol (1, 16, 39). Many antimicrobial peptides exert their toxicity by inserting into the membrane and forming channels, a process that is influenced by lipid/peptide interactions as well as the magnitude of the electrical membrane potential (1, 10, 16, 40). The first step in channel formation is an electrostatic attraction between the cationic peptide and the negatively-charged head groups of membrane phospholipids (Figure 5). Thus, a greater affinity of the cationic peptide for acidic phospholipids over the less negatively-charged zwitterionic lipids could be a factor in the specificity of these peptides for bacterial membranes (1, 16, 41). The presence of cholesterol may augment this specificity by interfering with the ability of the peptide to insert into the hydrophobic region of the membrane (42).

Another factor involved in channel formation is the electrical membrane potential, or $\Delta\psi$. The $\Delta\psi$ is the electrical charge difference across the cytoplasmic membrane and is oriented such that the interior (cytoplasm) of the cell carries a net negative charge. Channel formation by antimicrobial peptides has been shown to be voltage-dependent, indicating that peptide insertion into the membrane is an electrophoretic process (10, 40). It follows then that a greater $\Delta\psi$ value (more negative cell interior) would promote channel formation and increased cell killing. Indeed, bacteria in general are more susceptible to killing by antimicrobial peptides when tested under conditions where the magnitude of $\Delta\psi$ is expected to be high (5, 10). The observed influence of $\Delta\psi$ on peptide activity in bacteria has led some to propose that the decreased activity of certain antimicrobial peptides against mammalian cells could be partly due to a lower membrane potential in those cells (8, 41).

Modification of the cytoplasmic membrane into a state less susceptible to disruption by antimicrobial peptides may not be a viable option for most bacteria, although some possibilities do exist. A reduction in the acidic phospholipid content of the membrane could reduce the ability of peptides to bind the membrane, and ultimately disrupt its structure (16, 41). Indeed, the cytoplasmic membrane of *Proteus vulgaris,* which has a relatively high percentage of zwitterionic phosphatidylethanolamine and low amounts of the acidic phospholipids cardiolipin and phosphatidylglycerol, is less susceptible to permeabilization by magainins than the membranes of other Gram-negative bacteria that contain higher amounts of acidic phospholipids (16). Additional modifications could include adjusting the fatty acid content of the phospholipids and/or decreasing the $\Delta\psi$ component. In a manner analogous to that seen with palmitoylation of the outer membrane lipid A (27), an increased hydrophobic interaction of cytoplasmic membrane phospholipids, and reduced $\Delta\psi$, would serve to inhibit the ability of the peptide to insert into the membrane. However, the effect of these modifications may in themselves be detrimental to survival of the cell. Crucial metabolic processes are associated with the cytoplasmic membrane (e.g., energy transduction and nutrient transport) and the maintenance of appropriate membrane fluidity, surface charge and electrical membrane potential are required for optimal performance of these functions (39). Hence, physiological constraints may place strict limitations on modifications of this sort.

Concluding Remarks

The widespread distribution of antimicrobial peptides in nature and their broad spectrum of activity implies that resistance mechanisms in bacteria are more universal than that suggested by the limited studies published on this topic. Antimicrobial peptides may represent the next generation of antimicrobials to be used for the treatment of infectious disease. However, a fate similar to that seen with the classical antibiotics may follow the chemotherapeutic application of these peptides. The wide spread distribution of antimicrobial peptides in nature and a common theme in peptide activity suggests that a resistance mechanism evolving within a given microbe in a specific niche may have implications for medically relevant pathogens. Similar to that seen with the spread of antibiotic resistance, the acquisition of peptide-resistance genes from an organism unrelated to the recipient (i.e., via horizontal gene transfer) could conceivably be a means of acquiring peptide resistance (14). Additionally, the expression of genes encoding different peptide-resistance determinants is inducible and under regulatory control and may provide the potential for evolving multipeptide or peptide hyperresistant strains analogous to the multiantibiotic resistant forms arising today.

Acknowledgements

We thank Sella Garlich for preparation of the figures. Research in our laboratory on resistance to antimicrobial peptides is supported by grants GM54900 and AI42236 from the NIH and GROISM97Z0 from the Cystic Fibrosis Foundation. E. A. G. is an Associate Investigator of the Howard Hughes Medical Institute.

References

1. Hancock, R.E.W., Falla, T. and Brown, M. 1995. Cationic bactericidal peptides. Adv. Microbiol. Physiol. 37: 135-175.
2. Boman, H.G. 1995. Peptide antibiotics and their role in innate immunity. Annu. Rev. Immunol. 13: 61-92.
3. Huttner, K.M. and Bevins, C.L. 1999. Antimicrobial peptides as mediators of epithelial host defense. Pediatr. Res. 45: 785-794.
4. Baba, T. and Schneewind, O. 1998. Instruments of microbial warfare: bacteriocin synthesis, toxicity and immunity. Trends Microbiol. 6: 66-71.
5. Aspedon, A. and Groisman, E.A. 1996. The antibacterial action of protamine: evidence for disruption of cytoplasmic membrane energization in *Salmonella typhimurium*. Microbiol. 142: 3389-3397.
6. Westerhoff, H.V., Juretic, D., Hendler, R.W. and Zasloff, M. 1989. Magainins and the disruption of membrane-linked free-energy transduction. Proc. Natl. Acad. Sci. USA. 86: 6597-6601.
7. Skerlavaj, B., Romeo, D. and Gennaro, R. 1990. Rapid membrane permeabilization and inhibition of vital functions of Gram-negative bacteria by bactenecins. Infect. Immun. 58: 3724-3730.
8. Hancock, R.E.W. and Chapple, D.S. 1999. Peptide antibiotics. Antimicrobial Agents Chemother. 43: 1317-1323.
9. Spratt, B.G. 1994. Resistance to antibiotics mediated by target alterations. Science. 264: 388-393.
10. Cociancich, S., Ghazi, A., Hetru, C., Hoffmann, J.A. and Letellier, L. 1993. Insect defensin, in inducible antibacterial peptide, forms voltage-dependent channels in *Micrococcus luteus*. J. Biol. Chem. 268: 19239-19245.
11. Boman, H.G., Agerberth, B. and Boman, A. 1993. Mechanisms of action on *Escherichia coli* of cecropin P1 and PR-39, two antibacterial peptides from pig intestine. Infect. Immun. 61: 2978-2984.

12. Wu, M., Maier, E., Benz, R. and Hancock, R.E.W. 1999. Mechanism of interaction of different classes of cationic antimicrobial peptides with planar bilayers and with the cytoplasmic membrane of *Escherichia coli*. Biochem. 38: 7235-7242.

13. Groisman, E.A. 1994. How bacteria resist killing by host-defense peptides. Trends Microbiol. 2: 444-449.

14. Davies, J. 1994. Inactivation of antibiotics and the dissemination of resistance genes. Science. 264: 375-382.

15. Stumpe, S. and Bakker, E.P. 1997. Requirement of a large K^+-uptake capacity and of extracytoplasmic protease activity for protamine resistance of *Escherichia coli*. Arch. Microbiol. 167: 126-136.

16. Matsuzaki, K., Sugishita, K., Harada, M., Fujii, N. and Miyajima, K. 1997. Interactions of an antimicrobial peptide, magainin 2, with outer and inner membranes of Gram-negative bacteria. Biochim. Biophys. Acta. 1327: 119-130.

17. Ernst, R.K., Guina, T. and Miller, S.I. 1999. How intracellular bacteria survive: surface modifications that promote resistance to host innate immune responses. J. Infect. Dis. 179: S326-330.

18. Groisman, E.A., Kayser, J. and Soncini, F.C. 1997. Regulation of polymyxin resistance and adaptation to low-Mg^{2+} environments. J. Bacteriol. 179: 7040-7045.

19. Haynie, S.L., Crum, G.A. and Doele, B.A. 1995. Antimicrobial activities of amphiphilic peptides covalently bonded to a water-insoluble resin. Antimicrobial Agents Chemother. 39: 301-307.

20. Lehrer, R.I., Barton, A., Daher, K.A., Harwig, S.S.L., Ganz, T. and Selsted, M.E. 1989. Interaction of human defensins with *Escherichia coli*. J. Clin. Invest. 84: 553-561.

21. Rana, F.R., Macias, E.A., Sultany, C.M., Modzrakowski, M.C. and Blazyk, J. 1991. Interactions between magainin 2 and *Salmonella typhimurium* outer membranes: effects of lipopolysaccharide structure. Biochem. 30: 5858-5866.

22. Nikaido, H. 1994. Prevention of drug access to bacterial targets: permeability barriers and active efflux. Science. 264: 382-388.

23. Vaara, M. 1992. Agents that increase the permeability of the outer membrane. Microbiol. Rev. 56: 395-411.

24. Peschel, A., Otto, M., Jack, R.W., Kalbacher, H., Jung, G. and Götz, F. 1999. Inactivation of the *dlt* operon in *Staphylococcus aureus* confers sensitivity to defensins, protegrins, and other antimicrobial peptides. J. Biol. Chem. 274: 8405-8410.

25. Nikaido, H. and Vaara, M. 1985. Molecular basis of bacterial outer membrane permeability. Microbiol. Rev. 49: 1-32.

26. Hancock, R.E.W. 1984. Alterations in outer membrane permeability. Annu. Rev. Microbiol. 38: 237-264.

27. Guo, L., Lim, K.B., Poduje, C.M., Daniel, M., Gunn, J.S., Hackett, M. and Miller, S.I. 1998. Lipid A acylation and bacterial resistance against vertebrate antimicrobial peptides. Cell. 95: 189-198.

28. Guo, L., Lim, K.B., Gunn, J.S., Bainbridge, B., Darveau, R.P., Hackett, M. and Miller, S.I. 1997. Regulation of lipid A modifications by *Salmonella typhimurium* virulence genes *phoP-phoQ*. Science. 276: 250-253.

29. Gunn, J.S., Lim, K.B., Krueger, J., Kim, K., Guo, L., Hackett, M. and Miller, S.I. 1998. PmrA-PmrB-regulated genes necessary for 4-aminoarabinose lipid A modification and polymyxin resistance. Mol. Microbiol. 27: 1171-1182.

30. Soncini, F.C. and Groisman, E.A. 1996. Two-component regulatory systems can interact to process multiple environmental signals. J. Bacteriol. 178: 6796-6801.

31. García-Véscovi, E., Soncini, F.C. and Groisman, E.A. 1996. Mg^{2+} as an extracellular signal: environmental regulation of *Salmonella* virulence. Cell. 84: 165-174.

32. Ninfa, A.J. 1996. Regulation of gene transcription by extracellular stimuli. In: *Escherichia coli* and *Salmonella*: cellular and molecular biology. F. C. Neidhardt, R. Curtiss III, J. L. Ingraham, E. C. C. Lin, K. B. Low, B. Magasanik, W. S. Reznikoff, M. Riley, M. Schaechter, and H. E. Umbarger, eds. American Society for Microbiology, Washington, D.C. p. 1246-1262.

33. Groisman, E.A., Parra-Lopez, C., Salcedo, M., Lipps, C.J. and Heffron, F. 1992. Resistance to host antimicrobial peptides is necessary for *Salmonella* virulence. Proc. Natl. Acad. Sci. USA. 89: 11939-11943.

34. Shafer, W.M., Qu, X.-D., Waring, A.J. and Lehrer, R.I. 1998. Modulation of *Neisseria gonorrhoeae* susceptibility to vertebrate antibacterial peptides due to a member of the resistance/nodulation/division efflux pump family. Proc. Natl. Acad. Sci. USA. 95: 1829-1833.

35. Parra-Lopez, C., Baer, M.T. and Groisman, E.A. 1993. Molecular genetic analysis of a locus required for resistance to antimicrobial peptides in *Salmonella typhimurium*. EMBO J. 12: 4053-4062.

36. Lopez-Solanilla, E., Garcia-Olmedo, F. and Rodriguez-Palenzuela, P. 1999. Inactivation of the sapA to sapF locus of *Erwinia chrysanthemi* reveals common features in plant and animal bacterial pathogenesis. Plant Cell. 10: 917-924.

37. Devine, D.A., Marsh, P.D., Percival, R.S., Rangarajan, M. and Curtis, M.A. 1999. Modulation of antibacterial peptide activity by products of *Porphyromonas gingivalis* and *Prevotella* spp. Microbiol. 145: 965-971.

38. Stumpe, S., Schmid, R., Stephens, D.L., Georgiou, G. and Bakker, E.P. 1998. Identification of OmpT as the protease that hydrolyzes the antimicrobial peptide protamine before it enters growing cells of *Escherichia coli*. J. Bacteriol. 180: 4002-4006.

39. Kadner, R.J. 1996. Cytoplasmic membrane. In: *Escherichia coli* and *Salmonella*: Cellular and Molecular Biology. F. C. Neidhardt, R. Curtiss III, J. L. Ingraham, E. C. C. Lin, K. B. Low, B. Magasanik, W. S. Reznikoff, M. Riley, M. Schaechter, and H. E. Umbarger, eds. American Society for Microbiology, Washington, D. C. p. 58-87.

40. Kagan, B.L., Selsted, M.E., Ganz, T. and Lehrer, R.I. 1990. Antimicrobial defensin peptides form voltage-dependent ion-permeable channels in planar lipid bilayer membranes. Proc. Natl. Acad. Sci. USA. 87: 210-214.

41. Matsuzaki, K., Sugishita, K., Fujii, N. and Miyajima, K. 1995. Molecular basis for membrane selectivity of an antimicrobial peptide, magainin 2. Biochem. 34: 3423-3429.

42. Tytler, E.M., Anantharamaiah, G.M., Walker, D.E., Mishra, V.K., Palgunachari, M.N. and Segrest, J.P. 1995. Molecular basis for prokaryotic specificity of magainin-induced lysis. Biochem. 34: 4393-4401

From: *Development of Novel Antimicrobial Agents: Emerging Strategies*
ISBN 1-898486-23-9 © 2001 Horizon Scientific Press, Wymondham, UK.

4

Genetic Engineering of Novel Macrolide Antibiotics

Robert McDaniel and Leonard Katz

Abstract

Macrolides, such as erythromycin and tylosin, are a class of antibiotics that are used widely in human and veterinary medicine belonging to the family of natural products known as polyketides. These compounds are synthesized by large multienzyme complexes, polyketide synthases (PKSs), which contain a program at the genetic level encoding the structure of the compound. Recent advances in our understanding of PKSs as well as the development of molecular biology tools for the genetic engineering of PKSs has brought forth a number of strategies for creating novel polyketides by gene manipulation. Much of this development has been performed with the PKS that produces erythromycin and is reviewed here. Since the potential for creating structural diversity with PKSs is great, it is hoped that next-generation macrolides can be developed which overcome the increasing problem of resistant pathogens.

Introduction

The widespread use of antibiotics in human medicine and as supplements in animal feed has given rise to rapid spread of bacterial resistance (1). Of particular importance is the increase in the isolation of methicillin-resistant *Staphylococcus aureus* (MRSA), which, in 1992, was detected in >20% of hospitals surveyed in the United States and which also carried resistance to quinolones, macrolides, aminoglycosides and tetracyclines (2). Equally alarming is the world-wide rise of vancomycin-resistant enterococci (VRE), penicillin-resistant *Streptococcus pneumoniae* and multidrug-resistant *Mycobacterium tuberculosis* (3). The sudden and uncontrolled spread of resistance has underscored the need to discover novel antibacterial agents, for which there is no pre-existing resistance, or for the development of more potent analogs of existing compounds which can kill otherwise resistant pathogenic bacterial strains.

One class of antibiotics commonly used for the treatment of Gram-positive infections are the macrolides, exemplified by the compounds erythromycin (**1**) and its semi-synthetic analogs clarithromycin (**2**) and azithromycin (**3**) (Figure 1). These compounds act by binding to the 50S ribosomal subunit and stimulate dissociation of peptidyl-tRNA from the ribosome during the translocation process (4). Although binding of erythromycin to free 23S ribosomal RNA has not been demonstrated, resistance studies indicate that the drug interacts with the 2058-2062 nt region of the RNA. MLS (macrolide-linosamide-streptrogramin B) resistant

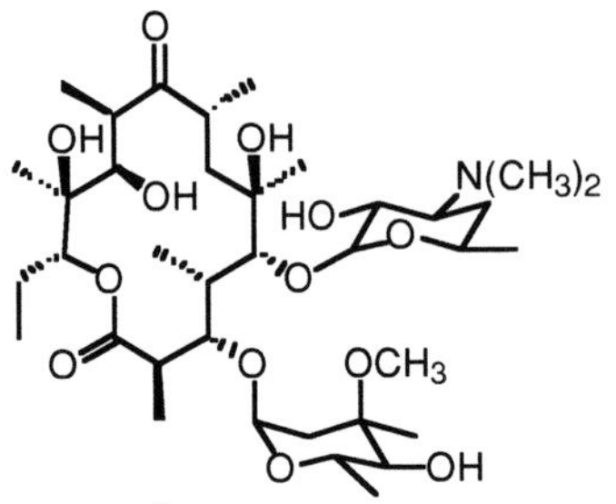

Figure 1. Structures of clinically relevant macrolides erythromycin (**1**), clarithromycin (**2**), azithromycin (**3**), and HMR 3647 (**4**).

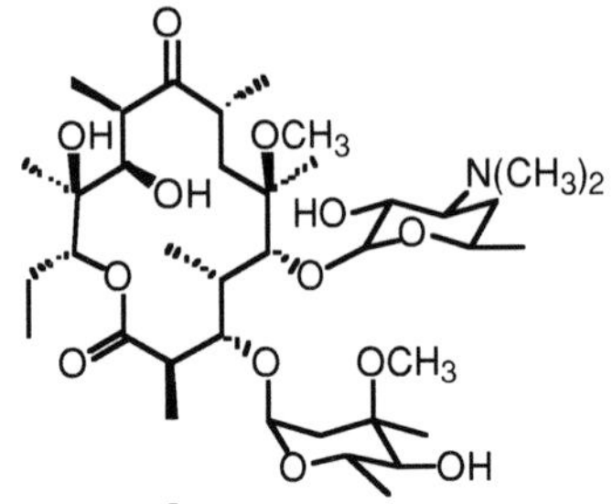

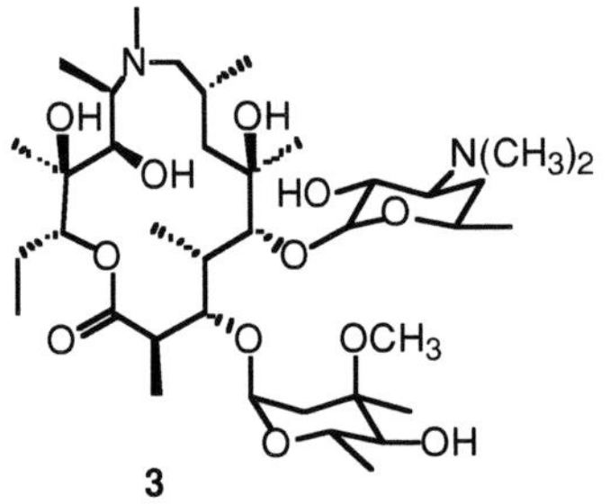

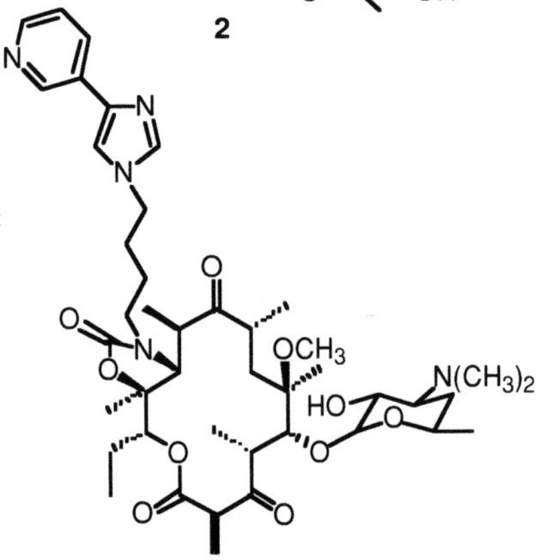

strains of *Staphylococcus, Streptococcus, Enterococcus*, etc. have been shown to carry a gene termed *erm*, acquired through plasmid-mediated or conjugal transfer, which encodes a methyltransferase that methylates the NH_2- group on the adenosine residue at position 2058 of the 23S rRNA conserved in all bacterial species. Mono- or dimethylated ribosomes exhibit decreased binding affinity for erythromycin and its derivatives (5). To date, more than 30 *erm* genes have been uncovered including those that are present in a number of actinomycete hosts that produce macrolide antibiotics - *ermE* in the erythromycin-producer *Saccharopolyspora erythraea* and *ermSF* in the tylosin producer *Streptomyces fradiae* (6,7). A second, clinically-important means of conferring resistance is the acquisition of a gene termed *mef* which encodes a membrane-localized protein that binds macrolides and assists in transporting them from the cell before they can accumulate in sufficient quantity to bind and inhibit their target (8). A number of *mef* genes have been reported in staphylococci and streptococci.

In addition to the continued search for novel macrolides from natural sources, two other approaches are currently being employed to generate novel macrolides with enhanced antibacterial properties. The first is chemical modification which has successfully produced the drugs clarithromycin and azithromycin, in clinical use since the early 1990s, and recently, the ketolide class, exemplified by the compound HMR-3647 (**4**), now in clinical development. These semisynthetic compounds are derived from erythromycin A produced by *Saccharopolyspora erythraea*. Although chemical modification has been and continues to be an important component of the discovery process for erythromycin and other antibiotics, the number of sites on these molecules that can be modified is limited. Since chemical modification has been ongoing since the discovery of erythromycin in 1952, at various

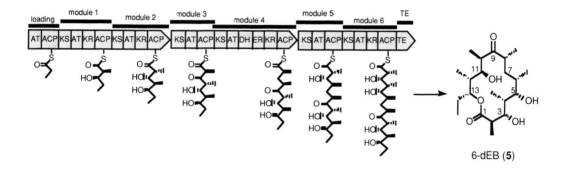

Figure 2. Modular organization of the deoxyerythronolide B synthase (DEBS). Structures attached as thioesters to the ACP domains of each module are shown to illustrate the pathway of synthesis. Completion of the synthesis produces the polyketide precursor of erythromycin, 6-deoxyerythronolide B (6-dEB, **5**). Abbreviations: KS – β-ketoacyl ACP synthase (ketosynthase); AT – acyltransferase; DH – dehydratase; ER – enoylreductase; KR – ketoreductase; KR° – inactive ketoreductase; ACP – acyl carrier protein; TE – thioesterase.

pharmaceutical companies, it is possible that the end point in production of new compounds employing this method may soon be reached.

A new approach, employing genetic manipulation to generate chemical diversity of erythromycin and other macrolides has been recently developed by several laboratories. This has been made possible by the cloning and sequencing of the genes involved in the biosynthesis of modular polyketides, a class of natural products to which macrolides belong. Access to these genes has led to an understanding of how the structures of these compounds are assembled. Along with the development of genetic engineering techniques for antibiotic-producing microorganisms, it has enabled the manipulation of genes for the generation of novel chemical entities. This chapter summarizes the progress made in this approach to novel chemical diversity.

Macrolide Biosynthesis and Polyketide Synthases

Polyketides are formed through cycles of condensation and β-keto modification of simple carboxylic acids in a manner similar to the synthesis of fatty acids (9). The enzymes that carry out these biochemical steps are called polyketide synthases (PKSs) to which the starter, extender and all acyl intermediates are tethered via thioester linkages. For most marolides, synthesis is initiated by a Claisen condensation between a starter carboxylic and an extender dicarboxylic acid (Figure 1). For erythromycin, the starter is propionyl-CoA and the extender is 2-methylmalonyl-CoA. Condensation is driven by decarboxylation of the extender unit followed by transfer of the starter residue to the extender unit resulting in the formation of a diketide. In both polyketide and fatty acid synthesis, the next step is reduction of the β-carbonyl of the diketide, which occurs after every condensation step in the latter but is not obligatory in polyketide synthesis. The first event is reduction of the β-carbonyl to an alcohol, followed by dehydration to yield an alkene, followed by enoylreduction to result in the fully saturated methylene group. At any given step in polyketide synthesis, all, none, or some of these reductive steps take place resulting in the appearance of the ketone, hydroxyl, alkene or methylene functionality, respectively at the site of the β-carbon. As will be described below, only one functionality is seen at the corresponding site in the final molecule. Once the reduction cycle is complete, the diketide undergoes another round of condensation, followed by reduction until the final (pre-determined) sized molecule is built. At the end of the condensation cycles, the molecule is released from the polyketide synthase through the action of a thioesterase, which can also enable the molecule to cyclize. The variation in the extent of reduction is best illustrated by examining the biosynthesis of erythromycin (Figure 2). The polyketide backbone is built from the condensation of a propionate starter and 6 successive methylmalonate extenders. The keto group at C-9 results from the lack of ketoreduction after the third condensation in synthesis of the backbone. The hydroxyls at C-3, 5, 11 and 13 result from ketoreduction after the first, second, fifth and sixth condensations. The methylene at C-7 results from a full cycle of ketoreduction, dehydration and enoylreduction after the fourth condensation. This pattern of events is programmed into the polyketide synthase that controls the condensation and reduction events. The hydroxyls at C-6 and 12 and the glycosidations at C-3 and C-5 are the results of post polyketide synthesis events.

The complexity of the biochemical events for the syntheses of polyketides such as erythromycin underscores the requirement for the programming of enzymatic machinery (i.e. the PKSs) to produce the correct structure. The PKSs that produce macrocyclic compounds, such as macrolides, rapamycin, or rifamycin, are organized in modules, with each module containing the activities corresponding to a ketosynthase (KS), acyltransferase (AT) and acyl carrier protein (ACP) as discrete functional domains within a multifunctional polypeptide (10-13). The AT domain selects and binds the correct extender, for example, in the case of the erythromycin PKS, a methylmalonyl CoA from the intracellular pools of the various CoA carboxylates and transfers it to the phosphopantotheine side chain of the ACP as a thioester. The KS domain contains a cysteine residue at the active center to which is transferred either a starter chain (if it is the first module) or a growing polyketide chain which is passed from the ACP of the loading domain or the previous module. Once the KS and ACP are loaded with their corresponding units, the KS catalyzes the Claisen condensation described above. Condensation results in the release of the polyketide chain from the KS domain and transfer to the β-carbon of the decarboxylated extender unit sitting on the ACP, resulting in the increase in length of the polyketide chain by two carbon units. If the extender unit is methylmalonate, the third carbon appears as a methyl side chain at the β-carbon. If the extender is malonate, the side chain is a proton.

Each PKS module also contains the domains that are necessary for any reductive activities to the β-keto group formed from the condensation. While the polyketide chain is attached to the ACP, the ketoreductase domain (KR) reduces the ketone to the hydroxyl function, the dehydratase domain (DH) eliminates water from the α,β carbon centers leaving a double bond and the enoylreductase domain (ER) reduces the ene function to yield the β-methylene group. In addition to the KS, AT and ACP domains, a PKS module may have no additional functional domains, only a KR, a KR and a DH or the full set of reduction domains KR, DH and ER. The order of the domains in the module, KS-AT-DH-ER-KR-ACP is found in all PKSs described. The presence of the domains in the module, therefore, will determine which extender is incorporated into the growing polyketide chain and the extent of reduction of the β-carbon formed at the given condensation. For macrolides, a thioesterase (TE) domain is usually found in the last module for chain termination and cyclization.

The first example of a modular PKS is deoxyerythronolide B synthase (DEBS) (10,11), the erythromycin PKS (Figure 2), named for the compound it synthesizes. It consists of six modules which are organized into three polypeptides and arranged in the order in which they are used in the synthesis of the polyketide chain. Upstream of module 1 is the loading domain which contains AT and ACP domains. At the end of module 6 is a TE which releases the heptaketide bound intermediate and lactonizes it to form 6-deoxyerythronolide B (**5**). 6-dEB is formed from a propionyl-CoA starter unit and six methylmalonyl extender units. The stereochemistry of the hydroxyl side chains is precisely controlled by the KR domain, but it is not yet known how the methyl stereocenters are determined by the PKS. It is known that, despite the fact that both the 2*R* and 2*S* enantiomer of methylmalonate appear to be present in the final molecule, only (2*S*)-methylmalonyl CoA was found to bind to all the AT domains. The general correspondence between polyketide structure, number of modules, predicted AT specificity, and presence of the correct ß-keto processing domains is maintained for all the other modular PKSs examined, although some variations occur.

Tools for Manipulating Polyketide Synthases

Before reviewing the genetic engineering strategies that have been used to create novel macrolides, it is worth discussing the advances in polyketide molecular biology that have allowed this field to progress so rapidly. Foremost is the ability to clone macrolide and other polyketide gene clusters from the producing organisms. This is simplified by the general observation that all the genes for biosynthesis (including those for post-PKS modification) and self-resistance are clustered within the genomes of actinomycetes. Thus, once a portion of the gene cluster is identified, usually by homology to another PKS or by complementation in mutant strains, the entire gene cluster can be cloned by chromosomal walking. Examples of some macrolide gene clusters in addition to erythromycin that have been identified include oleandomycin, tylosin, and picromycin/methymycin (14-17).

The second development is the ability to perform genetic manipulation in *Streptomyces* and related organisms. This is important because functional expression of PKSs in *E. coli* or other traditional hosts has proven extremely difficult. General techniques for transforming *Streptomyces* species have been developed and numerous plasmids, phages, and regulatory elements for gene expression are available (18). If the natural producing host is used, PKS genes can be engineered through a two-step homologous recombination procedure in the chromosome. This method has been used extensively in the erythromycin producing host, *Saccharopolyspora erythraea* (10,19). However, this process is relatively slow, labor intensive and often limited by the inability to introduce foreign DNA into the desired strain. In order to circumvent these difficulties, heterologous expression systems for PKSs have been developed which allow PKS genes to be efficiently manipulated in *E. coli* before introduction into a more "transformation friendly" host such as *Streptomyces coelicolor* or *Streptomyces lividans* for production of the polyketide (20,21). Both DEBS and the picromycin/methymycin PKS have been functionally expressed in this manner (22,23), allowing rapid construction of mutated and hybrid PKSs which produce novel macrolide aglycones. However, these organisms currently lack the genes for hydroxylation, glycosidation and methylation necessary to convert the aglycones into active macrolides. One solution that has been employed with analogs of 6-dEB is bioconversion of aglycones with a mutant of *S. erythraea* in which the PKS is disrupted, but all the downstream enzymes remain functional (24).

Genetic Strategies for Manipulating Macrolide PKSs

The discovery that the genetic architecture of the erythromycin PKS is essentially a linear template for the structure of the polyketide backbone opened doors to the possibility of manipulating macrolides and other natural products by genetic engineering. The one-to-one correspondence between the active sites of the PKS and each biosynthetic step of the pathway

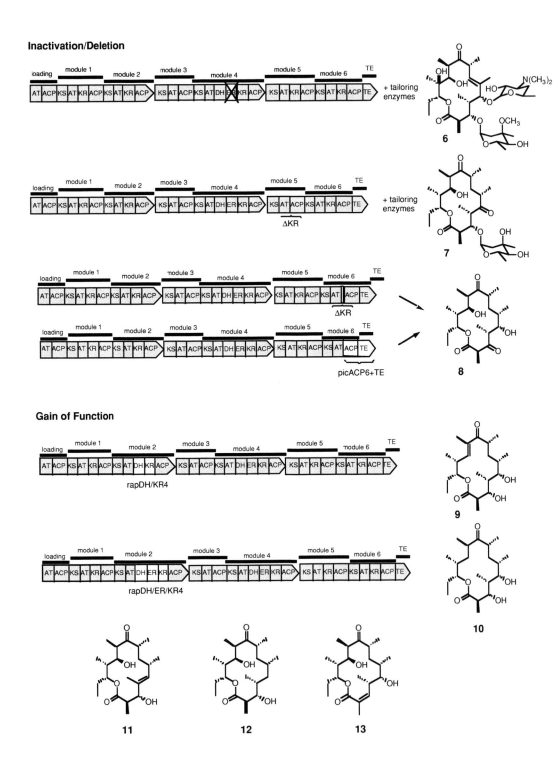

Figure 3. Modification of macrolides by controlling reduction cycle during biosynthesis. Examples of erythromycin or 6-dEB analogs that have been produced by inactivation, deletion, or substitution (gain-of-function) of the β-keto processing domains of DEBS are shown. Grey shading indicates DEBS, white shading indicates picromycin (pic) or rapamycin (rap) PKS domains.

suggests that any polyketide structure can be precisely controlled by engineering individual or groups of catalytic domains of the PKS. A number of different strategies have emerged since the discovery of the DEBS genes in 1990. Nearly all of the techniques have been developed using DEBS and include the removal, addition, substitution or site-specific mutagenesis of catalytic domains or modules to alter the alkyl side chains, oxidation state, macrolide ring size and even stereochemistry of 6-dEB.

At least two important criteria influence the success of PKS engineering. First, the PKS complexes must withstand any structural alterations that are introduced into the proteins themselves. Second, any altered intermediates that result from an earlier step genetic change must be processed by the downstream PKS domains. For example, a change incorporated in module 2 of DEBS results in an new intermediate for each of the remaining modules 3-6 and TE domain. It is unclear at this point which of these two are more critical considerations, however, the high degree of success thus far encountered with genetic manipulation indicates that PKSs are remarkably tolerant towards both.

Reduction Cycle

β-Keto processing domains, KR, DH and ER, offer perhaps the greatest degree of freedom for manipulating polyketide structures and are responsible for many of the functional groups (i.e. hydroxyls, carbonyls, and double bonds) present in macrolides. Since the activities encoded by these domains are optional for each module, they can be engineered by deletion, inactivation, replacement, or addition. In *S. erythraea*, for example, a mutation resulting in loss of function of the ER domain of module 4 led to production of **6** (19), while deletion of the KR domain in module 5 resulted in 7 (10) (Figure 3). Deletion of the KR activity in module 6 of DEBS has been accomplished by two different ways in the heterologous *S. lividans* system: by replacement of the KR domain with a short peptide linker (25), and by fusing the ACP and TE domains from the terminal module of the picromycin PKS downstream of the DEBS AT (23). Both PKSs produced the expected 3-keto 6-dEB analog (8). The former was the first attempt to produce a ketolide scaffold using DEBS. Ketolides are a new class of semisynthetic macrolides which are characterized by a ketone at carbon 3, and demonstrate excellent activity towards macrolide resistant pathogens (26). The second DEBS/Pic hybrid PKS represented an improvement over the initial construct since it eliminated some unwanted side products (23).

Introduction of β-keto processing domains, or gain of function mutagenesis, appears more challenging since the acyl chain intermediate must pass through a level of reduction that is the natural substrate for the downstream module. Thus, the additional reduction domain must compete for the same substrate as the downstream module. Gain of function was first demonstrated in a truncated trimodular DEBS system (27,28). The KR domain of module 2 was replaced with two cassettes from the rapamycin PKS (12), one encoding the KR and DH domains of module 4 (rapDH/KR4) and one encoding the KR, DH, and ER domains of module 1 (rapDH/ER/KR1). This led to compounds with a double bond or fully reduced methylene at the corresponding position. These same mutations were also introduced into the full DEBS system in *S. lividans* and resulted in 6-dEB analogs with a double bond at carbons C-10,11 (**9**), and a methylene at C-11 (**10**) (25). These cassettes were also used for gain of function mutagenesis in modules 5 and 6 of DEBS to produce compounds (**11-13**) (25).

Table 1. Summary of engineered loading domain and AT substitutions and polyketides produced.

PKS	Host	AT or LD substitution	Compounds
DEBS	S. lividans	AT1->rapAT2	13
DEBS	S. lividans	AT2->rapAT2	14
DEBS	S. lividans	AT3->rapAT2	15
DEBS	S. lividans	AT5->rapAT2	16
DEBS	S. lividans	AT6->rapAT2	17
DEBS	S. erythraea	AT1-> rapAT14, hygAT2, and venAT	18
DEBS	S. erythraea	AT2-> rapAT14, hygAT2, and venAT	19,20
DEBS	S. erythraea	AT4-> nidAT5	21
spiramycin	S. ambofaciens	LD->tylLD	22
DEBS	S. erythraea	LD->avrLD	23-28 + others

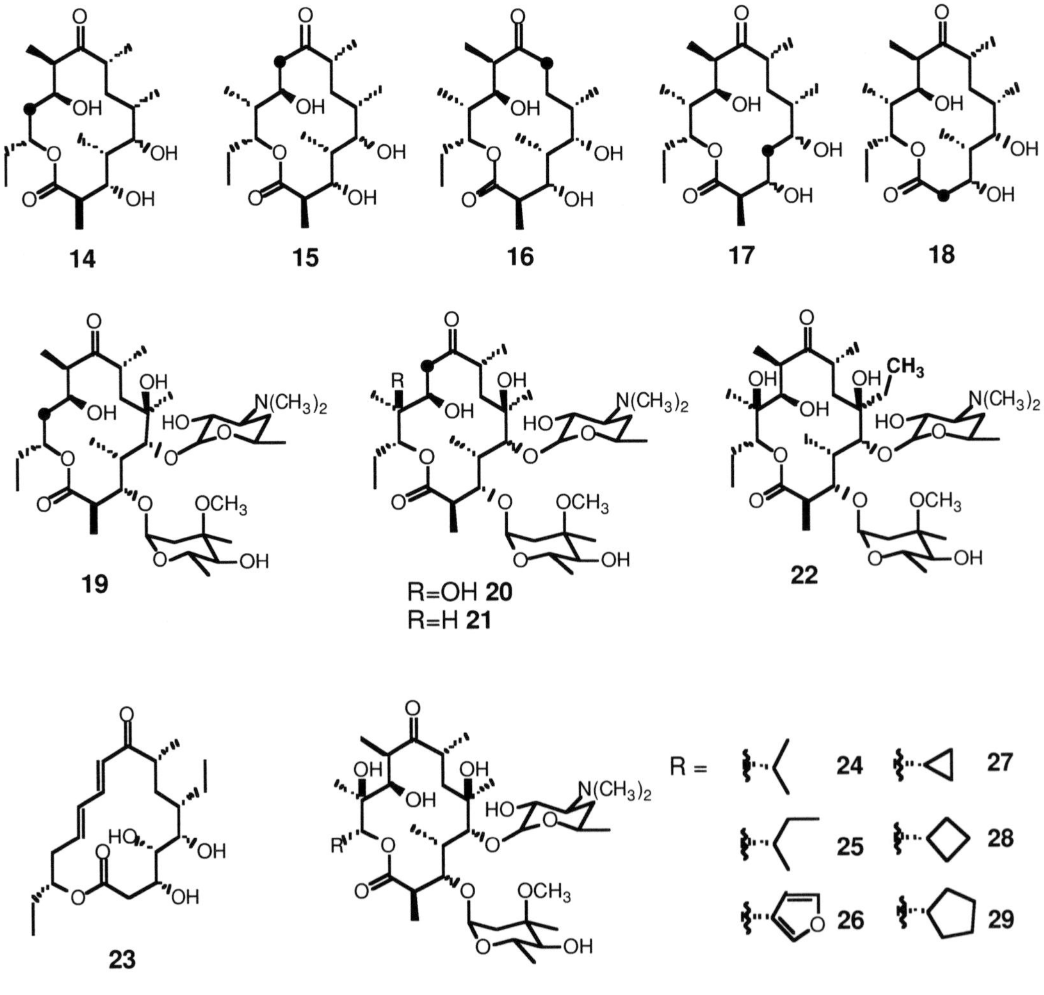

Figure 4. Engineering new side chains and starter units. These erythromycin, 6-dEB and platenolide analogs were generated from engineered DEBS or the spiramycin PKS in which either the loading domain or an AT domain was substituted. See Table 1 for a description of the genetic modifications and host organisms.

Side Chains and Starter Unit

Substitution of ATs and loading domains to generate alternative starter units and side chains on macrolides has been the subject of a number of studies (Table 1). There are multiple specificities encoded by the extender unit ATs of modular PKSs which can be used, including malonyl (no side chain at corresponding position), methylmalonyl (methyl side chain), ethylmalonyl (ethyl side chain), and potentially hydroxyl malonyl (hydroxyl side chain). In the *S. lividans*/DEBS system, exchanges of the methylmalonyl CoA-incorporating AT domains with the malonyl CoA-incorporating AT from module 2 of the rapamycin PKS, rapAT2, were performed in modules 1, 2, 3, 5 and 6 to produce desmethyl analogs of 6-dEB at carbons 2, 4, 8, 10, and 12 (Figure 4) (25,29). Side chains have also been engineered in *S. erythraea* to produce new erythromycin compounds. Three different malonyl AT domains, rapAT14 , hygAT2, and venAT, were used for exchange of AT1, each leading to the production of 12-desmethylerythromycin B (**19**) (30). Removal of the methyl group from C-12 by this substitution apparently rendered the compound inactive towards the hydroxylase that attaches the tertiary alcohol on C-12. Similarly, exchanges in module 2 of DEBS in *S. erythraea* with the same three AT domains led to the production of a mixture of 10-desmethylerythromycins A (**20**) and B (**21**) (30). Although removal of the methyl side chain at carbon C-6 of 6-dEB or erythromycin has not been reported, replacement with an ethyl group was accomplished by exchange of the AT domain in module 4 with the ethylmalonyl specific AT domain from module 5 of the niddamycin (nidAT5) PKS. In *S. erythraea* this substitution led to production of a low but detectable amount of 6-desmethyl-6-ethylerythromycin A (**22**) (31). In this strain, an increase in the supply of ethylmalonyl-CoA substrate was required and achieved by expressing a heterologous gene involved in butyryl-CoA metabolism.

Like extender units, nature utilizes a variety of starter units for polyketides but their degree of chemical diversity and complexity is much greater. Acetate and propionate units are commonly used to prime macrolide biosynthesis. However, more unusual starter units such as isobutyrate (for avermectin), 3,4-dihydroxycyclohexane carboxylate (rapamycin and FK506), and 3-amino-5-hydroxy benzoic acid (rifamycin) have been discovered in other macrocyclic polyketides. The PKSs from tylosin and spiramycin, which produce 16-membered macrolides, were used to demonstrate that starter units of polyketides could be controlled by substitution of entire loading domains (32). By exchanging the entire acetyl-CoA specific loading domain of the spiramycin PKS, which is comprised of a KS^Q decarboxylation domain, an AT, and an ACP, with the analogous propionate incorporating loading domain of the tylosin PKS, a new 16-membered macrolide aglycone, 16-methyl platenolide (**23**), was produced.

This approach was extended to incorporate branched-chain starter units using DEBS and the promiscuous loading domain from the avermectin PKS (33). When this substitution was constructed in *S. erythraea*, congeners of erythromycin with two new starter units, isobutyrate (**24**) and 2-methylbutyrate (**25**) were detected. The appeal of this recombinant PKS is further compounded by the knowledge that the avermectin loading domain has been used to generate over 40 starter unit analogs of avermectin by feeding carboxylic acid derivatives to a mutant *S. avermitilis* strain incapable of synthesizing the starter unit (34). One can imagine quickly increasing the number of erythromycin derivatives in an analogous manner. Subsequent studies with this strain have already led to several erythromycins derived from cycloalkyl and cycloaromatic starter units (**26-29**) (35).

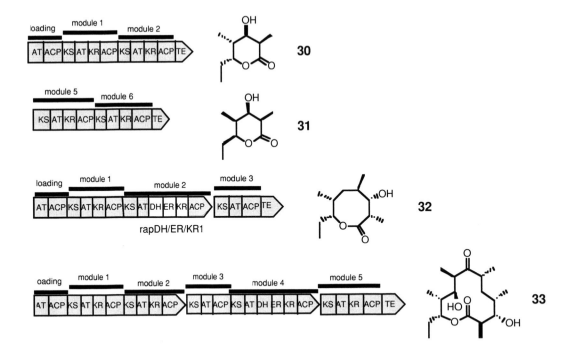

Figure 5. Controlling ring size. Deletion of DEBS modules has produced different lactone ring sizes, including an unusual 8-membered lactone.

Macrolide Ring Size

The chain length of macrolide polyketides is dictated by the number of modules present in the PKS. The 14-membered macrolides are derived from PKSs with 6 modules, 16-membered macrolides are synthesized from 7 modules and so forth. Since modules appear to function independently, the ring size of macrolides can potentially be controlled by deleting or inserting modules. Although the latter has not been demonstrated, module deletions have been performed with DEBS to generate smaller ring sizes (Figure 5). For example 6-membered lactones (**30**, **31**) have been generated using either the first two DEBS modules (36-38), or the last two DEBS modules (39). An 8-membered lactone (**32**) was produced from an engineered trimodular DEBS (28). Finally, deletion of the module 6 resulted in formation of a novel 12-membered macrolide (**33**) which is an analog of the methymycin aglycone (37).

Hybrid Pathways

The similarities among macrolide polyketide pathways suggests close evolutionary relationships among this class of PKS that may be advantageous for constructing hybrid pathways. For example, the aglycone of the 14-membered macrolide picromycin, narbonolide, differs from 6–dEB by the presence of a ketone at C-3, a double bond at C-10,11, and the lack of a methyl at C-10 (Figure 6). These differences in chemical structure are due to differences in the activities of only modules 2 and 6 of the PKSs. Recently, the entire picromycin PKS was functionally expressed using the same heterologous host/vector system as DEBS (23).

Experiments are currently underway to investigate the feasibility of generating novel polyketides through substitution or complementation of complete modules and subunits derived from related gene clusters but containing different activities. A recently reported hybrid DEBS in which all of module 2 was replaced with module 5 from the rifamycin PKS (13) to make high levels of 6-dEB suggests that this strategy should be successful (40).

Precursor Directed Biosynthesis

Precursor directed biosynthesis (Figure 7) is a powerful engineering approach which combines the advantages of modern synthetic chemistry with nature's complex polyketide biochemistry (24). It is based on the ability to engineer strains which are blocked at a particular step in polyketide biosynthesis or in the ability to produce a necessary precursor, and supply synthetically derived precursors which mimic intermediates but contain different substituents. A number of erythromycin analogs with different starter units were produced in this manner by introducing a mutation in the active site of the KS domain in module 1 of DEBS (KS1°). When this strain was fed N-acetylcysteamine thioesters of diketides, e.g. **34-37**, they were incorporated into module 2, and completely processed into 6-dEB or its analogs (**38-40**) (24,41). Since these diketides are apparently guided to module 2 by the functionality and stereochemical configuration of the alpha and beta carbons, the starter unit group or C-12 side chain could be precisely controlled by keeping this constant and varying the remaining portion of the molecule.

The ability to incorporate synthetic intermediates presents opportunities to generate molecules with diversity that surpasses what can be accomplished by genetic manipulation alone. In addition to supplying side chains or functional groups not encountered in nature, it can be used to introduce reactive functional groups that serve as chemical "handles" for post-PKS synthetic medicinal chemistry. The technique is not limited to engineering starter units, as intermediates can be incorporated into other modules. For example, feeding an unsaturated

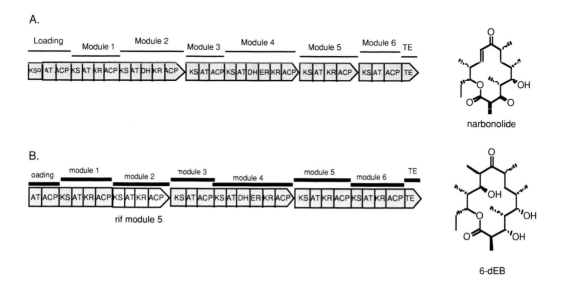

Figure 6. Hybrid pathways. A. The picromycin PKS has a modular organization similar to DEBS, but modules 2 and 6 encode slightly different activities. B. Module 5 from the rifamycin (rif) PKS encodes activities identical to module 2 of DEBS and leads to a functional hybrid PKS when substituted for DEBS module 2.

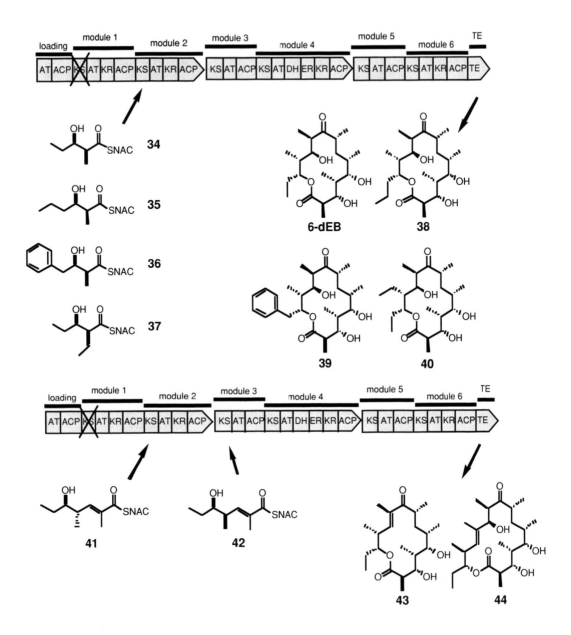

Figure 7. Precursor directed biosynthesis. A block in the 6-dEB pathway was created by a cys->ala mutation in the active site of the KS of module 1. Feeding diketides with different α and γ substitutions resulted in 6-dEB analogs in which the starter unit or side chain of carbon 10 was modified. The methyl stereochemistry of an unsaturated triketide dictated whether it was incorporated into module 2 or 3, leading to either a 14-membered or 16-membered macrolactone.

triketide (**42**) led to the corresponding 6-dEB analog with a double bond at C-10,11 (**43**). However, when the C-4 epimer was fed, **41**, it was unexpectedly incorporated into module 2 rather than module 3, resulting in production of a novel 16-membered macrolide similar to tylactone (**44**) (24). Finally, this method is an attractive means for constructing analogs in organisms for which genetic manipulation is difficult. Many derivatives of a single macrolide could be produced from a single engineered strain and a supply of different chemically synthesized precursors.

Combinatorial Biosynthesis: Prospects for Drug Discovery

Combinatorial chemistry is now a major component of most modern drug discovery programs. The examples reviewed above illustrate how genetic engineering can be used to control the chemistry of polyketides to generate novel macrolides. In light of this and the number of different polyketides that are present in nature, we can now consider the combinatorial manipulation of these enzymes for the generation of macrolide libraries and other "unnatural" natural product libraries.

An important milestone towards this achievement has been accomplished by constructing a combinatorial library of more than 100 6-dEB analogs (25). This library was constructed by engineering simultaneous changes in multiple modules of DEBS, employing a combination of the genetic engineering methods described above. DEBS constructs containing single, double, and triple domain substitutions produced corresponding macrolides with up to three modifications. Production of this library was significant because it demonstrated that multiple changes in a polyketide pathway can be affected by genetic manipulation. Thus, PKSs are tolerant enough to accept multiple heterologous domains and process substrates that have been significantly altered. Although the production levels of many of the engineered PKS were low, it demonstrated that the enzymes are capable of producing a large number of new macrolide aglycones that could serve as scaffolds for next-generation macrolide antibiotic development. Since these mutagenesis strategies appear to be complementary, the generation of new techniques should further increase the number of compounds that are accessible.

It should be emphasized here that genetic engineering of macrolides offers a complementary approach rather than a substitution for the traditional medicinal chemistry methods that have led to improved macrolides such as clarithromycin and azithromycin (42). Therefore, the design of novel macrolide scaffolds amenable to subsequent synthetic modification is likely to be of most utility in this field. Also, alterations in the polyketide macrolactone frequently render the new substrate inactive towards one or more of the glycosylation, hydroxylation or acylation enzymes which perform post-PKS tailoring (30,43). Since these reactions are usually required for optimal activity of the macrolide, synthetic derivatization, such as that demonstrated by the new ketolides, may be able to increase the potency in some cases.

Concluding Remarks

The ability to genetically manipulate polyketide structures presents new opportunities for drug engineers and could perhaps lead to a renaissance in development of macrolide antibiotics. Already a number of robust techniques have been described here which allow one to alter the structures of macrolides that would be extremely difficult by conventional means. As our understanding of PKSs increases and more advanced tools are developed for manipulation, it should be possible to further explore macrolide structure-activity and continue the improvements to this class of antibiotics.

References

1. Cohen, M.L. 1994. Antimicrobial resistance: prognosis for public health. Trends Microbiol. 2: 422-425.
2. Mylotte, J.M., Karuza, I. and Bentley, D.W. 1992. Methicillin-resistant *Staphylococcus aureus*: a questionnaire survey of 75 long-term care facilities in western New York. Infect. Control Hosp. Epidemiol. 13: 711-718.
3. Courvalin, P. 1990. Resistance of enterococci to glycopeptides. Antimicrobial Agents Chemother. 33: 2291-2296.
4. Brisson-Noel, A., Trieu-Cuot, P. and Courvalin, P. 1988. Mechanism of action of spiramycin and other macrolides. J. Antimicrobial Chemother. 22 (Supp. B): 13-33.
5. Weisblum, B. 1995. Erythromycin resistance by ribosome modification. Antimicrobial Agents Chemother. 39: 577-585.
6. Uchiyama, H. and Weisblum, B. 1985. N-Methyl transferase of *Streptomyces erythraeus* that confers resistance to the macrolide-lincosamide-streptogramin B antibiotics: amino acid sequence and its homology to cognate R-factor enzymes from pathogenic bacilli and cocci. Gene. 38: 103-110.
7. Kamimiya, S. and Weisblum, B. 1988. Translational attenuation control of *ermSF*, an inducible resistance determinant encoding rRNA N-methyltransferase from *Streptomyces fradiae*. J. Bacteriol. 170: 1800-1811.
8. Clancy, J., Petitpas, J., Dib-Hajj, F., Yuan, W., Cronan, M., Kamath, A.V., Bergeron, J. and Retsema, J.A. 1996. Molecular cloning and functional analysis of a novel macrolide-resistance determinant, *mefA*, from *Streptococcus pyogenes*. Mol. Microbiol. 22: 867-879.
9. O'Hagan, D. 1991. The polyketide metabolites, Ellis Horwood, Chichester, UK.
10. Donadio, S., Staver, M.J., McAlpine, J.B., Swanson, S.J. and Katz, L. 1991. Modular organization of genes required for complex polyketide biosynthesis. Science. 252: 675-679.
11. Cortés, J., Haydock, S.F., Roberts, G.A., Bevitt, D.J. and Leadlay, P.F. 1990. An unusually large multifunctional polypeptide in the erythromycin-producing polyketide synthase of *Saccharopolyspora erythraea*. Nature. 348: 176-178.
12. Schwecke, T., Aparicio, J.F., Molnar, I., Konig, A., Khaw, L.E., Haydock, S.F., Oliynyk, M., Caffrey, P., Cortes, J., Lester, J.B., Böhm, G.A., Staunton, J. and Leadlay, P.F. 1995. The biosynthetic gene cluster for the polyketide immunosuppressant rapamycin. Proc. Natl. Acad. Sci. USA. 92: 7839-7843.
13. August, P.R., Tang, L., Yoon, Y.J., Ning, S., Müller, R., Yu, T.-W., Taylor, M., Hoffmann, D., Kim, C.-G., Zhang, X., Hutchinson, C.R. and Floss, H.G. 1998. Biosynthesis of the ansamycin antibiotic rifamycin: deductions from the molecular analysis of the *rif* biosynthetic gene cluster *Amycolatopsis mediterranei* S699. Chem. Biol. 5: 69-79.
14. Swan, D.G., Rodriguez, A.M., Vilches, C., Mendez, C. and Salas, J.A. 1994. Characterisation of a *Streptomyces antibioticus* gene encoding a type I polyketide synthase which has an unusual coding sequence. Molec. Gen. Genet. 242: 358-362.
15. DeHoff, B.S., Sutton, K.L. and Rosteck, P.R. 1996. Sequence of *Streptomyces fradiae* tylactone synthase. GenBank accession #U78289.

16. Xue, Y., Zhao, L., Liu, H.-w. and Sherman, D.H. 1998. A gene cluster for the macrolide antibiotic biosynthesis in *Streptomyces venezuelae*: Architecture of metabolic diversity. Proc. Natl. Acad. Sci. USA. 95: 12111-12116.

17. Betlach, M.C., Kealey, J.T., Betlach, M.C., Ashley, G.A. and McDaniel, R. 1998. Characterization of the macrolide P-450 hydroxylase from *Streptomyces venezuelae* which converts narbomycin to picromycin. Biochemistry. 37: 14937-14942.

18. Hopwood, D.A., Bibb, M.J., Chater, K.F., Kieser, T., Bruton, C.J., Kieser, H.M., Lydiate, D.J., Smith, C.P., Ward, J.M. and Schrempf, H. 1985. Genetic Manipulation of Streptomyces: A Laboratory Manual, The John Innes Foundation, Norwich, UK.

19. Donadio, S., McAlpine, J.B., Sheldon, P.J., Jackson, M. and Katz, L. 1993. An erythromycin analog produced by reprogramming of polyketide synthesis. Proc. Natl. Acad. Sci. USA. 90: 7119-7123.

20. McDaniel, R., Ebert-Khosla, S., Hopwood, D. and Khosla, C. 1993. Engineered biosynthesis of novel polyketides. Science. 262: 1546-1557.

21. Ziermann, R. and Betlach, M.C. 1999. Recombinant polyketide synthesis in *Streptomyces*: Engineering of improved host strains. Biotechniques. 26: 106-110.

22. Kao, C.M., Katz, L. and Khosla, C. 1994. Engineered biosynthesis of a complete macrolactone in a heterologous host. Science. 265: 509-512.

23. Tang, L., Fu, H., Betlach, M.C. and McDaniel, R. 1999. Elucidating the mechanism of chain termination switching in the picromycin/methymycin polyketide synthase. Chem. Biol. 6: 553-558.

24. Jacobsen, J.R., Hutchinson, C.R., Cane, D.E. and Khosla, C. 1997. Precursor-directed biosynthesis of erythromycin analogs by an engineered polyketide synthase. Science. 277: 367-369.

25. McDaniel, R., Thamchaipenet, A., Gustafsson, C., Fu, H., Betlach, M., Betlach, M. and Ashley, G. 1999. Multiple genetic modifications of the erythromycin gene cluster to produce a library of novel "unnatural" natural products. Proc. Natl. Acad. Sci. USA. 96: 1846-1851.

26. Agouridas, C., Denis, A., Auger, J.-M., Benedetti, Y., Bonnefoy, A., Bretin, F., Chantot, J.-F., Dussarat, A., Fromentin, C., D'Ambrierès, S.G., Lachaud, S., Laurin, P., Le Martret, O., Loyau, V. and Tessot, N. 1998. Synthesis and antibacterial activity of ketolides (6-O-methyl-3-oxoerythromycin derivatives): A new class of antibacterials highly potent against macrolide-resistant and -susceptible respiratory pathogens. J. Med. Chem. 41: 4080-4100.

27. McDaniel, R., Kao, C.M., Fu, H., Hevezi, P., Gustafsson, C., Betlach, M., Ashley, G., Cane, D.E. and Khosla, C. 1997. Gain-of-function mutagenesis of a modular polyketide synthase. J. Am. Chem. Soc. 119: 4309-4310.

28. Kao, C.M., McPherson, M., McDaniel, R., Fu, H., Cane, D. and Khosla, C. 1997. Gain of function mutagenesis of the erythromycin polyketide synthase. 2. Engineered biosynthesis of an eight-membered ring tetraketide lactone. J. Am. Chem. Soc. 119: 11339-11340.

29. Liu, L., Thamchaipenet, A., Fu, H., Betlach, M. and Ashley, G. 1997. Biosynthesis of 2-nor-6-deoxyerythronolide B by rationally designed domain substitution. J. Am. Chem. Soc. 119: 10553-10554.

30. Ruan, X.R., Pereda, A., Stassi, D.L., Zeidner, D., Summers, R.G., Jackson, M., Shivakumar, A., Kakavas, S., Staver, M.J., Donadio, S. and Katz, L. 1997. Acyltransferase domain substitutions in erythromycin polyketide synthase yields novel erythromycin derivatives. J. Bacteriol. 179: 6416-6425.

31. Stassi, D.L., Kakavas, S.J., Reynolds, K.A., Gunawardana, G., Swanson, S., Zeidner, D., Jackson, M., Liu, H., Buko, A. and Katz, L. 1998. Ethyl-substituted erythromycin derivatives produced by directed metabolic engineering. Proc. Natl. Acad. Sci. USA. 95: 7305-7309.

32. Kuhstoss, S., Huber, M., Turner, J.R., Paschal, J.W. and Rao, R.N. 1996. Production of a novel polyketide through the construction of a hybrid polyketide synthase. Gene. 183: 231-236.

33. Marsden, A.F.A., Wilkinson, B., Cortés, J., Dunster, N.J., Staunton, J. and Leadlay, P.F. 1998. Engineering broader specificity into an antibiotic-producing polyketide synthase. Science. 279: 199-202.

34. Dutton, C.J., Gibson, S.P., Goudie, A.C., Holdom, K.S., Pacey, M.S., Ruddock, J.C., Bu'Lock, J.D. and Richards, M.K. 1991. Novel avermectins produced by mutational biosynthesis. J. Antibiot. 44: 357-365.

35. Pacey, M.S., Dirlam, J.P., Geldart, R.W., Leadlay, P.F., McArthur, H.A., McCormick, E.L., Monday, R.A., O'Connell, T.N., Staunton, J. and Winchester, T.J. 1998. Novel erythromycins from a recombinant *Sacharopolyspora erythraea* strain NRRL 238 pIG1. I. Fermentation, isolation and biological activity. J. Antibiot. 51: 1029-1034.

36. Kao, C.M., Luo, G., Katz, L., Cane, D.E. and Khosla, C. 1994. Engineered biosynthesis of a triketide lactone from an incomplete modular polyketide synthase. J. Am. Chem. Soc. 116: 11612-11613.

37. Kao, C.M., Luo, G., Katz, L., Cane, D.E. and Khosla, C. 1995. Manipulation of macrolide ring size by directed mutagenesis of a modular polyketide synthase. J. Am. Chem. Soc. 117: 9105-9106.

38. Cortés, J., Wiesmann, K.E.H., Roberts, G.A., Brown, M.J.B., Staunton, J. and Leadlay, P.F. 1995. Repositioning of a domain in a modular polyketide synthase to promote specific chain cleavage. Science. 268: 1487-1489.

39. Jacobsen, J., Cane, D.E. and Khosla, C. 1998. Spontaneous priming of a downstream module in 6-deoxyerythronolide B synthase leads to polyketide biosynthesis. Biochemistry. 37: 4928-4934.

40. Gokhale, R.S., Hunziker, D., Cane, D. and Khosla, C. 1999. Mechanism and specificity of the terminal thioesterase domain from the erythromycin polyketide synthase. Chem. Biol. 6: 117-125.

41. Jacobsen, J.R., Cane, D.E. and Khosla, C. 1998. Dissecting the evolutionary relationship between 14-membered and 16-membered macrolides. J. Am. Chem. Soc. 120: 9096-9097.

42. Chu, D.T.W. 1995. Recent developments in 14- and 15-membered macrolides. Exp. Opin. Invest. Drugs. 4: 65-94.

43. Andersen, J.F., Tatsuta, K., Gunji, H., Ishiyama, T. and Hutchinson, C.R. 1993. Substrate specificity of 6-deoxyerythronolide B hydroxylase, a bacterial cytochrome P450 of erythromycin A biosynthesis. Biochemistry. 32: 1905-1913.

From: *Development of Novel Antimicrobial Agents: Emerging Strategies*
ISBN 1-898486-23-9 © 2001 Horizon Scientific Press, Wymondham, UK.

5

Granulysin, A Novel Mediator of Antimicrobial Activity of Cytolytic T-cells

Steffen Stenger, Alan M. Krensky and Robert L. Modlin

Abstract

Cytolytic T-cells play an important role in immunity against many intracellular pathogens. One effector mechanism of cytolytic T-lymphocyts (CTL) is the lysis of infected target cells, thereby exposing the microbes to the hostile extracellular environment. Here, we characterize a novel effector mechanism of CTL which not only results in lysis of the target cell but simultaneously to the death of the intracellular invader. The lethal hit is delivered by the combined action of two components of cytolytic granules of CD8+ CTL. One is the well known lytic protein perforin; the other is the recently discovered antimicrobial protein granulysin. In this review we will summarize current knowledge about granulysin with special focus on its functional role as an antibacterial effector molecule of the adaptive immune response.

Introduction

Cytolytic T-lymphocyte (CTL) activity has been detected in multiple T-cell subsets, principally $CD8^+$ T-cells which recognize 9-10 amino acid peptide antigens presented by MHC class I molecules, but also $CD4^+$ T-cells which recognize 15 amino acid peptides presented by MHC class II molecules. TCR $\gamma\delta^+$ T-cells have also been shown to lyse bacteria-pulsed targets (1). Recently a novel subset of human CTL has been characterized, which recognizes mycobacterial lipid and lipoglycan antigens presented by the non polymorphic molecule CD1 (2). CTL were initially described as central effector cells in the immune response to viruses. Work over the past decade has expanded the spectrum of pathogens which can induce a CTL-response to include intracellular bacteria and parasites. For example, β2-microglobulin knockout mice ($\beta_2 m^{-/-}$) generally fail to transport MHC class I or class I-like molecules to the cell surface and therefore cannot generate $CD8^+$ T-cells, NK1.1 T-cells or CD1-restricted T-cells. Infection of β2m knockout mice with *Trypanosoma cruzi* (3), *Listeria monocytogenes* (4,5) or *Mycobacterium tuberculosis* (6) resulted in an exacerbated course of infection. These studies indicate a substantial role for CD8+ T-cells in complementing other components of the immune system to mount an efficient and long-lasting immune response to intracellular microbes. It should be pointed out, however, that a critical role for CD8+ T cells does not hold for all intracellular pathogens (for example *Leishmania major* infection) (7,8). Several effector functions by which CD8+ T-cells contribute to protective immunity have been suggested:

1) Secretion of interferon γ (IFN-γ), which is a major activator of macrophages.
2) Killing of bacteria in the extracellular environment by the secretion of a bactericidal factor (e.g. *Pseudomonas aeruginosa*, *Escherichia coli* and *Staphylococcus aureus* (9,10) or mechanisms yet to be defined (e.g. *Candida albicans* (11), *Cryptococcus neoformans* (12), *Schistosoma mansoni* (13), *Entamoeba histolytica* (14) or *Toxoplasma gondii* (15)).
3) CTL-mediated lysis of the target cells, resulting in the release of the pathogen into the extracellular environment. The microbe could then be taken up at low multiplicities of infection by freshly recruited, possibly lymphokine-activated macrophages which have the capacity to kill them (16).
4) CTL mediated lysis and simultaneous killing of the intracellular pathogen. This mechanism, initially described by DeLibero et al. in the context of killing of intracellular *Mycobacterium bovis BCG* (16) has recently been characterized in more detail and will be the focus of this chapter.

Two Functionally Distinct Subsets of Human CTL

In the course of studying the ability of CD1-restricted CTL to recognize and kill *M. tuberculosis*-infected target cells, we identified two distinct phenotypes of CTL associated with two different cytolytic mechanisms (17). Cytotoxicity by CD4-, CD8- double negative (DN) lines was markedly inhibited by blocking antibodies to Fas or to FasL. In contrast, cytotoxicity by CD8+ CTL lines was not affected by blocking of Fas or FasL, suggesting that they kill targets via the granule exocytosis pathway. The selective usage of these pathways by the two subsets was confirmed by the finding that DN, but not CD8+ CTL expressed mRNA for FasL. In contrast CD8+, but not DN, CTL expressed perforin mRNA and secreted granzyme A. Importantly, lysis induced by these phenotypically distinct subsets of human CTL also resulted in a different fate for the intracellular mycobacteria. Only lysis by CD8+ CTL utilizing the granule exocytosis pathway killed the bacteria efficiently, while DN CTL had no impact on the viability of the pathogen. The granule exocytosis pathway of cytotoxicity involves directed and regulated secretion of the lytic granule constituents into the infected target cell (18). To determine whether killing of the bacteria was mediated by one or more components of the cytolytic granules we performed experiments, in which CTL were degranulated by preincubation with strontium, a methodology previously shown to be useful for degranulating lymphocytes. Degranulation inhibited the lysis mediated by CD8+ but not DN T cells. Consequently, the antimycobacterial activity of CD8+ CTL was completely abrogated by preincubation of the cells with strontium (17). These experiments suggested that CTL that kill infected cells via the granule-exocytosis pathway release granular effector molecules with the capacity to directly kill the intracellular microbial pathogen. In order to uncover the molecular mechanism underlying our observations we considered the biology of cytotoxic granules and candidate molecules for antibacterial activity involved in the granule exocytosis pathway.

Granule Exocytosis Pathway

Each activated CTL contains in the order of 30-50 lytic granules in the cytoplasm which are the hallmark of specialized killer cells (19). Within five minutes after ligation of the T cell receptor the microtubule organising center and the Golgi complex reorient to face the bound target. Subsequently the lytic granules concentrate at the area of membrane contact. In this way the lymphocyte assumes polarity with respect to its target. As soon as this polarity has been established some of the granules fuse with the plasma membrane, releasing their contents into the space between between the killer cell and the target (20). Such regulated secretion within a restricted area of membrane contact ensures a high degree of target specificity during killing.

Several proteins which are expressed specifically in CTL and are injected into the target cell were of interest for our observation of CTL-mediated killing of *M. tuberculosis*. The most abundant and well known granular protein is perforin (21-23). Perforin is homologous, both structurally and functionally, to the membrane attack complex of the complement system (24). In its monomeric form perforin inserts into the targert membrane, where the monomers subsequently polymerize into pore structures. The perforin pores, ranging from 5 to 20 nm in internal diameter, are large enough to allow free passage of water and ions across the plasma membrane of the target cell (21). In contrast to the terminal complement cascade which induces unorganised destruction of cells, perforin-mediated lysis causes programmed cell death (apoptosis) of the target. A second group of molecules involved in lysis by the granule-exocytosis pathway are the granzymes (25). To date, seven mouse (granzymes A-G) and four human granzymes have been identified (26). All are highly homologous to each other and share the ability to act as serine esterases, whereby the esterases differ in their substrate specificity. Granzymes are important players in CTL-mediated lysis and induce apoptosis by activating the caspase cascade. A third molecule, termed granulysin, is a more recently described member of the family of granular proteins with lytic activity. Its distribution, structure and functional diversity have only recently been discovered and will therefor be discussed in greater detail.

Granulysin, A Novel Effector Molecule of Cytolytic T-cells

519 was discovered as a cDNA by subtractive hybridization searching for genes expressed by T-lymphocytes 3-5 days after antigen activation (27). A recombinant form of this protein was lytic against tumor cell targets (27). The protein was renamed granulysin to reflect its subcellular localization and lytic function. The granulysin gene was localized to chromosome 2 in a region containing other lymphocyte-specific genes, including CD8α (28). There are several naturally occuring variants of human granulysin, resulting from three differently spliced forms of mRNA and from proteolytic processing of the primary polypeptide products. The most abundant transcript, referred to as 520, encodes for a protein of 145 amino acids that contains a putative 15-residue signal sequence. The proteins encoded by the 519, 520 and 522 mRNAs were referred to as P519, P520, P522, respectively, and for simplicity, were collectively referred to as 519. After cleavage of the predicted hydrophobic leader sequence a protein of 15 kDa is produced. Antisera raised against either the entire 15 kDa protein recognized 15 and 9 kDa proteins in CTL lysates. Peptide using defined antisera showed that the 9 kDa protein is a processed product of the 15 kDa form by the cleavage at both the amino- and carboxy termini (29). The 9 kDa granulysin is basic (arginine-rich) and has an overall charge of +11. The experimentally determined molecular weights differed slightly

from theoretical calculations, suggesting that a small number of polymorphisms may be present in the gene coding for granulysin (29). Pulse chase experiments indicated that, while the 15kDa protein is produced rapidly and has a shorter half-life, the 9 kDa protein is produced more slowly and is relatively stable. The 9 kDa form is a processed product of the 520 cDNA (30). Similar to perforin and granzymes, granulysin is expressed late after T-cell activation. The protein is highly expressed about 3-5 days following activation by either alloantigen, mitogen or IL-2, which reflects the time frame during which T cells are capable of carrying out their defined functions (29). Granulysin mRNA has been detected in some lymphocyte tumor lines and NK cell lines. The gene appears not to be expressed in non-hematopoietic cell-types. Granulysin itself has lytic activity against tumor cell and immortalized targets, including YAC-1, K562, and JY, but does not lyse fresh human red blood cells (29).

Granulysin is a member of a larger group of proteins, referred to as saposin-like proteins (**s**phingolipid **a**ctivator **p**rotein **li**ke **p**rotein=SAPLIP). These proteins share a common predicted structure and conduct a variety of functions in association with lipids. Most members of this group interact with lipids and are small (6-10 kDa) proteins processed from larger precursors (31). The carboxy-terminal region of granulysin contains a SAPLIP domain. SAPLIP proteins share a common structure based primarily on the conserved positioning of cysteines and a series of hydrophobic residues. The predicted structure of family members, including granulysin, suggests the presence of four, relatively amphipathic helices held together as a bundle of disulfide bonds. Members of the SAPLIP family contain six half-cysteines that are involved in three intramolecular disulfide bonds. These bonds allow formation of an extremely stable structure that is both heat- and protease-resistant (25,32). Saposins play a role in the hydrolysis of sphingolipids by particular lysosomal enzymes, acting as a catalyst for the reaction by association with either the lipid, the enzyme, or both. Other members of the SAPLIP family include pulmonary surfactant protein B, lipid hydrolases (acyloxyacyl hydrolase, acid sphingomyelinase), sphingolipid hydrolase activators (saposin A,B,C,D) and most intriguingly several granular peptides from eukaryotic organisms:

1) Amoebapores (Amoebapores A, B, C) are a family of lytic peptides found within granules of the parasite *Entamoeba histolytica*. These small proteins, which are lytic against both eukaryotic and bacterial cells, act by polymerizing in the target cell membranes causing the formation of pores that lead to death of the target cell. Amoebapores kill a broad spectrum of bacterial prey (33).

2) *Caenorhabditis (C.) elegans* contains several distantly related members of the gene family of saposin-like proteins. The recombinant protein (T07C4.4) of *C. elegans* was found to have antibacterial activity against *E. coli* suggesting that these amoebapore homologues may play a role in antibacterial mechanisms of *C. elegans* (34).

3) The liver fluke *Fasciola hepatica* produces a protein that consists of only a single saposin-like domain and a signal peptide. The saposin-like domain of one representative protein, which is most closely related to NK-lysin and granulysin, has been expressed in *E. coli* and the recombinant protein was shown to have a circular dichroism spectrum consistent with the helix bundle structure characteristic of saposin-like domains (34).

4) NK-lysin is a tumorolytic protein that is present in porcine cytoplasmic granules of T and NK cells. NK-lysin has antibacterial activity against *Escherichia coli, Bacillus megaterium, Acinetobacter calcoaceticus* and *Streptococcus pyogenes* (35).

The region of granulysin coding for the 9 kDa form corresponds to the 9 kDa functional region of NK-lysin. These two 9 kDa proteins share 35% amino acid identity and 66%

similarity. Interestingly, the regions of NK-lysin and granulysin lacking the sequence coding for the 9 kDa proteins share an even higher identity of 59% and 70% similarity, which suggests an important role for this region in protein folding and processing (29).

Recent studies have begun to characterize the lytic activity of granulysin in more detail. It was demonstrated that treatment of Jurkat cells with granulysin induced cell shrinkage, chromatin condensation, nuclear fragmentation and phosphatidylserine translocation, all hallmarks of apoptosis (36). Programmed cell death was associated with a sixfold increase in the ceramide/sphingomyelin ratio, implicating the activation of sphingomyelinases characteristic for SAPLIP proteins. Saposins are not pore-forming proteins, but are proteins that interact with lipid membranes and activate lipid degrading enzymes (specifically, glucosylceramidases and sphingomyelinidases). The immediate consequence of the activation of these enzymes is the increase in intracellular ceramide content. As ceramide has been proposed as a mediator of apoptotic processes (37), this suggests that granulysin might induce cell lysis through activation of sphingomyelinases and an increase in cellular ceramide content. Studies with general inhibitors of caspase activity also implicate the caspase cascade in the mechanism of apoptosis induced by recombinant granulysin (36).

Antimicrobial Activity of Cytolytic T-cells Mediated by Granulysin

Based on the homology to T07.C4.4 (*C. elegans*) (34), amoebapores (33) and NK-lysin (35), all of which have antibacterial activity, we hypothesized that granulysin might be the effector molecule responsible for the antimycobacterial activity of CD8+ CTLs described above. Granulysin was detected in CD8+ CTL, colocalizing with perforin in cytotoxic granules, but was absent from DN CTL (38). Therefore the expression of granulysin by T-cell lines correlates with the ability to kill intracellular *M. tuberculosis*. Recombinant granulysin (38) had a wide range of antimicrobial activity. It showed a dose-dependent growth inhibition of Gram-positive and Gram-negative bacteria, causing a greater than thousand-fold reduction in CFU of *Salmonella typhimurium*, *L. monocytogenes*, *Escherichia coli*, and *Staphylococcus aureus*. Granulysin also killed fungi and parasites, including *Cryptococccus neoformans*, *Candida albicans* and *Leishmania major*. The broad antimicrobial spectrum of granulysin is reminiscent of defensins, which are nonspecifically released from cytoplasmic granules of polymorphonuclear leukocytes to kill phagocytized pathogens. However, the mode of action of granulysin is clearly distinct from defensins, as it is secreted upon antigen activation of CD8+ CTL and is therefore a weapon of the specific immune response.

M. tuberculosis is one of the most resistant pathogens to the antimicrobial mechanisms of the immune system. Nitrogen radicals are the only effector molecules among the armamentarium of mononuclear phagocytes that have been convincingly shown to be involved in protective immunity to *M. tuberculosis* (39-43). Even oxygen radicals, which are produced by macrophages upon activation with interferon γ and show antibacterial activity against a variety of intracellular pathogens, do not contribute to resistance against virulent strains of *M. tuberculosis*. Identifying antimycobacterial effector molecules of the specific cellular immune response therefore poses a great challenge to researchers and we were interested whether granulysin would exert such activity. Granulysin was very efficient in killing extracellular *M. tuberculosis*. However, no antibacterial activity was detected when recombinant granulysin was added to *M. tuberculosis*-infected cells. One explanation for the inability of granulysin to kill intracellular as compared to extracellular *M. tuberculosis* was the possible failure of the recombinant protein to gain access to the intracellular compartment in which mycobacteria reside. In fact, although granulysin lyses tumor targets, it possesses relatively weak lytic activity against *M. tuberculosis*-infected macrophages. To mimic the chain of events possibly occurring in vivo, infected macrophages were treated with a combination of granulysin and perforin, which colocalize in the cytotoxic granules of CD8+ T-cells and are released together upon antigen activation. Perforin alone, which forms pores in the membrane of various hematopoetic targets (44) showed no antibacterial activity, whereas the combination of both molecules resulted in a dramatic decrease in the viability of intracellular mycobacteria. These data indicated that granulysin could kill intracellular *M. tuberculosi* if perforin, or possibly other pore-forming molecules of T-cell granules, provided access to the intracellular compartment.

Scanning electron microscopy (SEM) revealed that granulysin induces discrete lesions and marked distortions in the bacterial surface topography of *M. tuberculosis*. Mycobacteria, singly and in clusters, were found to contain multiple small protrusions that were almost entirely absent from control samples. The reduction in viable CFU in conjunction with evidence of bacterial surface alteration indicate that granulysin is directly cytotoxic to *M. tuberculosis* and other bacteria. The ability of members of the saposin-like family of proteins (e.g. granulysin) to activate lipid degrading enzymes in lipophilic membranes might explain the ultrastructural lesions in the cell wall of *M. tuberculosis,* which is particularly rich in lipids.

These findings suggest a novel pathway by which antigen-specific T cells directly contribute to the death of microbial pathogens, specifically microorganisms residing in intracellular compartments. The role of CD8+ CTL in protection against intracellular pathogens might therefore not merely be to lyse target cells and disperse the intracellular pathogens, but in addition, to deliver granulysin, a lethal weapon in the arsenal of CTL which can reduce the viability of intracellular invaders. The presence of granulysin in NK cells and of related peptides in cytoplasmic granules of *Entamoeba histolytica* (the amoebapores), suggests that the saposin-like protein family represents an ancient yet highly conserved form of antimicrobial host defense, likely contributing to innate immune responses. The importance of this pathway is reflected by the presence of granulysin and other family members in CTL, indicating that the adaptive immune response has evolved to include antimicrobial peptides for effective immunity.

Concluding Remarks

There is a continuing need for new antimicrobial agents, particularly those that are effective at killing pathogens resistant to conventional antibiotics, e.g. *M. tuberculosis.* Granulysin is a protein involved in the natural pathways of resistance to microbial infection. The structure and activity of granulysin suggests that its action is unrelated to that of conventional antibiotics. It may therefore prove exceptionally useful as a clinical drug. The native forms of human granulysin provide a basis for further therapeutic development of the polypeptides having modified biological and chemical properties. Specifically, it would be desirable to design genetically engineered variants of granulysin, which lack cytolytic activity against eukaryotic cells while retaining the beneficial antibacterial effect agains pathogenic prokaryotic organisms.

References

1. Munk, M.E., Gatrill, A.J. and Kaufmann, S.H.E. 1990.Target cell lysis and IL-2 secretion by gamma/delta T lymphocytes after activation with bacteria. J. Immunol. 145:2434-2439.

2. Porcelli, S., Brenner, M.B., Greenstein, J.L., Balk, S.P., Terhorst, C. and Bleicher, P.A. 1989. Recognition of cluster of differentiation 1 antigens by human CD4-CD8-cytolytic T lymphocytes. Nature. 341:447-450.

3. Rottenberg, M.E., Bakhiet, M., Olsson, T., Kristensson, K., Mak, T., Wigzell, H. and Orn, A. 1993. Differential susceptibilities of mice genomically deleted of CD4 and CD8 to infections with *Trypanosoma cruzi* or *Trypanosoma brucei*. Infect. Immun. 61:5129-5133.

4. Gazzinelli, R.T., Hakim, F.T., Hieny, S., Shearer, G.M. and Sher, A. 1991. Synergistic role of CD4+ and CD8+ T lymphocytes in IFN-gamma production and protective immunity induced by an attenuated *Toxoplasma gondii* vaccine. J. Immunol. 146:286-292.

5. Ladel, C.H., Flesch, I.E., Arnoldi, J. and Kaufmann, S.H. 1994. Studies with MHC-deficient knock-out mice reveal impact of both MHC I- and MHC II-dependent T cell responses on *Listeria monocytogenes* infection [published erratum appears in J. Immunol. 1995 Apr 15;154(8):4223]. J. Immunol. 153:3116-3122.

6. Flynn, J.L., Goldstein, M.M., Triebold, K.J., Koller, B. and Bloom, B.R. 1992. Major histocompatibility complex class I-restricted T cells are required for resistance to *Mycobacterium tuberculosis* infection. Proc. Natl. Acad. Sci. USA. 89:12013-12017.

7. Wang, Z.-E., Reiner, S.L., Hatam, F., Heinzel, F.P., Bouvier, J., Turck, C.W. and Locksley, R.M. 1993.Targeted activation of CD8+ cells and infection of 2-microglobulin deficient mice fail to confirm a primary protective role for CD8+ cells in experimental leishmaniasis. J. Immunol. 151:2077-2086.

8. Huber, M., Timms, E., Mak, T.W., Rollinghoff, M. and Lohoff, M. 1998. Effective and long-lasting immunity against the parasite Leishmania major in CD8-deficient mice. Infect. Immun. 66:3968-3970.

9. Markham, R.B., Goellner, J. and Pier, G.B. 1984. *In vitro* T cell-mediated killing of *Pseudomonas aeruginosa*. I. Evidence that a lymphokine mediates killing. J. Immunol. 133:962-968.

10. Markham, R.B., Pier, G.B., Goellner, J.J. and Mizel, S.B. 1985. *In vitro* T cell-mediated killing of *Pseudomonas aeruginosa*. II. The role of macrophages and T cell subsets in T cell killing. J. Immunol. 134:4112-4117.

11. Beno, D.W., Stover, A.G. and Mathews, H.L. 1995. Growth inhibition of *Candida albicans* hyphae by CD8+ lymphocytes. J. Immunol. 154:5273-5281.

12. Levitz, S.M. and Dupont, M.P. 1993. Phenotypic and functional characterization of human lymphocytes activated by interleukin-2 to directly inhibit growth of *Cryptococcus neoformans in vitro*. J. Clin. Invest. 91:1490-1498.

13. Ellner, J.J., Olds, G.R., Lee, C.W., Kleinhenz, M.E. and Edmonds, K.L. 1982. Destruction of the multicellular parasite *Schistosoma mansoni* by T lymphocytes. J. Clin. Invest. 70:369-378.

14. Salata, R.A., Cox, J.G. and Ravdin, J.I. 1987. The interaction of human T-lymphocytes and *Entamoeba histolytica*. Parasite Immunol. 9:249-261.

15. Khan, I.A., Smith, K.A. and Kasper, L.H. 1988. Induction of antigen-specific parasiticidal cytotoxic T cell splenocytes by a major membrane protein (P30) of *Toxoplasma gondii*. J. Immunol. 141:3600-3605.

16. DeLibero, G., Flesch, I., Kaufmann, S.H.E. 1988. Mycobacteria-reactive Lyt-2+ T cell lines. Eur. J. Immunol. 18:59-66.

17. Stenger, S., Mazzaccaro, R.J., Uyemura, K., Cho, S., Barnes, P.F., Rosat, J.P., Sette, A., Brenner, M.B., Porcelli, S.A. , Bloom, B.R. and Modlin, R.L. 1997. Differential effects of cytolytic T cell subsets on intracellular infection. Science. 276:1684-1687.

18. Griffiths, G.M. 1997. Protein sorting and secretion during CTL killing. Sem. Immunol. 9:109-115.

19. Timonen, T., Ortaldo, J.R. and Herberman, R.B. 1981. Characteristics of human large granular lymphocytes and relationship to natural killer and K cells. J. Exp. Med. 153:569-582.

20. Yannelli, J.R., Sullivan, J.A., Mandell, G.L. and Engelhard, V.H. 1986. Reorientation and fusion of cytotoxic T lymphocyte granules after interaction with target cells as determined by high resolution cinemicrography. J. Immunol. 136:377-382.

21. Podack, E.R. and Dennert, G. 1983. Assembly of two types of tubules with putative cytolytic function by cloned natural killer cells. Nature. 302:442-445.

22. Masson, D. and Tschopp, J. 1985. Isolation of a lytic, pore-forming protein (perforin) from cytolytic T- lymphocytes. J. Biol. Chem. 260:9069-9072.

23. Henkart, P.A., Millard, P.J., Reynolds, C.W. and Henkart, M.P. 1984. Cytolytic activity of purified cytoplasmic granules from cytotoxic rat large granular lymphocyte tumors. J. Exp. Med. 160:75-93.

24. Young, J.D., Cohn, Z.A. and Podack, E.R. 1986. The ninth component of complement and the pore-forming protein (perforin 1) from cytotoxic T cells: structural, immunological and functional similarities. Science. 233:184-190.

25. Peters, P.J., Borst, J., Oorschot, V., Fukuda, M., Krahenbuhl, O., Tschopp, J., Slot, J.W. and Geuze, H.J. 1991. Cytotoxic T lymphocyte granules are secretory lysosomes, containing both perforin and granzymes. J. Exp. Med. 173:1099-1109.

26. Smyth, M.J., O'Connor, M.D. and Trapani, J.A. 1996. Granzymes: a variety of serine protease specificities encoded by genetically distinct subfamilies. J. Leukoc. Biol. 60:555-562.

27. Jongstra, J., Schall, T.J., Dyer, B.J., Clayberger, C., Jorgensen, J., Davis, M.M. and Krensky, A.M. 1987. The isolation and sequence of a novel gene from a human functional T cell line. J. Exp. Med. 165:601-614.

28. Donlon, T.A., Krensky, A.M. and Clayberger, C. 1990. Localization of the human T lymphocyte activation gene 519 (D2S69E) to chromosome 2p12——q11. Cytogenet. Cell Genet. 53:230-231.

29. Pena, S.V., Hanson, D.A., Carr, B.A., Goralski, T.J. and Krensky, A.M. 1997. Processing, subcellular localization and function of 519 (granulysin), a human late T cell activation molecule with homology to small, lytic, granule proteins. J. Immunol. 158:2680-2688.

30. Hanson, D.A., Kaspar, A.A., Poulain, F.R. and Krensky, A.M. 1999. Biosynthesis of granulysin, a novel cytolytic molecule. Mol. Immunol. 36 (7): 413-422.

31. O'Brien, J.S. and Kishimoto, Y. 1991. Saposin proteins: structure, function and role in human lysosomal storage disorders. FASEB J. 5:301-308.

32. Andersson, M., Curstedt, T., Jornvall, H. and Johansson, J. 1995. An amphipathic helical motif common to tumourolytic polypeptide NK- lysin and pulmonary surfactant polypeptide SP-B. FEBS Lett. 362:328-332.

33. Leippe, M., Tannich, E., Nickel, R., van der Goot, G., Pattus, F., Horstmann, R.D. and Muller-Eberhard, H.J. 1992. Primary and secondary structure of the pore-forming peptide of pathogenic *Entamoeba histolytica*. EMBO J. 11:3501-3506.

34. Banyai, L. and Patthy, L. 1998. Amoebapores homologs of *Caenorhabditis elegans*. Biochem. Biophys. Acta. 1429: 259-264.

35. Andersson, M., Gunne, H., Agerberth, B., Boman, A., Bergman, T., Sillard, R., Jornvall, H., Mutt, V., Olsson, B., Wigzell, H. 1995. NK-lysin, a novel effector peptide of cytotoxic T and NK cells. Structure and cDNA cloning of the porcine form, induction by interleukin 2, antibacterial and antitumour activity. EMBO J. 14:1615-1625.

36. Gamen, S., Hanson, D.A., Kaspar, A., Naval, J., Krensky, A.M. and Anel, A. 1998. Granulysin-induced apoptosis. I. Involvement of at least two distinct pathways. J. Immunol. 161:1758-1764.

37. Obeid, L., Linardic, C., Karolak, L. and Hannun, Y. 1993. Programmed cell death induced by ceramide. Science. 259:1769-1771.

38. Stenger, S., Hanson, D.A., Teitlebaum, R., Dewan, P., Niazi, K.R., Froelich, C.J., Ganz, T., Thoma-Uszynski, S., Melian, A., Bogdan, C., Porcelli, S.A, Bloom, B.R., Krensky, A.M. and Modlin, R.L. 1998. An antimicrobial activity of cytolytic T cells mediated by granulysin. Science. 282:121-125.

39. Denis, M. 1991. Interferon-gamma-treated murine macrophages inhibit growth of tubercle bacilli via the generation of reactive nitrogen intermediates. Cell Immunol. 132:150-157.

40. Chan, J., Xing, Y., Magliozzo, R.S. and Bloom, B.R. 1992. Killing of virulent *M. tubeculosis* by the reactive nitrogen intermediates produced by activated murine macrophages. J. Exp. Med. 175:1111-1122.

41. Nicholson, S., da Gloria Bonecini-Almeida, M., Lapa e Silva, J.R., Nathan, C., Xie, Q.-W., Mumford, R., Weidner, J.R., Calaycay, J., Geng, J., Boechat, N., Linhares, C., Rom, W. and Ho, J.L. 1996. Inducible nitric oxide synthase in pulmonary alveolar macrophages from patients with tuberculosis. J. Exp. Med. 183:2293-2302.

42. MacMicking, J.D., North, R.J., LaCourse, R., Mudgett, J.S., Shah, S.K. and Nathan, C.F. 1997. Identification of nitric oxide synthase as a protective locus against tuberculosis. Proc. Natl. Acad. Sci. USA. 94:5243-5248.

43. Bonecini-Almeida, M.G., Chitale, S., Boutsikakis, I., Geng, J., Doo, H., He, S. and Ho, J.L. 1998. Induction of *in vitro* human macrophage anti-*Mycobacterium tuberculosis* activity: requirement for IFN-gamma and primed lymphocytes. J. Immunol. 160:4490-4499.

44. Duke, R.C., Persechini, P.M., Chang, S., Liu, C.C., Cohen, J.J. and Young, J.D. 1989. Purified perforin induces target cell lysis but not DNA fragmentation. J. Exp. Med. 170:1451-1456.

From: *Development of Novel Antimicrobial Agents: Emerging Strategies*
ISBN 1-898486-23-9 © 2001 Horizon Scientific Press, Wymondham, UK.

6

The Immunostimulatory Properties of Bacterial DNA

David S. Pisetsky

Abstract

Depending on base sequence, DNA can cause powerful immunostimulation and serve as a "danger signal" to activate host defense. In bacterial DNA, immunostimulation results from sequences of 6 bases that occur much more commonly in prokaryotic than eukaryotic DNA. These sequences, known as CpG motifs or immunostimulatory sequences (ISS), center on an unmethylated CpG dinucleotide and lead to the activation of B cells and macrophages as well as the production of cytokines such as IL-12, TNF-α and IFN-α/β. These cytokines can lead to Th1 cell predominance. In addition to CpG motifs, other DNA sequences such as runs of dG have immunostimulatory activity. The properties of DNA as an immunomodulator are also influenced by backbone chemistry since phosphorothioate oligonucleotides can cause potent immune stimulation. The immunostimulatory properties of DNA allow the design of novel therapeutic agents that can augment host defense during infection as well as influence the balance of TH1/TH2 responses.

Introduction

DNA is a complex macromolecule whose immunological properties depend on base sequence. Although mammalian DNA is immunologically inert, bacterial DNA exerts a broad range of activities that could modulate host defense. These activities affect multiple cell types and, in their totality, resemble those of endotoxin or lipopolysaccharide. Because of its capacity as an immune stimulator, bacterial DNA may serve as a "danger signal" to activate host defense during infection and promote the generation of protective immune responses (1-2).

The immunostimulatory properties of DNA are relevant not only to events during infection but to emerging strategies for vaccination and immunotherapy. Indeed, synthetic DNA molecules represent powerful agents that could serve as adjuvants as well as immunomodulators to shift the balance of cellular immune responses. This shift may be beneficial during certain infectious diseases as well as autoimmune and inflammatory conditions where disturbances in Th1/Th2 cell numbers and function may be pathogenic. In this review, the immunological properties of bacterial DNA will be reviewed as they pertain to host defense as well as new nucleic acid-based therapies. As this review will indicate, DNA has potent and unexpected immune properties and, as such, may play a pivotal role in the regulation of humoral and cellular responses.

Immunostimulatory Properties of Bacterial DNA

The immunostimulatory properties of DNA were discovered initially during investigation of the anti-tumor properties of an extract of *Mycobacterium bovis* BCG organisms. This extract, called MY-1, could promote rejection of tumors transplanted into mice. By biochemical fractionation, the active component of this extract was DNA, with DNase but not RNase treatment leading to its elimination. As shown by both *in vivo* as well as *in vitro* experiments, the inhibitory properties of the extract resulted from induction of NK cell activity rather than any direct cytotoxic effect on tumor cells. The induction of NK cell activity in turn depended on stimulation of interferon (IFN) production (3, 4).

As subsequently demonstrated in *in vitro* experiments, the enhancement of NK cell activity as well as IFN production was not confined to mycobacterial DNA. Rather, DNA from many bacterial species showed immunostimulatory capacity (5). In contrast, DNA from mammalian sources was void of activity. Since prior experiments had shown that a variety of polyanions could stimulate immune responses, the activity of bacterial, but not mammalian DNA, indicated the immune stimulation was not simply a function of a charged backbone (6). Rather, immune stimulation required some features unique to prokaryotic DNA.

In a remarkable set of experiments, the immunostimulatory component of bacterial DNA was identified by analysis of the *in vitro* activity of DNA of defined sequences. Using sequences derived from cloned mycobacterial DNA, Tokunaga *et al.* (7) showed that immune stimulation resulted from a sequence motif that centers on an unmethylated CpG dinucleotide. This motif has the general structure of two 5' purines, an unmethylated CpG sequence, and two 3' pyrimidines; depending on the bases that flank the CpG motif, this 6 base structure can form a palindrome (8). This sequence is now known as either a CpG motif or an immunostimulatory sequence (ISS).

Because of marked quantitative differences in the content of CpG motifs in bacterial and mammalian DNA, this sequence can serve as signal of foreignness. CpG motifs occur much more commonly in bacterial DNA than mammalian DNA for at least two reasons. The first concerns DNA methylation. Thus, in mammalian DNA, cytosine is commonly methylated presumably as a mechanism for regulating gene transcription in differentiated cells. The second reason concerns a phenomenon known as CpG suppression. In mammalian DNA, the CpG doublet occurs much less commonly than predicted by the overall base composition of DNA (9, 10). While the basis for structural differences in eukaryotic and prokaryotic DNA are not known, the combination of CpG suppression and cytosine methylation provides a motif that, in code-like fashion, can signal the presence of foreign nucleic acids (2).

As shown using either natural DNA or synthetic oligonucleotides containing an ISS, the immune activity of bacterial DNA encompasses B cells, T cells as well as macrophages and dendritic cells. Thus, foreign DNA can stimulate the production of IFN-α/β, IFN-γ, IL-12, IL-6, TNF-α as well as chemokines. In addition, bacterial DNA is mitogenic for B cells, causing proliferation and antibody production; the effects of foreign DNA on T cells, however, reflect the influence of cytokines such as IFN-α/β rather than a direct stimulatory effect (11-14). While bacterial DNA can stimulate B cell activation and cytokine production in both human and mouse, it is mitogenic only for murine B cells. Human B cells do not directly respond to bacterial DNA *in vitro*, although they may respond to other nucleic acids (15).

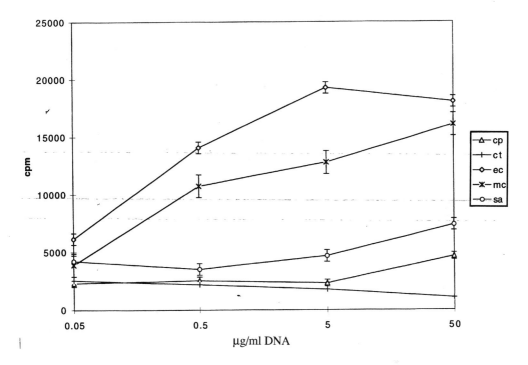

Figure 1. Stimulation of mitogenesis by bacterial DNA. Murine spleen cells were stimulated by various doses of bacterial DNA and mitogenesis assessed 48 hours later by thymidine incorporation. The following DNA were tested: cp (*Clostridium perfringens*); ec (*E. coli*); mc (*Micrococcus lysodeikticus*); sa (*Staphyloccus aureus*); and ct (calf thymus). Reproduced with permission.

Antibodies to Bacterial DNA

While bacterial DNA could act either alone or in concert with endotoxin (or other macromolecules) to exert systemic effects in infection, there are few data to substantiate such a role at this time. The major evidence that bacterial DNA exerts any immunomodulatory effects during ordinary encounters with microorganisms, however, relates to the expression of anti-DNA antibodies. As shown by ELISA assays with single stranded DNA, normal human sera (NHS) commonly display antibodies directed to DNA from some, but not all, bacterial species. These antibodies bind exclusively to bacterial DNA antigen and differ from SLE anti-DNA autoantibodies that show broad reactivity to DNA independent of species origin. The anti-DNA antibodies in NHS are primarily IgG2. This isotype pattern is similar to that of antibodies to bacterial carbohydrates. This finding suggests that antibodies to foreign DNA arise by a T independent mechanism that may reflect cytokine induction without direct involvement of antigen-specific T cells (16-17).

Structure-function Relationships of Immunostimulatory DNA

Studies on the structure-function relationships of immunostimulatory DNA have involved three types of DNA preparations: natural bacterial DNA; synthetic phosphodiester oligonucleotides; and synthetic phosphorothioate oligonucleotides. Phosphorothioates are DNA derivatives in which one of the non-bridging oxygens in the phosphodiester backbone is substituted by a sulfur atom. These derivatives are used commonly as antisense reagents and differ from phosphodiesters in various physical-chemical properties (18,19).

Analysis of immune stimulation by natural DNA indicates that, while all bacterial DNA display CpG motifs, they differ in their potency in activating either murine B cells or macrophages. Among bacterial DNA studied, stimulation varies markedly, although activity for B cells and macrophages generally show a similar pattern (Figure 1). The differences in the immunostimulatory capacity of DNA from various bacterial species may relate to the content of CpG motifs; because of the limited amount of sequence information on DNA from certain species, however, precise determination of this content is not possible (5, 20). Another explanation for differences in the potency of bacterial DNA may be the arrangement of CpG motifs along the DNA molecule as well as the presence of other sequences which could be inhibitory.

While theoretically possible, differences in the binding and uptake of natural bacterial DNA to cells does not appear to explain the variation in activity. Thus, stimulation of murine spleen cells with different bacterial DNA complexed to cytofectin agents leads to the same rank order of stimulation as these DNA alone. Since cytofectins promote DNA uptake by cells, the similarity in the rank order of DNA with and without these agents argues that DNA show similar uptake irrespective of base composition or sequence. In this regard, the augmentation of cell responses by cytofectins differs for B cells and macrophages, suggesting that uptake mechanisms may differ among cell types (20).

Among phosphodiester oligonucleotides, CpG motifs have attracted the most interest as the source of immunostimulation. The general rules for an ISS (two 5' purines, an unmethylated CpG motif, two 3' pyrimidines) provide the basis for a large number of active sequences. Depending on the exact sequence of bases as well as the surrounding bases, however, ISS show considerable differences in their potency as immunostimulators (21). These differences could result from a variety of properties such as conformation, stability and interaction with proteins, including intracellular target(s) involved in the signal transduction cascade.

As shown in studies on cytokine production, base sequence may influence the activity of an ISS because of interactions with a membrane receptor. This receptor interaction could result from increased uptake into cells via receptor mediated endocytosis; alternatively, binding to a surface receptor could cause a signaling event that contributes to the cellular response. The best evidence for the role of a membrane receptor in the response to oligonucleotides comes from the analysis of the activities of oligonucleotides composed of an ISS (AACGTT) embedded in the context of 5' and 3' flanking sequences of a single base. In *in vitro* cultures with murine cell populations, the extent of interferon stimulation by the oligonucleotides varied dramatically, with an ISS in the context of dG flanks producing the greatest stimulation of macrophage cytokine response; among this set of compounds, an ISS with dA flanks produced minimal stimulation, arguing that flanking sequences as well as an ISS contribute to the overall response (22).

The explanation for the effects of dG flanks comes from observations on the specificity of the Type A macrophage scavenger receptor (MSR). This receptor is part of a larger family of so-called scavenger receptors whose function is to bind to foreign or structurally altered self macromolecules, possibly for degradation and disposal. This receptor was originally defined by its interaction with acetylated (Ac) or oxidized (Ox) low density lipoproteins (LDL) and shows broad specificity for polyanions. Thus, the MSR binds, in addition to AcLDL and OxLDL, dextran sulfate, fucoidin, lipopolysaccharide and dG-rich DNA. Because of dG's ability to base pair with itself, DNA rich in this base can form an alternative structure known as quadruplex DNA. Quadruplex DNA is a four strand array comprised of four polynucleotide chains aligned in either the parallel or anti-parallel orientation (23, 24).

When presented with dG sequences that bind to the MSR, an ISS may have a higher level of DNA uptake and ability to cause cell activation. Indeed, ligands of the scavenger receptor (e.g., dextran sulfate) can block the response of an ISS containing runs of dG sequence, suggesting that receptor interaction is the first step in macrophage activation by an ISS (22). While the effect of dG runs on stimulation by an ISS may result from increased binding and uptake, binding to the MSR may also transduce a signal that promotes the cell response. The ability of MSR to transduce signals has been the subject of many studies, with results varying depending on the stimulating ligand as well as macrophage cell population.

The role of the Type A MSR in the response to natural DNA may differ from the response to synthetic oligonucleotides. Several studies have now demonstrated that the MSR mediates the *in vivo* clearance of DNA from the circulation as well as the *in vitro* response of macrophages to DNA stimulation (25-27). These findings suggest that natural DNA may behave differently than synthetic oligonucleotides and bind to the MSR even in the absence of extended dG sequences that can form quadruplexes. Alternatively, natural DNA may display sufficient amounts of a quadruplex or related structure to allow MSR interaction. While the explanation for the differences in properties of natural DNA and synthetic oligonucleotides requires further investigation, it appears nevertheless that cell surface interactions may influence the response to DNA.

The Properties of Phosphorothioates

Although phosphorothioates are commonly used as substitutes or models for immunostimulatory DNA, they differ significantly from phosphodiesters in their *in vitro* and *in vivo* properties. These differences reflect the interplay between base sequence and the phosphorothioate backbone. Because of charge differences conferred by the sulfur moiety, phosphorothioates differ from phosphodiesters in their melting temperature, resistance to nuclease digestion, uptake into cells and *in vivo* distribution (18, 19). These differences, while facilitating the use of phosphorothioates as antisense agents, may alter their behavior as immunomodulators.

Analysis of the activity of phosphorothioates oligonucleotides indicates a profound influence of backbone chemistry on immune stimulation. For human B cells, phosphorothioate oligonucleotides cause activation and antibody production whereas phosphodiester oligonucleotides as well as natural DNA are inactive (15). For murine B cells, phosphorothioate oligonucleotides display different structure-function relationships than phosphodiesters, with the phosphorothioate backbone itself leading to immune stimulation. Even in the absence of a CpG motif, certain phosphorothioates are mitogenic. Among single base phosphorothioates, dG compounds show particular mitogenic activity, providing additional evidence that quadruplex structures may have unique activities (28, 29).

The effects of phosphorothioates on cytokine production are more complex and involve both stimulatory as well as inhibitory activities. In general, phosphorothioates with an ISS appear more active than comparable phosphodiesters, perhaps reflecting greater stability and half-life *in vivo* or *in vitro*. Among single base phosphorothioates, however, dG oligonucleotides can inhibit the production of IFN-γ induced either by Concanavalin A or the combination of a phorbol ester and calcium ionophore. Other single base compounds can also inhibit IFN-γ production, suggesting that inhibition of cytokine production may be a property of the phosphorothioate backbone depending on sequence (30). Together these findings suggest that, with phosphorothioate oligonucleotides, immune effects will represent the balance between activating sequences (e.g., CpG motifs) and inhibitory sequences (e.g., dG runs) that differentially affect B cells and macrophages.

Signal Transduction Pathways

Studies on signal transduction by immunostimulatory DNA have utilized a variety of different systems to delineate mechanisms for cell activation. As stimulators, natural DNA, synthetic phosphodiesters and synthetic phosphorothioates have been utilized essentially interchangeably despite evidence that these compounds may differ in biochemical and immunological properties. As target cells, cell lines as well as mononuclear cell preparations from peripheral sources have been studied.

A key issue in the pathways of cell activation by oligonucleotides concerns the requirement for cell uptake. This issue has been approached primarily by assessing the activity of oligonucleotides attached to beads to prevent intracellular uptake. For murine cells, oligonucleotides on bead surfaces are inactive, suggesting that interaction with a surface receptor is insufficient to cause B cell mitogenesis (12). This result does not exclude the possibility that extracellular oligonucleotides can transduce at least a partial signal. In contrast, phosphorothioates on beads can activate human B cells, suggesting a role of surface receptors in activation as well as important mouse-human differences in mechanisms of responsiveness (15).

The identity of DNA receptors has been a matter of considerable investigation. In addition to the macrophage scavenger receptor, CD4, Mac1 and a variety of other binding molecules can bind oligonucleotides depending on sequence (31, 32). Since phosphorothioates show higher non-specific protein binding than phosphodiesters, these compounds may be particularly likely to interact with surface molecules and cause activation by receptor cross-linking (33). The reason human and murine cells differ in their response to extracellular oligonucleotides may reflect the binding affinity of the membrane receptors and their triggering function.

In the murine system, where intracellular uptake of DNA appears required, cell activation is linked to endocytosis since antimalarial agents such as chloroquine can block cell activation. These agents prevent acidification of endocytic vesicles and their function. The steps following endocytosis are less clear although signal transduction critically involves generation of reactive oxygen species as well as activation of stress kinases (34, 35). The coupling mechanisms for these activating events are less clear, although it is likely that they involve intracellular DNA binding proteins with specificity for CpG motifs. The mechanisms by which certain DNA sequences (e.g., dG runs) can stimulate cytokine production have not been investigated.

Therapeutic Uses of Immunostimulatory DNA

The immunological properties of bacterial DNA are relevant to therapeutic manipulations in at least three clinical settings: antisense therapy; DNA vaccination; and immunomodulation by ISS. Antisense agents are short oligonucleotides complementary to sequences in the messenger RNA (mRNA) for a specific protein. When bound to sense mRNA, antisense compounds prevent translation of the protein and therefore can be therapeutic by blocking the production, for example, of an essential viral protein, oncogenic protein or inflammatory mediator. Although a variety of backbone chemistries have been explored for this activity, phosphorothioates have emerged as the most popular (19). Depending on its sequence, an antisense compound could display immunostimulatory activity, with effects on B or T cells as well as macrophages complicating its use. Judicious selection of sequences could prevent some of these effects, although the phosphorothioate backbone itself may be prone to unintended positive or negative influences.

Of gene-based modalities, DNA vaccination represents a potentially powerful approach for eliciting protective immunity against a wide range of microorganisms. This approach involves immunization with plasmid DNA vectors ("naked DNA") that encode, under the influence of a strong promoter, a foreign protein to be targeted for protective immunity mediated by either B or T cells. Since the plasmids contain bacterial DNA sequences, including CpG motifs, they can exert immune effects including B cell activation as well as induction of cytokines that lead to a Th1 cell predominance. As shown using plasmids varying in sequence as well as methylation state, the encoded ISS are critical to the adjuvant activity of the DNA vectors. Thus, elimination of an ISS or methylation of CpG motifs diminishes vaccine potency (36). In this regard, addition of an ISS to an active vector may confer only limited benefits because high levels of cytokine production can diminish promoter activity and in vivo protein expression.

Another use of ISS oligonucleotides is immunomodulation. Both phosphodiester and phosphorothioate compounds can induce directly or indirectly cytokines such as IL-12, IFN-α/β and IFN-γ and can therefore promote the induction of Th1 cell responses. These properties have many potential uses: serve as adjuvants to enhance response to conventional vaccines; stimulate host defense in the setting of infection; and shift the host responses to a Th1 mode to promote protective immune responses (37, 38). Thus far, the *in vivo* use of oligonucleotides in animal models has not elicited hazardous side effects although, at certain doses, these compounds can cause inflammatory reactions as well as severe complications such as septic shock (39).

While oligonucleotides hold great promise as immunomodulators, many aspects of this approach remain unknown. These aspects include the spectrum of *in vivo* activity in man as well as other species; optimal sequences for activity in man; optimal routes of administration; and the possible development of tolerance to the activating effects of these agents. Furthermore, the occurrence of antibodies to bacterial DNA in normal humans emphasizes the immunogenicity of foreign DNA and suggests that exposure to foreign nucleic acids could cause anti-DNA induction. The pathogenicity of such antibodies will likely depend on host factors, with anti-DNA autoantibody induction representing at least a theoretical complication of the therapeutic use of oligonucleotides.

Concluding Remarks

Depending on base sequence and backbone chemistry, DNA can exert powerful immunostimulatory activities that suggest a role in host defense. These activities also provide the basis of novel strategies that could be useful in the prevention and treatment of human disease. Future research will define the mechanisms by which DNA activates the immune systems in normal and aberrant settings and delineate structures with the greatest therapeutic potential.

References

1. Pisetsky, D.S. 1996. The immunologic properties of DNA. J. Immunol. 156: 421-423.
2. Pisetsky, D.S. 1996. Immune activation by bacterial DNA: a new genetic code. Immunity. 5: 303-310.

3. Tokunaga, T., Yamamoto, H., Shimada, S., Abe, H., Fukuda, T., Fujisawa, Y., Furutani, Y., Yano, O., Kataoka, T., Sudo, T., Makiguchi, N. and Suganuma, T. 1984. Antitumor activity of deoxyribonucleic acid fraction from *Mycobacterium bovis* BCG. I. Isolation, physicochemical characterization, and antitumor activity. J. Natl. Cancer Inst. 72: 955-962.

4. Shimada, S., Yano, O. and Tokunaga, T. 1986. *In vivo* augmentation of natural killer cell activity with a deoxyribonucleic acid fraction of BCG. Jpn. J. Cancer Res. 77: 808-816.

5. Yamamoto, S., Yamamoto, T., Shimada, S., Kuramoto, E., Yano, O., Kataoka, T. and Tokunaga, T. 1992. DNA from bacteria, but not from vertebrates, induces interferons, activates natural killer cells and inhibits tumor growth. Microbiol. Immunol. 36: 398-997.

6. Vogt, W., Ruhl, H., Wagner, B. and Diamantstein, T. 1973. Stimulation of DNA synthesis in mouse lymphoid cells by polyanions *in vitro*. II. Relationship between adjuvant activity and stimulation of DNA synthesis by polyanions. Eur. J. Immunol. 3: 493-496.

7. Tokunaga, T., Yano, O., Kuramoto, E., Kimura, Y., Yamamoto, T., Kataoka, T. and Yamamoto, S. 1992. Synthetic oligonucleotides with particular base sequences from the cDNA encoding proteins of *Mycobacterium bovis* BCG induce interferons and activate natural killer cells. Microbiol. Immunol. 36: 55-66.

8. Yamamoto, S., Yamamoto, T., Kataoka, T., Kuramoto, E., Yano, O. and Tokunaga, T. 1992. Unique palindromic sequences in synthetic oligonucleotides are required to induce INF and augment INF-mediated natural killer activity. J. Immunol. 148: 4072-4076.

9. Bird, A.P. 1987. CpG islands as gene markers in the vertebrate nucleus. Trends Genet. 3: 342-346.

10. Hergersberg, M. 1991. Biological aspects of cytosine methylation in eukaryotic cells. Experientia. 47: 1171-1185.

11. Messina, J.P., Gilkeson, G.S. and Pisetsky, D.S. 1991. Stimulation of *in vitro* murine lymphocyte proliferation by bacterial DNA. J. Immunol. 147: 1759-1764.

12. Krieg, A.M., Yi, A.-K., Matson, S., Waldschmidt, T.J., Bishop, G.A., Teasdale, R., Koretzky, G.A. and Klinman, D.M. 1995. CpG motifs in bacterial DNA trigger direct B-cell activation. Nature. 374: 546-549.

13. Klinman, D.M., Yi, A.-K., Beaucage, S.L., Conover, J. and Krieg, A.M. 1996. CpG motifs present in bacterial DNA rapidly induce lymphocytes to secrete interleukin 6, interleukin 12, and interferon γ. Proc. Natl. Acad. Sci. USA. 93: 2879-2883.

14. Halpern, M.D., Kurlander, R.J. and Pisetsky, D.S. 1996. Bacterial DNA induces murine interferon-γ production by stimulation of interleukin-12 and tumor necrosis factor-α. Cell. Immunol. 167: 72-78.

15. Liang, H., Nisioka, Y., Reich, C.F., Pisetsky, D.S. and Lipsky, P.E. 1996. Activation of human B cells by phosphorothioate oligodeoxynucleotides. J. Clin. Invest. 98: 1119-1129.

16. Robertson, C.R., Gilkeson, G.S., Ward, M.M. and Pisetsky, D.S. 1992. Patterns of heavy and light chain utilization in the antibody response to single-stranded bacterial DNA in normal human subjects and patients with systemic lupus erythematosus. Clin. Immunol. Immunopath. 62: 25-32.

17. Pisetsky, D.S. 1998. Antibody responses to DNA in normal immunity and aberrant immunity. Clin. Diag. Lab. Immunol. 5: 1-6.

18. Stein, C.A., Subasinghe, C., Shinozuka, K. and Cohen, J.S. 1988. Physicochemical properties of phosphorothioate oligodeoxynucleotides. Nucl. Acids Res. 16: 3209-3221.

19. Wagner, R.W. 1994. Gene inhibition using antisense oligodeoxynucleotides. Nature. 372: 333-335.

20. Neujahr, D.C., Reich, C.F. and Pisetsky, D.S. 1999. Immunostimulatory properties of genomic DNA from different bacterial species. Immunobiol. 200: 106-119.
21. Pisetsky, D.S. and Reich, C.F. III. 1998. The influence of base sequence on the immunological properties of defined oligonucleotides. Immunopharm. 40: 199-208.
22. Kimura, Y., Sonehara, K., Kuramoto, E., Makino, T., Yamamoto, S., Yamamoto, T., Kataoka, T. and Tokunaga, T. 1994. Binding of oligoguanylate to scavenger receptors is required for oligonucleotides to augment NK cell activity and induce IFN. J. Biochem. 116: 991-994.
23. Krieger, M. 1994. Structures and functions of multiligand lipoprotein receptors: macrophage scavenger receptors and LDL receptor-related protein (LRP). Annu. Rev. Biochem. 63:601-637.
24. Pearson, A.M., Rich, A. and Krieger, M. 1993. Polynucleotide binding to macrophage scavenger receptors depends on the formation of base-quartet-stabilized four-stranded helices. J. Biol. Chem. 268: 3546-3554.
25. Kawabata, K., Takakura, Y. and Hashida, M. 1995. The fate of plasmid DNA after intravenous injection in mice: involvement of scavenger receptors in its hepatic uptake. Pharm. Res. 12: 825-830.
26. Takagi, T., Hashiguchi, M., Mahato, R.I., Tokuda, H., Takakura, Y. and Hashida, M. 1998. Involvement of specific mechanism in plasmid DNA uptake by mouse peritoneal macrophages. Biochem. Biophys. Res. Commun. 245: 729-733.
27. Wloch, M.K., Pasquini, S., Ertl, H.C.J. and Pisetsky, D.S. 1998. The influence of DNA sequence on the immunostimulatory properties of plasmid DNA vectors. Hum. Gene Ther. 9:1439-1447.
28. Pisetsky, D.S. and Reich, C. 1993. Stimulation of *in vitro* proliferation of murine lymphocytes by synthetic oligodeoxynucleotides. Molec. Biol. Rep.18: 217-221.
29. Monteith, D.K., Henry, S.P., Howard, R.B., Flournoy, S., Levin, A.A., Bennett, C.F. and Crooke, S.T. 1997. Immune stimulation a class effect of phosphorothioate oligodeoxynucleotides in rodents. Anti-Cancer Drug Des. 12: 421-432.
30. Halpern, M.D. and Pisetsky, D.S. 1995. *In vitro* inhibition of murine IFN-γ production by phosphorothioate deoxyguanosine oligomers. Immunopharm. 29: 47-52.
31. Yakubov, L., Khaled, Z., Zhang, L.-M., Truneh, A., Vlassov, V. and Stein, C.A. 1993. Oligodeoxynucleotides interact with recombinant CD4 at multiple sites. J. Biol. Chem. 268: 18818-18823.
32. Benimetskaya, L., Loike, J.D., Khaled, Z., Loike, G., Silverstein, S.C., Cao, L., El Khoury, J., Cai, T.-Q. and Stein, C.A. 1997. Mac-1 (CD11b/CD18) is an oligodeoxynucleotide-binding protein. Nature Med. 3: 414-420.
33. Brown, D.A., Kang, S.-H., Gryaznov, S.M., DeDionisio, L., Heidenreich, O., Sullivan, S., Xu, X. and Nerenberg, M.I. 1994. Effect of phosphorothioate modification of oligodeoxynucleotides on specific protein binding. J. Biol. Chem. 269: 26801-26805.
34. Yi, A.-K., Tuetken, R., Redford, T., Waldschmidt, M., Kirsch, J. and Krieg, A.M. 1998. CpG motifs in bacterial DNA activate leukocytes through the pH-dependent generation of reactive oxygen species. J. Immunol. 160: 4755-4761.
35. Yi, A.-K. and Krieg, A.M. 1998. Cutting Edge: Rapid induction of mitogen-activated protein kinases by immune stimulatory CpG DNA. J. Immunol. 161:4493-4497.
36. Tighe, H., Corr, M., Roman, M. and Raz, E. 1998. Gene vaccination: plasmid DNA is more than just a blueprint. Immunol. Today. 19:89-97.
37. Lipford, G.B., Bauer, M., Blank, C., Reiter, R., Wagner, H. and Heeg, K. 1997. CpG-containing synthetic oligonucleotides promote B and cytotoxic T cell responses to protein antigen: a new class of vaccine adjuvants. Eur. J. Immunol. 27: 2340-2344.

38. Weiner, G.J., Liu, H.-M., Wooldridge, J.E., Dahle, C.E. and Krieg, A.M. 1997. Immunostimulatory oligodeoxynucleotides containing the CpG motif are effective as immune adjuvants in tumor antigen immunization. Proc. Natl. Acad. Sci. USA. 94: 10833-10837.

39. Sparwasser, T., Miethke, T., Lipford, G., Borschert, K., Hacker, H., Heeg, K. and Wagner, H. 1997. Bacterial DNA causes septic shock. Nature. 386: 336-337.

From: *Development of Novel Antimicrobial Agents: Emerging Strategies*
ISBN 1-898486-23-9 © 2001 Horizon Scientific Press, Wymondham, UK.

7

Bacteria-Mediated DNA Transfer for Gene Therapy and Genetic Vaccination

Siegfried Weiss and Trinad Chakraborty

Abstract

Transfer of eukaryotic expression plasmids to mammalian cells has recently been achieved using live attenuated bacteria. These successes have encouraged the use and generation of bacterial vector delivery systems that use local, mucosal and systemic routes of infection to deliver the desired gene directly to the cell type or organ of interest. Intrinsic properties of invasive bacteria such as their tropism for cell types or cell to cell spread are currently understood in great detail and provide the necessary basis for the design of novel vehicles. Finally, the ability of bacteria to harbor very large plasmids makes them very attractive as vehicles for gene therapy.

   Escherichia coli rendered artificially invasive and *Listeria monocytogenes* were used *in vitro* to transfer reporter genes into various types of cells from different species. Transfer could also be observed *in vivo* with *L. monocytogenes* albeit at low frequency. *Shigella flexneri*, *Salmonella typhimurium* and *Salmonella thyphi* were used as vehicles to transfer plasmids for genetic immunisation *in vivo*. Immune responses against the antigen encoded by the expression plasmid could be detected in all cases. Thus, bacteria represent a simple and versatile carrier system for genetic immunisation that should provide the flexibility required for the various vaccination problems. The low production cost of such vaccines makes them attractive candidates for mass application.

Introduction

The possibility to transfer eukaryotic expression plasmids *in vivo* into mammalian hosts has extended the panel of treatments of disease by an extremely promising variant. Disorders caused by monogenic defects might be cured by transferring a complementing gene into appropriate precursor cells (1). Similarly, vaccination against a broad range of infectious diseases could possibly be achieved by transferring the genetic information encoding protective antigens into inductive cells of the immune system rather than the antigens themselves (2, 3). The use of expression plasmids in this case allows the co-expression of immune modulatory molecules. Thus, it appears feasible that such a vaccination strategy will not only be used in prophylactic immunization but also in immune therapy of cancer, autoimmune disease or allergies.

A major obstacle hampering the success of gene therapy and genetic vaccination so far is the lack of efficient vehicles that transfer the plasmids into appropriate cells. Currently, the commonly used vectors consist of replication-defective viruses or plasmids transferred either as 'naked' DNA or encapsulated into various inert polymer matrices or liposomes. Thus, plasmids will be only transferred into cells that are in the vicinity of the site of injection. This might be sufficient for vaccination purposes because antigen presenting cells will transport antigens that they express themselves or antigens that they take up from apoptotic neighbouring cells into lymph nodes or the spleen to stimulate T cells (4, 5). For gene therapy such an approach is more limiting since appropriate targetting of the DNA into immature precursor or stem cells is a prerequisite. Therefore, continuous administration of curative vectors would be required.

Bacteria have been considered as alternative vehicles for gene delivery since they offer several advantages over other delivery systems. Many bacteria disseminate from their port of entry and display regional and local tropism to certain target organs. Current studies aimed at understanding the mechanisms underlying these dispositions suggest that it will be possible to introduce targeted specificity. Systemic spread, in particular cell to cell spread, by some organisms allow to reach layers of tissue that are not accessible to non-replicating vehicles. In addition, it is now possible to generate attenuating mutations into any bacterial species and indeed several live attenuated bacteria are currently approved for use in animals and humans (6-8). The advent of bacterial genome sequencing provides a further level of sophistication in the generation of well-characterized vehicles and the ability of bacteria to harbour megaplasmids add to the attraction of these vehicles. Finally, bacterial vehicles are controllable by commonly used antibiotics which provide an additional safety aspect to these systems. Here we will summarize experiments in which bacteria have been shown to efficiently transfer eukaryotic expression plasmids *in vitro* and *in vivo* for the purpose of gene therapy and genetic vaccination.

Principle of Bacteria-mediated Gene Transfer

The basic idea for the transfer of an eukaryotic expression plasmid by a bacterial carrier to the mammalian host cells (Figure 1) was as follows: bacteria such as *Shigella* or *Listeria* that carry an expression plasmid invade the host cell by bacterial-induced phagocytosis. These bacteria express virulence factors that allow them to escape from the phagolysosome into the cytosol of the host cell. In this cellular compartment bacterial lysis must be induced either through metabolic deficiency (auxotrophy) or by the use of an appropriate antibiotic. The liberated plasmids will find their way into the nucleus of the host cell like in any other transfection. This will result in transient expression or stabile integration of the plasmid into the host genome depending on the particular experimental set up. In the meantime, however, it has become clear that escape of bacteria from the vacuole to the cytosol is not a prerequisite for DNA transfer suggesting additional mechanisms for transfer of the DNA to the cell nucleus. The way of transfer appears to be dependent on the particular bacterium host cell combination.

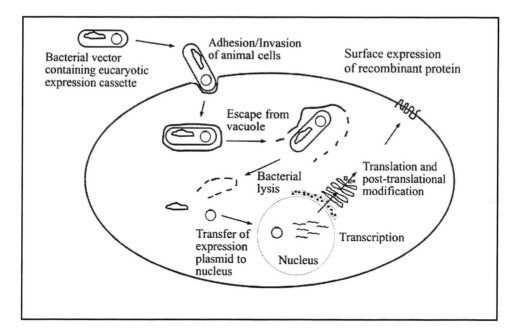

Figure 1. Schematic representation of bacteria-mediated expression plasmid transfer into host cells. Phagocytosis is induced by the recombinant bacteria which subsequently will evade from the vacuole and colonize the cytosol. Lysis of the bacteria will result from an attenuating auxotrophy or treatment with antibiotics. Plasmids will be liberated, transferred to the nucleus and proteins encoded by the plasmid will be expressed. Depending on the bacterial vehicle this could lead to complementation of a dysfunction or to the induction of an immune response.

Bacteria-mediated Gene Transfer for Gene Therapy

Experiments assessing transfer of expression plasmids for gene therapy are to the best of our knowledge restricted mainly to *in vitro* studies and two bacterial species - recombinant *Escherichia coli* and *Listeria monocytogenes* - so far.

E. coli that were auxotroph for a component of cell wall synthesis were transformed with the 200 kb virulence plasmid of *Shigella flexneri* which contains the genes responsible for invasion and phagosomal escape of this pathogen (9). Alternatively, the auxotroph *E. coli* were transformed with a plasmid encoding the invasin gene (Inv) of *Yersinia pseudotuberculosis* (10). Since the products of this gene only support invasion but not the escape into the cytosol, additional recombinants were established where inv was co-expressed with listeriolysin (LLO), a pore forming hemolysin of *L. monocytogenes*.

Such recombinant *E. coli* were used to transfer eukaryotic expression cassettes encoding β-galactosidase (β-gal) or the green fluorescent protein (GFP) as reporter proteins. Using this system, gene transfer and stable transfectants were obtained with several cell lines mainly of epithelial origin. Interestingly, plasmid transfer was also observed when listeriolysin was not co-expressed, which suggests that the bacteria do not need to escape from the phagocytic vesicles. Co-expression of LLO only enhanced the DNA transfer under certain conditions. We have performed similar experiments using *L. monocytogenes* as transfer vehicle (Hense *et al.*, submitted; Krusch *et al.*, to be published). As vector a shuttle plasmid was used that contained an origin of replication (ori) for plasmid construction in *E. coli* and an additional ori for propagation in the Gram-positive *L. monocytogenes*. The expression cassette was derived from pCMVβ (Clontech) that was driven by the human Cytomegalovirus (hCMV) immediate early promoter and contained SV40 derived splice donor and acceptor sites. Intracellular bacteria were killed by addition of antibiotics to the culture after allowing sufficient time for the bacteria to invade the host cells and escape from the vacuoles. DNA transfer was observed with several cell lines including cells of epithelial, endothelial and parenchymal origin. Interestingly, no or only inefficient transfer was obtained with cells of macrophage origin. Also, primary macrophages could not be transfected using *L. monocytogenes* as transfer vehicle. This is in contrast to the experiments described above where recombinant *E. coli* were capable of transferring plasmids into the macrophage-like cell line J774, suggesting that the bacterial carrier has a strong influence on the mechanism of DNA transfer.

Transfer of expression plasmids from the bacterial carrier to the host cell was proved by several experiments. Staining patterns of transfectants were clearly different when compared to cells that were infected with *L. monocytogenes* expressing the reporter protein under the control of a prokaryotic promoter. In addition, host cell transcription of the reporter gene could be shown by the detection of splice products of the reporter gene. Furthermore, the reporter protein was synthesized in the presence of tetracycline, an antibiotic that blocks bacterial protein biosynthesis. Finally, stable transfectants could be established in which the plasmid had integrated into the genome of the host cell. Integration occurred with a frequency that is also observed in other types of transfection regimens (1×10^{-5}).

Detailed analysis in our antibiotics driven system revealed that with *L. monocytogenes*, for plasmid transfer to take place it was necessary for the bacteria to invade the cells and to escape from the vacuole. Mutants that lack these properties were not able to support the DNA transfer. Interestingly, DNA transfer was also observed when bacteria were not killed by antibiotics. Under these circumstances many host cells died probably due to overload with bacteria. However, a few survivors displayed expression of the reporter gene. Obviously, in some cells bacteria lysed spontaneously or were killed by host cell defence mechanisms leading to cell survival and DNA transfer. This suggests that also in natural conditions plasmid transfer from the bacteria to the host takes place since many commensal or pathogenic bacteria carry plasmids.

Using recombinant *L. monocytogenes* harbouring an eukaryotic expression plasmid encoding GFP together with an antibiotics treatment regimen, we have tried to expand the DNA transfer system to an *in vivo* model. As target organ liver was chosen since *L. monocytogenes* has a natural tropism for this organ. It was possible to control the infection by antibiotics early enough to avoid a specific immune response even when high doses of *L. monocytogenes* were used. However, induction of inflammation after intravenous administration of the bacteria seemed to be effective in curbing infection for detectable DNA transfer to take place. It is well established that neutrophilic granulocytes kill hepatocytes that are infected with *L. monocytogenes*. We, therefore, depleted this cell population using a specific antibody. This pre-treatment of recipients resulted in a few hepatocytes clearly expressing the reporter gene after infection with *L. monocytogenes* carrying a GFP encoding expression plasmid. These findings demonstrate that DNA transfer by bacterial carriers for gene therapy is in principle possible *in vivo*. However, several optimization steps will be required until this system is developed as a serious alternative to the commonly used transfer vehicles.

Bacteria-mediated Gene Transfer for Genetic Vaccination

Bacteria have been widely used as antigen carriers. Commensal bacteria as well as attenuated pathogens expressing heterologous antigens have been shown to elicit protective immune responses in prophylactic and therapeutic situations (11-13). The ability of bacteria to induce local inflammation at their port of entry or in target organs provides a local adjuvant effect. In addition, most of these bacteria have a mucosal port of entry, thus providing an attractive way of administration for mass vaccination via the oral or nasal route. So far only a few bacterial species have been tested for delivery of DNA vaccines but the results are extremely promising.

Expression Plasmid Delivery by Attenuated Shigella flexneri

S. flexneri are capable of escape from the phagolysosome. It was the first bacterial species that has been published to be able to transfer DNA *in vitro* into established cell lines (14). Due to the lack of an appropriate animal infection model *in vivo,* DNA transfer by such bacteria into the cornea of guinea pigs was visualized first. In the mouse model these bacteria have been used to nasally immunize mice against the model antigen β-gal (15) or against two protective antigens of measles virus [haemagglutinin (HA) and nucleoprotein (NP); (16)]. T cell responses could readily be demonstrated in the spleen of mice immunized by at least two nasal administrations. Since in the case of the antigens of the measles virus cytotoxic as well as T_H1 type helper T cells were induced in mice immunized in such a way the response was considered to be protective. The presence of specific antibodies against the antigens could also be demonstrated in the serum. However, the titers were low and in the case of measles virus antigens it was considered doubtful whether the titre would be sufficient to protect an individual from infection with this virus. Nevertheless, histological examination of lung tissue after nasal administration of the attenuated Shigella expression plasmid carrier showed negligible tissue damage (16), thus rendering this DNA carrier system a very promising one. Since *Shigella flexneri* usually does not disseminate from the mucosa to deep organs this carrier might be particularly suited for induction of a local immune response.

L. monocytogenes as Carrier for DNA Vaccination

Interestingly, *L. monocytogenes* has not only been considered as a carrier for expression plasmids for gene therapy but also as delivery system for genetic vaccination. This is based on the fact that *L. monocytogenes* when used as carrier for heterologous antigens elicits an extremely strong cytotoxic T cell response that results in therapeutic reactions even against established tumors. Thus, the combination of DNA vaccines with the adjuvant effect of *L. monocytogenes* seems to be a favourable combination. The use of this bacterial species in gene therapy in which the aim is to avoid any immune response and as carrier for DNA vaccination in which the aim is to efficiently induce an immune response appears to be mutually exclusive, but only at first sight. The judicious use of appropriate mutant strains and different administration protocols could favour one or the other outcome of a DNA transfer. To date, only *in vitro* experiments on *L. monocytogenes*-mediated DNA transfer for genetic vaccination have been reported. DNA delivery was achieved by introducing a lysin gene derived from a bacteriophage under the control of a promoter that is activated only when the bacterium is inside a host cell. With this system, DNA transfer into macrophages has been reported (17). As presented above, the efficiency of transfer was very low into these macrophage cell lines. However the transfectants were capable of presenting antigen to CD8 T cells. More recently it has been shown that also dendritic cells could be transfected by *L. monocytogenes*. Thus, all conditions for the induction of an effective immune response, at least for the generation of cytotoxic T cells, are fulfilled.

Transfer of Expression Plasmids Using Attenuated Salmonella

The use of *Salmonella* as vehicle for expression plasmids for genetic vaccination offers several unique advantages over other transfer systems. Firstly, live-attenuated strains of *Salmonella* exist and have been used both in animal husbandry and as vaccines (6-8). Indeed, it is now possible to generate attenuated mutant Salmonella that specifically induce Th-1 or Th-2 driven responses (18). In mice such strains have successfully been used as antigen carriers expressing various heterologous proteins (19, 20). In fact, mice represent an excellent model for testing efficiency of *Salmonella*-based carrier strains. Furthermore, the mucosal port of entry of these bacteria allows either oral or nasal administration of vaccine candidate strains (21, 22). The physiology and virulence properties of Salmonella have been under great scrutiny in the past few years and Salmonella remains one of the best studied pathogens today (23). Most conveniently recombinant plasmids constructed in *E. coli* can be directly transferred into *Salmonella* without any further manipulation. The sequence of the whole genome of these bacteria will soon be available providing even further possibilities to improve on current vaccine carrier strains. Finally the technologies for large scale production of *Salmonella* strains already exist. For all these reasons *Salmonella* as carrier in combination with DNA immunisation should have a great potential for vaccination and immunotherapies. The bacterial carrier should have an amplifying effect on the immune response against the antigens encoded by the expression plasmids.

We transformed *S. thyphimurium* that were metabolically attenuated (*aroA*⁻) with eukaryotic expression plasmids encoding two virulence factors of *L. monocytogenes* (LLO and the membrane protein ActA) or β-gal. The transformants were administered orally to mice and the specific immune response was monitored (24). Results showed that efficient induction of all three arms of the specific immune system took place by such vaccination. Already after a single dose cytotoxic T cells, helper T cells and specific antibodies could be detected. Multiple doses improved the responses. To see whether such an immune response would protect mice against a pathogen we immunized mice with *Salmonella* that were carrying a LLO encoding plasmid. Such mice were completely protected against a lethal challenge with *L. monocytogenes* when multiple administrations of the vaccine were applied. A single administration already protected more than half of the mice.

Similar results were obtained in a murine tumor model in which an aggressive fibrosarcoma had been transfected with β-gal to serve as surrogate tumor specific antigen (25). Most of the mice that were orally vaccinated with *S. typhimurium* that contained an eukaryotic expression plasmid encoding β-gal were protected completely against the tumor and tumor growth was significantly retarded in the few mice in which the tumor could be established.

The immune response elicited by this vaccination where DNA transfer was required was far superior over a vaccination with a similar *Salmonella* strain that constitutively expressed the antigen itself due to a prokaryotic promoter. This was taken as first evidence that transfer of expression plasmids into cells of the host had indeed taken place *in vivo*. Direct evidence for plasmid transfer from the bacterial carrier to the host cell was obtained *in vitro* by using primary peritoneal macrophages as target cells. It could be shown that the transcription and translation of the reporter gene was due to the machinery of the host since the mRNA derived from the expression plasmid was spliced and translation was not inhibited by the presence of tetracycline. Interestingly, so far we have not been able to transfect any other cells (including macrophage cell lines) with this method.

The efficiency of induction of T cells and antibodies and particularly the observation that already a single dose results in an almost maximal response renders *Salmonella*-mediated oral DNA vaccination a very promising way of genetic immunisation provided a long lasting immunological memory is induced. To test this, mice were immunised once or four times with *Salmonella* bearing plasmids that encoded β-gal under the control of either an eukaryotic or a prokaryotic promoter. Mice were then rested and analysed 8 months after administration of the single dose or 6 months after the fourth dose. Efficient T cell memory responses could be demonstrated in mice immunized by multiple or single doses of *Salmonella* bearing the eukaryotic expression plasmids while the administration of transformants that constitutively express the antigen did not induce any detectable immunological memory under these conditions. Thus, *Salmonella*-mediated genetic vaccination fulfils all the criteria for a successful vaccination strategy.

Nasal vaccination has been often found to be superior to oral application and might result in a general mucosal immune reaction. Therefore, we compared oral and nasal administration of *Salmonella* harbouring a eukaryotic expression plasmid that encoded β-gal. Systemic T cell responses were induced by both routes although nasal application was clearly inferior to oral administration since multiple doses were required to obtain a significant response. This is most likely due to the 500-fold lower number of bacteria that could be applied nasally due to toxicity of higher numbers of bacteria under these circumstances.

Total antibodies specific for β-gal from serum, lung lavage, saliva and the gut lumen were also determined in these mice. Vaccination via both routes induced systemic antibodies whereas mucosal antibody was specific to the way of application i.e. oral administration induced antibodies in the gut and not in the lung and nasal administration induced antibodies in the lung but not in the gut. This seems to reflect the property of *Salmonella* to remain only shortly in the mucosa associated lymphoid tissue and to quickly disseminate into deeper organs such as spleen and liver. However by generating mutant strains of the bacterial carrier that remain at their site of entry efficient mucosal immunisation might be achieved.

Salmonella-mediated genetic vaccination could also be achieved by intraperitoneal injection. In this case an attenuated vaccine strain of *S. typhi* (Ty21a) was used that carried an expression plasmid encoding the NP of measles virus. Specific cytotoxic T cells could be induced by such treatment. This indicates that plasmid transfer for DNA vaccination is not restricted to *S. typhimurium*. This is important since veterinary DNA vaccines might require different *Salmonella* species for efficient mucosal vaccination. It also demonstrates that various routes of application can successfully be used for *Salmonella*-mediated DNA vaccination.

Originally, it was suggested that the orally administered attenuated recombinant *Salmonella* invade the host from the gut lumen via M cells. Subsequently they are taken up by macrophages. In these cells, they lyse due to the attenuation and the plasmids they contain are transferred into the nucleus. In turn, the antigen is expressed and the immune response is started by these cells. In order to learn more about the cells and mechanisms involved in this type of vaccination we orally administered *Salmonella* that carried a β-gal encoding expression plasmid. After various periods of time cells from Peyer's Patches, mesenteric lymph nodes and spleen were removed and tested for their capacity to stimulate CD4 and CD8 T cells specific for β-gal. Surprisingly, already after eight hours cells from the Peyer's Patches were able to stimulate CD8 as well as CD4 T cells. Thus, plasmid transfer had taken place very quickly after administration. Similar activities could be revealed in lymph nodes and spleen. This stimulation capacity diminished with time but was detectable for more than a month in spleen and mesenteric lymph nodes. The decline was more rapid in Peyer's Patches. After one month stimulatory activity was barely detectable in the latter organs. This observation now explains why even a single dose was so efficient in inducing an immune reaction since antigen presenting cells are present for a considerable length of time. It also provides an explanation as to why the immune response is mainly systemic under these conditions.

Preliminary experiments revealed that the cell-type responsible for this stimulation is most likely a dendritic cell whereas macrophages might not take part in the induction of T cells or at a very low level. Whether dendritic cells express the antigen or whether they acquire antigen by phagocytosis of apoptotic macrophages that express the antigen remains to be determined. However, since an early stimulatory capacity for MHC class II restricted CD4 T cells is observed that are usually directed against exogenous antigens, the second possibility appears to be more likely.

Concluding Remarks

The principle of a transfer of eukaryotic expression plasmids by bacteria for gene therapy and genetic vaccination is well established. The fact that the vehicle and the plasmid are different entities allows the manipulation of either one of them to optimize the system to the particular problem in question. The complex biology of the carrier might provide the potential to change particular properties by appropriate selection e. g. dissemination versus retention at the port of entry or tropism to particular cell types and organs. At the same time the plasmids might be manipulated to achieve certain responses and to avoid others. Thus, it is foreseeable that a very versatile system for DNA transfer will be established. This in combination with the low cost of production of such carriers are cogent reasons to use these carrier systems for vaccination and immunotherapies in the near future.

Acknowledgements

We would like to thank Sabine Schiller for expert secretarial assistance and M. Hense, A. Darji and S. zur Lage for providing unpublished data. Work in the laboratories of both authors was supported in part by the BMBF and the DFG.

References

1. Engel, B.C. and Kohn, D.B. 1999. Stem cell directed gene therapy. Front-Biosci. 4: 26-33.
2. Donnelly, J.J., Ulmer, J.B., Shiver, J.W. and Liu, M.A. 1997. DNA vaccines. Annu. Rev. Immunol. 15: 617-648.
3. Lai, W.C. and Bennett, M. 1998. DNA vaccines. Crit. Rev. Immunol. 18: 449-484.
4. Porgador, A., Irvine, K.R., Iwasaki, A., Barber, B.H., Restifo, N.P. and Germain, R.N. 1998. Predominant role for directly transfected dendritic cells in antigen presentation to CD8+ T cells after gene gun immunization. J. Exp. Med. 188: 1075-1082.
5. Akbari, O., Panjwani, N., Garcia, S., Tascon, R., Lowrie, D. and Stockinger, B. 1999. DNA vaccination: transfection and activation of dendritic cells as key events for immunity. J. Exp. Med. 189: 169-178.
6. Germanier, R. and Fürer, E. 1975. Isolation and characterization of *gal* E mutant Ty 21a of *Salmonella typhi*: A candidate strain for a live, oral typhoid vaccine. J. Infect. Dis. 131: 553-558.
7. Hassan, J.O. 1996. Effect of vaccination of hens with an avirulent strain of *Salmonella typhimurium* on immunity of progeny challenged with wild-type *Salmonella* strains. Infect. Immun. 64: 938-944.

8. Fox, B.C. 1997. Safety and efficacy of an avirulent live *Salmonella cholerae*suis vaccine for protection of calves against *S. dublin* infection. Am. J. Vet. Res. 58: 265-271.

9. Courvalin, P., Goussard, S. and Grillot, C.-C. 1995. Gene transfer from bacteria to mammalian cells. C. R. Acad. Sci. III 318: 1207-1212.

10. Grillot Courvalin, C., Goussard, S., Huetz, F., Ojcius, D.M. and Courvalin, P. 1998. Functional gene transfer from intracellular bacteria to mammalian cells. Nat. Biotechnol. 16: 862-866.

11. Dougan, G., Chatfield, S., Roberts, M., Charles, I., Comerford, S., Li, L.J. and Fairweather, N. 1990. Bacterial pathogens - a route to oral drug delivery. Biochem. Soc. Trans. 18: 746-748.

12. Fairweather, N.F., Chatfield, S.N., Charles, I.G., Roberts, M., Lipscombe, M., Li, L.J., Strugnell, D., Comerford, S., Tite, J. and Dougan, G. 1990. Use of live attenuated bacteria to stimulate immunity. Res. Microbiol. 141: 769-773.

13. Paterson, Y. and Ikonomidis, G. 1996. Recombinant *Listeria monocytogenes* cancer vaccines. Curr. Opin. Immunol. 8: 664-669.

14. Sizemore, D.R., Branstrom, A.A. and Sadoff, J.C. 1995. Attenuated Shigella as a DNA delivery vehicle for DNA-mediated immunization. Science. 270: 299-302.

15. Sizemore, D.R., Branstrom, A.A. and Sadoff, J.C. 1997. Attenuated Shigella as a DNA delivery vehicle for DNA-mediated immunization. Vaccine. 15: 804-807.

16. Fennelly, G.J., Khan, S.A., Abadi, M.A., Wild, T.F. and Bloom, B.R. 1999. Mucosal DNA vaccine immunization against measles with a highly attenuated *Shigella flexneri* vector. J. Immunol. 162: 1603-1610.

17. Dietrich, G., Bubert, A., Gentschev, I., Sokolovic, Z., Simm, A., Catic, A., Kaufmann, S.H., Hess, J., Szalay, A.A. and Goebel, W. 1998. Delivery of antigen-encoding plasmid DNA into the cytosol of macrophages by attenuated suicide *Listeria monocytogenes*. Nature Biotechnol. 16: 181-185.

18. VanCott, J.L., Chatfield, S.N., Roberts, M., Hone, D.M., Hohmann, E.L., Pascual, D.W., Yamamoto, M., Kiyono, H. and McGhee,, J.R. 1998. Regulation of host immune responses by modification of *Salmonella* virulence genes. Nature Med. 4: 1247-1252.

19. Hormaeche, C.E. 1991. Live attenuated *Salmonella* vaccines and their potential as oral combined vaccines carrying heterologous antigens. J. Immunol. Methods. 142: 113-120.

20. Hess, J., Gentschev, I. Miko, D., Welzel, M., Ladel, C., Goebel, W. and Kaufmann, S.H. 1996. Superior efficacy of secreted over somatic antigen display in recombinant *Salmonella* vaccine induced protection against listeriosis. Proc. Natl. Acad. Sci. USA. 93: 1458-1463.

21. Rudin, A., Johansson, E.L., Bergquist, C. and Holmgren, J. 1998. Differential kinetics and distribution of antibodies in serum and nasal and vaginal secretions after nasal and oral vaccination of humans. Infect. Immun. 66: 3390-3396.

22. Velin, D., Hopkins, S. and Kraehenbuhl, J. 1998. Delivery systems and adjuvants for vaccination against HIV. Pathobiology. 66: 170-175.

23. Jones, B.D and Falkow, S. 1996. Salmonellosis: host immune responses and bacterial virulence determinants. Annu. Rev. Immunol.14: 533-561.

24. Darji, A., Guzman, C.A., Gerstel, B., Wachholz, P., Timmis, K.N., Wehland, J., Chakraborty, T. and Weiss, S. 1997. Oral somatic transgene vaccination using attenuated *S. typhimurium*. Cell. 91: 765-775.

25. Paglia, P., Medina, E., Arioli, I., Guzman, C.A. and Colombo, M.P. 1998. Gene transfer in dendritic cells, induced by oral DNA vaccination with *Salmonella typhimurium*, results in protective immunity aginst a murine fibrosarcoma. Blood. 92: 3172-3176.

From: *Development of Novel Antimicrobial Agents: Emerging Strategies*
ISBN 1-898486-23-9 © 2001 Horizon Scientific Press, Wymondham, UK.

8

Potential Mucosal Adjuvants For Human Use

Mari Ohmura, Raymond J. Jackson, Yoshifumi Takeda
and Jerry R. McGhee

Abstract

There have been no effective and safe adjuvants for use in humans since aluminium compounds were approved by the US Federal Drug Administration. At present new adjuvants approved for use such as MDP and ISCOMs are limited to veterinary vaccines. Recent advances at the cellular and molecular levels of the immune system have led to the clinical application of certain cytokines for both immunotherapeutic and conventional vaccines. While not yet in widespread use, the cytokines IL-2 and IL-12 hold potential promise for human adjuvant applications. A purified saponin, QS-21 is another promising adjuvant candidate that has proven safe in Phase I and Phase II human clinical trials. Finally, considerable effort has been focused on developing detoxified derivatives of bacterial enterotoxins. Thus, mutants of cholera toxin (CT) produced by *Vibrio cholerae* and of the heat labile enterotoxin (labile toxin; LT) produced by enterotoxigenic *Escherichia coli,* which are non-toxic but which retain adjuvanticity have been constructed. Among these are CT E112K and S61F, which harbor single amino acid substitutions in the enzymatically active A subunit and are promising and excellent adjuvant candidates for mucosal vaccination. These molecules are currently undergoing pre-clinical evaluation as potential mucosal adjuvants for use in humans. Thus, several potentially safe and effective mucosal adjuvants with the ability to redirect the immune system to a Th1 or Th2-type response are on the horizon.

Introduction

Adjuvants are substances which enhance humoral and/or cellular immune responses when given with vaccine antigen. Many antigens (vaccines) including proteins, are often poorly immunogenic. Therefore, adjuvants are needed to enhance the immunogenicity of these vaccines. To date, the only adjuvants approved by the US Federal Drug Administration (FDA) for use with parenteral vaccines are aluminium compounds (aluminium hydroxide and aluminium phosphate) or Alum; however, Alum is a relatively weak adjuvant (1-3). The protection of mucosal surfaces such as the respiratory, gastrointestinal and urogenital tracts is mediated by the production of secretory IgA (S-IgA) antibodies (Abs) which can be effectively induced only when vaccine is administered by a mucosal route (4). Novel adjuvants are needed which can be administered by mucosal routes such as nasal, orogastrointestinal or urogenital tracts for induction of protective S-IgA Abs at mucosal surfaces. It is of interest that oral administration of enterotoxins such as cholera toxin (CT) or heat labile enterotoxin (LT) does not induce tolerance to these molecules and abrogates oral tolerance to co-administered proteins, despite the fact that feeding of large amounts of protein generally induces a state of systemic unresponsiveness (oral tolerance) (5, 6).

The development of new adjuvants which are more potent is necessary for the next generation of vaccines. Although a number of characteristics for an ideal adjuvant can be described, the most important issue is safety (which includes lack of carcinogenicity, teratogenicity and abortogenicity)(1). Other attributes of an adjuvant should be that it: (1) is biodegradable and biocompatible, (2) does not induce frank inflammatory reactions, (3) does not cross-react with human antigens, (4) induces humoral and/or cell-mediated immune responses, (5) elicits systemic and mucosal immunity, (6) is of defined chemical composition, (7) is stable during manufacture, on a shelf, and preferably at room temperature to avoid the cold chain, (8) is inexpensive, (9) is manufactured with the same reproducibility on a large scale, and finally (10) can be administered with different types of vaccines through several routes including mucosal delivery.

Many adjuvant candidates were previously reported which had high potency for adjuvanticity but displayed toxicity incompatible with clinical use (7). In fact, several of these adjuvant candidates induce inflammatory responses. The most potentially promising adjuvant candidates would be determined after taking into consideration their potency for adjuvanticity, a low level of side effects, and the type of immune response required for a particular vaccine to be effective. Thus, for toxoid vaccines (i.e., tetanus, diphtheria, cholera, etc.) a strong serum or mucosal Ab response (T-helper 2 (Th2) type) would be required, while viral or intracellular bacterial pathogen vaccines would require vigorous cellular mediated immune (CMI) responses (T-helper (Th1) 1 type).

Development of Nontoxic Adjuvants

Current studies are attempting to develop new adjuvant candidates which are non-toxic and which have adjuvant activity for human use. This brief review is not intended to be an all inclusive discussion of adjuvants but is focused on the characteristics of several current and potential human adjuvant candidates.

Mineral Compounds

Alum ($AlK(SO_4)_2.12H_2O$) has been widely used as an adjuvant in humans since the 1930's. It is considered to be safe even though it is known to induce IgE Ab responses. Immune responses are skewed toward the Th2 type and thus Alum is not suitable for induction of directed CMI-type responses (1, 8). Not all proteins and peptides can be adsorbed to Alum. Aluminium compounds have not always enhanced the immunogenicity of vaccines and have failed to augment the protection provided by vaccines against whooping cough and typhoid fever. Although other salts of metals such as calcium phosphate, cerium nitrate, zinc sulfate, colloidal iron hydroxide, and calcium chloride also increases the antigenicity of the toxoids, Alum has been shown to be superior to these other compounds as adjuvants (9).

Biodegradable Microspheres

Biodegradable microspheres are composed of the biodegradable and biocompatible co-polymer poly (DL-lactide-co-glycolide) (DL-PLG) into which antigen is incorporated. Controlled release of antigen molecules at a rate of degradation which is determined by the ratio of lactide to glycolide enables induction of protective immune responses with a minimal number of immunizations and establishment of long-term immune responses (10-13). Microspheres increase antigen stability in environments such as the digestive tract by serving to protect antigen from proteases and low pH. Particle-size influences adjuvanticity. When given by the oral route, particles of 10 μm (or smaller) in diameter are more effective than particles >10 μm, presumably due to enhanced uptake by M cells in the gastrointestinal tract (14). Microspheres induce CMI including CD4$^+$ Th1 and CD8$^+$ cytotoxic T lymphocytes (CTLs) as well as Ab responses after both systemic or mucosal administration (15, 16). For instance, a synthetic peptide from the measles virus representing a cytotoxic T cell epitope, when encapsulated in microspheres, induced strong CTL responses after a single intraperitoneal immunization of mice (17). Furthermore, immunization of mice with a short CTL epitope from the circumsporozoite protein of *Plasmodium berghei* encapsulated in microspheres enhanced specific CTL responses comparable to those obtained with the incomplete Freund's adjuvant formulation (18). On the other hand, the Ab response induced through immunization with encapsulated staphylococcal enterotoxin B toxoid (SEB) was protective against the weight loss and splenic V beta 8$^+$ T cell expansion induced by intravenous injection of SEB (19). The continuous release of encapsulated QS-21 with HIV-1 recombinant gp120 at the injection site provided higher immune responses than was induced by bolus injection (20). Production problems, exposure of antigens to organic solvents, and limited uptake by mucosal routes are difficulties yet to be overcome before microspheres can be considered practical for mucosal immunization.

Liposomes

Liposomes consist of phospholipid bilayers separated by aqueous compartments that have been widely used *in vivo* and *in vitro* as carriers of antigens and adjuvants. Liposomes are poorly immunogenic; however, the incorporation of molecules such as drugs and antigens protects these molecules from degradation, and increases antigen presentation to macrophages (21). It was reported that liposomes acted as adjuvants with a number of antigens such as diphtheria toxoid (22), adenovirus type 5 (23) and served as a carrier for the adjuvant lipid A constituent of endotoxin (24). The clearance of adenovirus type 5 incorporated into liposomes was markedly prolonged (25). The use of the cationic liposome, DOTAP (N-(1-(2,3-dioleoyloxy)propyl)-N,N,N-trimethylammonium methylsulfate) originally used for DNA transfection of mammalian cells (26, 27), reduced the dose of antigen and adjuvant required, i.e., tetanus toxoid (TT) and mutant cholera toxin (mCT), presumably due to an improvement in stability of TT and mCT against digestive enzymes in the gastrointestinal tract (28). Both humoral and cell-mediated immunity are involved in the enhancement of immune responses by liposomes (29, 30). One liposomal formulation given orally to humans induced both mucosal and systemic antibody responses (31).

Immunostimulating Complexes (ISCOMs)

ISCOMs consist of approximately 40 nm cage-like particles of cholesterol, Quil A and amphipathic antigen (32). Quil A, a partially purified saponin extract derived from the bark of the *Quillaja saponaria* tree, consists of at least 23 different saponin derivatives some of which are hemolytic and toxic (33). Therefore, ISCOMs are limited to use in veterinary vaccines. The efficacy of adjuvanticity of ISCOMs was investigated in a variety of species including mice, rabbits and monkeys using human vaccines such as influenza (34), measles (35), rabies (36), gps40 from EB-virus (37) and gp120 from HIV (38). While viral envelope proteins readily incorporate into ISCOMs, soluble hydrophilic protein antigens generally do not readily incorporate into the ISCOM complex. ISCOMs stimulated humoral and cell-mediated immune responses and induced CD8$^+$ CTLs (34, 39, 40). The ability of ISCOMs to induce MHC class I restricted CTLs should play an important role in protection from viral pathogens.

QS-21

QS-21 is one of the numerous saponins found in Quil A. This triterpene glycoside has been purified by high pressure liquid chromatography to greater than 98 % purity (33) and its structure elucidated (41). The molecule has a molecular weight of 1990 Da, is water soluble and thus can readily be formulated into aqueous vaccines.

Most pre-clinical studies in the mouse employed doses of 10-20 µg mixed with a variety of protein antigens delivered subcutaneously (sc) or intramuscularly (im). These studies (see ref. 42 for a review) indicated that QS-21 induced strong Th1 type immune responses which were associated with vigorous antigen-specific CTLs and serum IgG1, IgG2b and IgG2a Ab responses. Recent studies (43) have shown that QS-21 delivered intranasally with a DNA plasmid vaccine encoding the gp160 envelope protein of HIV enhanced both humoral anti-gen-specific Ab primarily of the IgG2a isotype, intestinal S-IgA Ab responses in addition to CTLs. Our studies have clearly shown that QS-21 can also be delivered by the oral route and that Th2 (doses < 50 µg) or mixed Th1 and Th2-type (dose > 100 µg) Ab responses are elicited against tetanus toxoid (TT) (44). Moreover, mucosal (intestinal and vaginal) Abs which were induced were IL-12 dependent (44). QS-21 delivered with TT by the nasal route also enhanced serum and mucosal antibody responses. The overall enhancement of the im-mune response when QS-21 was employed as an adjuvant with TT was comparable to those when cholera toxin (CT) was used as an adjuvant.

QS-21 acts as an adjuvant in a variety of higher mammals and it has been advanced to several human Phase I and Phase II clinical trials (for a review see ref. 45). QS-21 as adjuvant is currently being assessed in vaccines for melanoma, breast cancer, prostate cancer, HIV, streptococcal pneumonia, malaria, influenza, herpes and hepatitis B. QS-21 has been used in these studies at doses of 10 to 200 µg subcutaneously (s.c.) or intramuscularly (i.m.) with the appropriate antigen. While most of these trials were designed to assess safety, a study using a GM2 ganglioside-keyhole limpet haemocyanin conjugate vaccine (GM2 KLH) and QS-21 demonstrated adjuvanticity. In terms of safety, QS-21 has been injected into over 1,500 hu-man subjects with no serious adverse events, although the 200 µg dose administered s.c. or i.m. did produce local erythema (up to 20 cm), induration and tenderness. No ulceration or subcutaneous nodules were observed and systemic symptoms were mild. To our knowledge, no human trials employing QS-21 as a mucosal adjuvant have been conducted. It would appear from the data accumulated to date that QS-21 is a promising candidate for use as an adjuvant in both systemic and mucosally administered human vaccines.

Cytokines

Cytokines are secreted proteins, generally of low to moderate molecular weight, that act as signals for communication between different cell types or serve an autocrine function. A single cytokine may be pleotrophic and thus exhibit a variety of biological responses depend-ent upon the particular cell phenotype. The cytokine network together with hormones and growth factors act via membrane receptors inducing gene transcription which may result in cell activation, trafficking, differentiation or proliferation. The immune system has been shown to consist of distinct CD4$^+$ T helper subsets distinguished by the cytokine patterns they secrete following stimulation. Th1 CD4$^+$ cells secrete IL-2, IFN-γ and LT-α upon stimulation, while Th2 CD4$^+$ T cells secrete IL-4, IL-5, IL-6, IL-10 and IL-13 (46). Since the outcome of an immune response is dependent upon the stimulation of Th cell subsets, various cytokines have been used with both conventional and immunotherapeutic vaccines to tip the balance in favor of the desired immune response. The stimulation of the Th1 cell subset with resultant generation of CTL activity is effective for combating most intracellular viral and bacterial infections as well as carcinomas. Activation of this T cell subset also promotes neutralizing complement fixing antibody as well as antibody dependent effector cells. Activation of Th2 type immune responses is important for most extracellular bacterial infections, as well as neutralization of exotoxins. The above is a generalization and there is some overlap in responses. Thus a Th1 response may prove protective for exotoxins. However, a strong Th2 response generally does not promote significant CTL activity. We will not attempt to review here all cytokines that have shown adjuvant activity with various vaccines but will use IL-2 and IL-12 as an illustration for their potential use and concomitant pitfalls for cytokines as potential adjuvants.

Human peripheral blood lymphocytes when stimulated with T cell mitogens were found to secrete a factor (IL-2) which promoted T cell growth (47). Subsequent data indicated that this molecule acted via an IL-2 receptor found primarily on T cells, B cells, natural killer (NK) cells and lymphokine activated killer (LAK) cells. The ability to expand the T cell population including NK and LAK cells by administering recombinant IL-2 led to a flurry of clinical testing of this cytokine for a variety of diseases. Metastatic malignant melanoma was an early target for IL-2 adjuvant effects (48). However, response rates of only 15-20 % were obtained with a variety of regimens and there is no consensus for optimal dose and treatment schedule. Metastatic renal cell carcinomas have been treated by subcutaneous injection of IL-2 with similar response rates (15-20 %) as malignant melanomas (49). Subcutaneous administration generally resulted in lower toxic responses than intravenous infusion. Certain leukemias, particularly chronic myelogenous leukemia, have responded to treatment with IL-2 (50). Recent studies of HIV$^+$ subjects receiving subcutaneous IL-2 treatment have shown an increase in CD4$^+$ cells, prolonged reduction of levels of viremia, and the cells regained the ability to secrete IL-2, IL-4 and IFN-γ *in vitro* (51). The improvement in immune status may prove beneficial in the suppression of various secondary infections. Two recent studies point to the potential future use of IL-2 for immunotherapy. A tetravalent guanylhydrazone (CN1-1493) has been found to block the toxic effects of fatal doses of IL-2 in an animal model. CN1-1493 is an inhibitor of macrophage activation and thus blocks nitric oxide and tumor necrosis factor production implicated in IL-2 toxicity. Moreover, combination therapy with IL-2 led to the regression of 10 of 10 hepatomas in these animals. Thus, if tolerated in humans, this compound could lead to more aggressive treatment with IL-2 (52). The recent discovery that two commercially available recombinant IL-2 preparations (Chiron IL-2 and Hoffmann La Roche IL-2) differ in their biological and clinical efficacy (1 IU HLR IL-2 = 3 to 5 IU of Chiron IL-2) will lead to a re-examination of a number of clinical trials in light of the source of IL-2 employed in a particular study (53).

IL-12 is a heterodimeric cytokine secreted primarily by activated macrophages, dendritic cells and possibly other antigen-presenting cells (54, 55). IL-12 in turn, stimulates NK and T cells to produce IFN-γ directing Th0 cell precursors into a Th1 pathway. Numerous studies have shown that IL-12 has strong cytotoxic T cell adjuvant activity to a variety of antigens when delivered by the parental route. Interestingly, when delivered by a mucosal (nasal) route with antigen, it was recently shown to induce S-IgA Ab responses and a mixed Th1 and Th2 antigen-specific response (56). *In vitro* studies with human lymphocytes infected with HIV (57) demonstrating that IL-12 restored HIV-specific CTL activity provided additional impetus to conduct Phase I clinical trials in humans. A Phase I dose escalation trial was conducted on patients with advanced cancer to determine toxicity, maximum tolerated dose, biological and antineoplastic efficacy for IL-12 given by the intravenous route. Patients were given a single injection, rested 2 weeks followed by once daily injections for 5 days every 3 weeks. This study established that 500 ng/kg was the maximum tolerated dose. As with most cytokines, the *in vivo* half life was relatively short (5-10 hrs). Biological effects included a rise in serum neopterin (an indication of activated monocytes and macrophages (58) and a dose-dependent increase in IFN-γ. Failure of the disease to progress or partial, transient responses to IL-12 were found in six of 40 patients (59). In a Phase II trial, the 500 ng/kg dose was given successively (without the 2 week interim) and of the 17 renal cancer patients receiving this regimen, 12 patients required hospitalization and 2 died of acute toxicity (60). It was later determined in mice and non-human primates, that a single dose followed by a rest before giving consecutive doses of IL-12 attenuated the IFN-γ response and acute toxicity. While IL-12 is still considered a promising potential adjuvant, this study reinforces the fact that adjuvants are a balance between toxicity and adjuvanticity (61) and carefully controlled clinical trials must be carried out for all potential adjuvants. Mucosal delivery of IL-12 may prove less toxic than i.v. infusion and is an effective adjuvant (56). Whether such a strategy will prove effective for certain human conventional and immunotherapeutic vaccines remains to be investigated.

Bacterial Products as Potential Adjuvants

Bordetella pertussis Compounds (Pertussigen)

Killed whole organisms of *Bordetella pertussis* have strong adjuvant effects for immune responses to co-administered diphtheria and tetanus toxoids, and has been given as the diphtheria-tetanus-pertussis (DTP) vaccine since 1927 (62). However, due to adverse events such as erythema, swelling, fever, and much less frequently convulsions and hypotonic hyporesponsive episodes, the development of a more purified pertussis vaccine associated with a lower frequency of adverse events was considered to be necessary. Acellular pertussis vaccines contain inactivated pertussis toxin which is detoxified either by treatment with a chemical or by using molecular genetic techniques and may contain one or more other bacterial components. Though the acellular pertussis vaccine contains substantially less endotoxin than whole-cell pertussis vaccines, the adjuvanticity of the cellular form was thought to be due to the presence of pertussis toxin (PT) and LPS in the vaccine (63, 64). Since PT itself is too toxic to be administered to humans, a recent genetically detoxified PT, 9K-129G was generated and results suggested that it has mucosal adjuvanticity when intranasally administered with antigen (65). Indeed, when tetanus toxoid (TT) plus PT 9K-129G was given by the nasal route to mice, IgG1 and IgA but not anti-TT IgE were induced and this immune response was protective against a tetanus toxin challenge (65).

Lipopolysaccharide (LPS)

LPS has a number of biological activities including adjuvant activity. However, LPS administration results in pyrogenicity and toxicity (66-68). LPS consists of three covalently linked regions: O-specific polysaccharide, core oligosaccharide and lipid A. One of the chemically modified forms of lipid A, monophosphorylipid A (MPL^R) was not only relatively less toxic and less pyrogenic than native LPS, but retained many of the biological activities including adjuvanticity (69-71). Another attempt for detoxification of LPS was the removal of fatty acid chains from lipid A. Removal of the fatty acid chains linked to hydroxyl groups of 3-hydroxytetradecanoyl residues on lipid A resulted in loss of most of the biological activity including toxicity, but did not affect adjuvanticity (72). These derivatives induce CTL activity and Th1-type helper cells. MPL^R and a mycobacterial cell wall extract (CWS) used in combination (Detox^{TM}) is currently undergoing clinical trials as an adjuvant in immunotherapy of cancer and other terminal diseases.

Muramyl Dipeptide (MDP)

MDP (N-acetyl-muramyl-L-alanyl-D-isoglutamine) is the smallest water-soluble component derived from mycobacterial cell walls and is one of the active components in Freund's complete adjuvant. Although the mechanism of adjuvanticity of MDP is unknown, MDP itself could induce IL-1 production which activates macrophages and T cells. MDP itself is unacceptable for use as an adjuvant for humans due to its pyrogenicity (7, 73). Derivatives of MDP have been generated such as murametide, threonyl-MDP and murabutide which retain adjuvanticity with less toxicity, and have been evaluated in clinical trials (74). MDP enhances the serum Ab responses to influenza vaccine. Derivatives of MDP increase cell-mediated immunity when given parenterally. It was demonstrated that immunization with murabutide and TT in humans showed an adjuvant effect, although these compounds have been associated with adverse reactions in a clinical trial (75).

Lipopeptides

The molecular structure of one (P3C-SSNA) of the synthetic lipopeptides is N-palmitoyl-S-(2,3-bis(palmitoyloxy)-(2RS) propyl)-(R)-cystenyl-(S)-asparaginyl-(S)-alanine). Lipopeptides are non-toxic when injected. These lipopeptide analogs induced humoral immune responses comparable to those obtained by complete Freund's adjuvant and could decrease the quantity of vaccine required (76). Furthermore, conjugation of lipopeptides with a CTL epitope induced CTLs which eliminated virus-infected cells.

Derivatives of Cholera Toxin and Heat-Labile Enterotoxin of E. coli

Cholera toxin (CT) is a bacterial protein toxin produced by *Vibrio cholerae* very similar to the heat labile toxin (LT) produced by enterotoxigenic *Escherichia coli* (77). These toxins are known to be strong adjuvants when given by the oral, intranasal or parenteral routes. Moreover, both toxins exhibit long-term memory to co-administered protein as well as to themselves (78, 79). However, both holotoxins cause watery diarrhea in humans which is mediated by the A subunit of the molecule. This region is responsible for catalysis of ADP-ribosylation of the G protein, Gsα (77). Therefore, many studies have been focused on altering the A subunit of the molecule to eliminate Gsα ADP ribosyltransferase but to retain adjuvanticity.

Several chimeric fusion proteins have been constructed with the CTB/A2 subunits (80) or the A1 subunit (81) of CT. In the case of the former construct the CTB/A2 subunits were fused to a 42 kDa segment of the saliva binding region (SBR) of Ag I/II from *Streptococcus mutans*. The goal was to elicit both serum and more importantly S-IgA in the oral cavity for

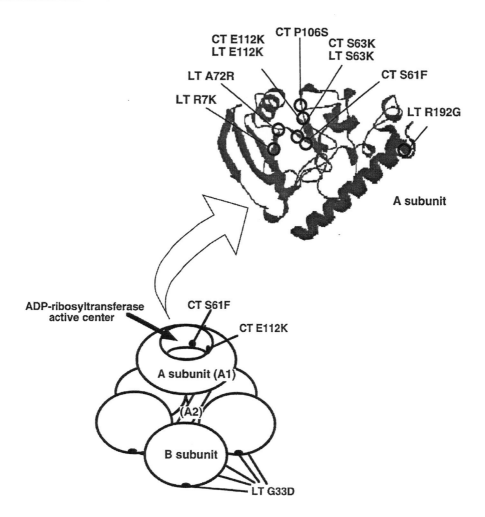

Figure 1. Amino acid substitutions in mutants of CT and LT in attempts to develop nontoxic adjuvants.

protection against dental caries. While this construct was shown to elicit salivary and intestinal S-IgA when given by the nasal or intragastric routes, SBR immunogenicity was greatly enhanced when given with CT. This suggests that this construct acts more as a carrier (binding to mucosal surfaces facilitating antigens entry) than as an adjuvant. The second fusion protein was generated by genetic fusion of the CTA1 C-terminal subunit of cholera toxin to two Ig-binding domains (DD) of staphylococcal protein A. This protein CTA1-DD (37 kDa) retained approximately 25 % of the ADP ribosyltransferase activity compared to holotoxin. The protein was found to bind to all Ig , CTA1-DD when given i.v. or i.n. elicited an adjuvant effect (although somewhat lower than CT) at a dose 10 fold higher than that of CT. The authors argue that ADP-ribosylation is the dominating adjuvant component of CT. However, this view is still controversial as discussed below.

We have summarized a number of individual amino acid substitutions in derivatives of LT and CT (for instance, LT R7K, CT S61F designate that arginine at position 7 of the LT A subunit was substituted for lysine and that serine at position 61 of the CT A subunit was substituted for phenylalanine, respectively) (Figure 1). LT mutants S61F (82) and E112K (83) were the first attempts to inhibit ADP-ribosyltransferase activity and diarrheagenicity. Although a number of CT and LT mutants have been created using site-directed mutagenesis by predicting the ADP-ribosyltransferase active center and NAD binding site by computer analysis of the crystallographic structure of these molecules, some of them were unstable or retained toxic activity. However, several molecular derivatives could be candidates for the purpose of developing vaccines against cholera or traveller's diarrhea by CT or LT since they induced neutralizing antibodies against the A subunit in murine models (84-86).

The first report of a mucosal adjuvant of an LT mutant suggested that neither recombinant cholera toxin B subunit nor a non-toxic mutant of LT E112K exhibit adjuvanticity when given orally and concluded that ADP-ribosyltransferase activity was necessary for adjuvanticity of CT or LT (87). However, in later studies LT R7K (88), S63K (89), A72R (90), CT P106S, S63K (91) all retained their adjuvanticity when given nasally, suggesting ADP-ribosyltransferase activity is not necessary for their adjuvanticity. The effectiveness of another LT mutant R192G as a mucosal adjuvant has been shown by a combination of the oral and parenteral routes (92). Mice nasally given rotavirus 2/6 virus like particles composed of proteins VP2 and VP6 (2/6-VLPs) with LT R192G were totally protected from rotavirus challenge. Specific humoral and cellular immune responses via Th1-type responses were induced when killed *Salmonella dublin* with LT R192G were given orally and the responses were protective against lethal oral challenge using wild-type *S. dublin* (93, 94). LT mutant G33D which lacks the ability to bind to the receptor GM-1 was able to serve as an effective nasal adjuvant (95) although it did not induce IgG or IgA directed against co-administered antigen when given orally (96). CT S61F and E112K showed adjuvanticity comparable to that of native CT when used at doses of 5-10 fold that of CT via the parenteral or intranasal routes (97, 98). Protein antigen plus CT S61F induced high serum titers of antigen-specific IgG Abs, largely of the IgG1 and IgG2b subclasses, as well as high antigen-specific IgA Ab responses in mucosal secretions with increased numbers of antigen-specific IgG and IgA Ab forming cells in mucosal tissues (97). Moreover, nasal immunization of pneumococcus surface protein A plus CT S61F showed protection against a fatal challenge with *Pneumococci* (99). The beneficial aspects of intranasal administration when compared with parenteral immunization were lower dose of antigen required to induce Ab responses and the induction of not only serum IgG Ab responses, but also mucosal S-IgA Ab responses in mucosal effector tissues.

As the LT or CT mutants may be less stable in the gastrointestinal tract than wild type toxins, the use of the cationic liposome DOTAP could effectively induce immune responses and decrease the dose of antigen and adjuvant when protein antigen plus CT E112K was orally administered to mice (28). Furthermore, adjuvanticity for CT E112K was induced via up-regulation of mainly B7-2 on APCs and through preferential inhibition of Th1-type $CD4^+$ cell responses (100). The two CT mutants, E112K and S61F elicit $CD4^+$ Th2 type responses via IL-4 production which enhances the responses of IgG1 and IgG2b subclasses and IgE responses when given nasally or orally.

Concluding Remarks

Despite considerable efforts to develop other adjuvant candidates, no other adjuvant since Alum has yet reached widespread use in human vaccination. Several cytokines including IL-2 and IL-12 are being tested for immunotherapeutic activity in humans and studies should be extended to investigate their efficacy as mucosal adjuvants. The proven safety of the purified saponin QS-21 when delivered parenterally and its ability to act as a mucosal adjuvant in animal models should spur additional trials for its use in human mucosal vaccinations. The development of mutated derivatives of LT or CT which are devoid of ADP-ribosyltransferase activity (toxicity) could be promising candidates for use as both parenteral and mucosal adjuvants. Pre-clinical evaluation of the safety and efficacy of these molecules for use in humans as mucosal adjuvants is in progress. Thus, an array of different molecules have been identified as potential mucosal adjuvants and only the rigors of clinical testing will prove which are suitable for human use.

Acknowledgements

We thank Drs. P. N. Boyaka and F.W. van Ginkel for helpful discussions of this manuscript. The work in our laboratory is supported by NIH grants AI-18958, AI-43197, AI-35932, DK-44240, AI-65298, AI-65299 and DE-09837 and by Aquila Biopharmaceuticals, Inc.

References

1. Gupta, R.K., Rost, B.E., Relyveld, E. and Siber, G.R. 1995. Adjuvant properties of aluminium and calcium compounds. Vaccine Design, The Sub-unit and Adjuvant Approach. Plenum Press, New York. p. 229-248.
2. Bomford, R. 1980. The comparative selectivity of adjuvants for humoral and cell-mediated immunity. I. Effect on the antibody response to bovine serum albumin and sheep red blood cells of Freund's incomplete and complete adjuvants, alhydrogel, *Corynebacterium parvum*, *Bordetella pertussis*, muramyl dipeptide and saponin. Clin. Exp. Immunol. 39 (2): 426-434.
3. Vanselow, B. A., I. Abetz, and K. Trenfield. 1985. A bovine ephemeral fever vaccine incorporating adjuvant Quil A: a comparative study using adjuvants Quil A, aluminium hydroxide gel and dextran sulphate. Vet. Rec. 117 (2): 37-43.
4. Mestecky, J., and J. R. McGhee. 1987. Immunoglobulin A (IgA): molecular and cellular interactions involved in IgA biosynthesis and immune response. Adv. Immunol. 40: 153-245.
5. Elson, C. O., and W. Ealding. 1984. Cholera toxin feeding did not induce oral tolerance in mice and abrogated oral tolerance to an unrelated protein antigen. J. Immunol. 133 (6): 2892-2897.
6. Clements, J. D., N. M. Hartzog, and F. L. Lyon. 1988. Adjuvant activity of *Escherichia coli* heat-labile enterotoxin and effect on the induction of oral tolerance in mice to unrelated protein antigens. Vaccine. 6 (3): 269-277.
7. Riveau, G., K. Masek, M. Parant, and L. Chedid. 1980. Central pyrogenic activity of muramyl dipeptide. J. Exp. Med. 152 (4): 869-877.
8. Brewer, J. M., M. Conacher, A. Satoskar, H. Bluethmann, and J. Alexander. 1996. In interleukin-4-deficient mice, alum not only generates T helper 1 responses equivalent to Freund's complete adjuvant, but continues to induce T helper 2 cytokine production. Eur. J. Immunol. 26 (9): 2062-2066.
9. Glenny, A. T., G. A. H. Buttle, and M. F. Stevens. 1931. Rate of disappearance of diphtheria toxoid injected into rabbits and guinea pigs: toxoid precipitated with alum. J. Pathol. Bacteriol. 34: 267.
10. Miller, R. A., J. M. Brady, and D. E. Cutright. 1977. Degradation rates of oral resorbable implants (polylactates and polyglycolates): rate modification with changes in PLA/PGA copolymer ratios. J. Biomed. Mater. Res. 11 (5): 711-9.

11. O'Hagan, D. T., D. Rahman, J. P. McGee, H. Jeffery, M. C. Davies, P. Williams, S. S. Davis, and S. J. Challacombe. 1991. Biodegradable microparticles as controlled release antigen delivery systems. Immunology. 73 (2): 239-42.

12. Eldridge, J. H., J. K. Staas, D. Chen, P. A. Marx, T. R. Tice, and R. M. Gilley. 1993. New advances in vaccine delivery systems. Semin. Hematol. 30 (4 Suppl 4): 16-24; discussion 25.

13. Gupta, R. K., J. Alroy, M. J. Alonso, R. Langer, and G. R. Siber. 1997. Chronic local tissue reactions, long-term immunogenicity and immunologic priming of mice and guinea pigs to tetanus toxoid encapsulated in biodegradable polymer microspheres composed of poly lactide-co-glycolide polymers. Vaccine. 15 (16): 1716-1723.

14. Eldridge, J. H., C. J. Hammond, J. A. Meulbroek, J. K. Staas, R. M. Gilly, and T. R. Tice. 1990. Controlled vaccine release in the gut-associated tissues. 1. Orally administered biodegradable microspheres target the Peyer's patches. J. Control Release. 11: 205-214.

15. Maloy, K. J., A. M. Donachie, D. T. O'Hagan, and A. M. Mowat. 1994. Induction of mucosal and systemic immune responses by immunization with ovalbumin entrapped in poly(lactide-co-glycolide) microparticles. Immunology. 81 (4): 661-667.

16. Moore, A., P. McGuirk, S. Adams, W. C. Jones, J. P. McGee, D. T. O'Hagan, and K. H. Mills. 1995. Immunization with a soluble recombinant HIV protein entrapped in biodegradable microparticles induces HIV-specific CD8$^+$ cytotoxic T lymphocytes and CD4$^+$ Th1 cells. Vaccine. 13 (18): 1741-1749.

17. Partidos, C. D., P. Vohra, D. Jones, G. Farrar, and M. W. Steward. 1997. CTL responses induced by a single immunization with peptide encapsulated in biodegradable microparticles. J. Immunol. Methods. 206 (1-2): 143-19051.

18. Men, Y., H. Tamber, R. Audran, B. Gander, and G. Corradin. 1997. Induction of a cytotoxic T lymphocyte response by immunization with a malaria specific CTL peptide entrapped in biodegradable polymer microspheres. Vaccine. 15 (12-13): 1405-1412.

19. Eldridge, J. H., J. K. Staas, J. A. Meulbroek, T. R. Tice, and R. M. Gilley. 1991. Biodegradable and biocompatible poly(DL-lactide-co-glycolide) microspheres as an adjuvant for staphylococcal enterotoxin B toxoid which enhances the level of toxin-neutralizing antibodies. Infect. Immun. 59 (9): 2978-2986.

20. Cleland, J. L., A. Lim, A. Daugherty, L. Barron, N. Desjardin, E. T. Duenas, D. J. Eastman, J. C. Vennari, T. Wrin, P. Berman, K. K. Murthy, and M. F. Powell. 1998. Development of a single-shot subunit vaccine for HIV-1. 5. programmable *in vivo* autoboost and long lasting neutralizing response. J. Pharm. Sci. 87 (12): 1489-1495.

21. Alving, C. R., J. N. Verma, M. Rao, U. Krzych, S. Amselem, S. M. Green, and N. M. Wassef. 1992. Liposomes containing lipid A as a potent non-toxic adjuvant. Res. Immunol. 143 (2): 197-198.

22. Allison, A. G., and G. Gregoriadis. 1974. Liposomes as immunological adjuvants. Nature. 252 (5480): 252.

23. Kramp, W. J., H. R. Six, S. Drake, and J. A. Kasel. 1979. Liposomal enhancement of the immunogenicity of adenovirus type 5 hexon and fiber vaccines. Infect. Immun. 25 (2): 771-773.

24. Schuster, B. G., M. Neidig, B. M. Alving, and C. R. Alving. 1979. Production of antibodies against phosphocholine, phosphatidylcholine, sphingomyelin, and lipid A by injection of liposomes containing lipid A. J. Immunol. 122 (3): 900-905.

25. Kramp, W. J., H. R. Six, and J. A. Kasel. 1982. Postimmunization clearance of liposome entrapped adenovirus type 5 hexon. Proc. Soc. Exp. Biol. Med. 169 (1): 135-139.

26. Wattiaux, R., M. Jadot, F. Dubois, S. Misquith, and S. Wattiaux-De Coninck. 1995. Uptake of exogenous DNA by rat liver: effect of cationic lipids. Biochem. Biophys. Res. Commun. 213 (1): 81-87.

27. Meyer, K. B., M. M. Thompson, M. Y. Levy, L. G. Barron, and F. C. Szoka, Jr. 1995. Intratracheal gene delivery to the mouse airway: characterization of plasmid DNA expression and pharmacokinetics. Gene Ther. 2 (7): 450-460.

28. Yamamoto, M., S. Yamamoto, M. Ohmura, M.-N. Kweon, J. VanCott, M. Noda, J. Peterson, K. Fujihashi, Y. Takeda, H. Kiyono, and J. R. McGhee. A nontoxic mutant of cholera toxin enhances immunity to oral vaccines in mice. (submitted).

29. Lawman, M. J., P. T. Naylor, L. Huang, R. J. Courtney, and B. T. Rouse. 1981. Cell-mediated immunity to herpes simplex virus: induction of cytotoxic T lymphocyte responses by viral antigens incorporated into liposomes. J. Immunol. 126 (1): 304-308.

30. Sanchez, Y., I. Ionescu-Matiu, G. R. Dreesman, W. Kramp, H. R. Six, F. B. Hollinger, and J. L. Melnick. 1980. Humoral and cellular immunity to hepatitis B virus-derived antigens: comparative activity of Freund complete adjuvant alum, and liposomes. Infect. Immun. 30 (3): 728-733.

31. Childers, N. K., S. M. Michalek, D. G. Pritchard, and J. R. McGhee. 1990. Mucosal and systemic responses to an oral liposome-Streptococcus mutans carbohydrate vaccine in humans. Reg. Immunol. 3 (6): 289-296.

32. Morein, B., B. Sundquist, S. Hoglund, K. Dalsgaard, and A. Osterhaus. 1984. Iscom, a novel structure for antigenic presentation of membrane proteins from enveloped viruses. Nature. 308 (5958): 457-460.

33. Kensil, C. R., U. Patel, M. Lennick, and D. Marciani. 1991. Separation and characterization of saponins with adjuvant activity from Quillaja saponaria Molina cortex. J. Immunol. 146 (2): 431-437.

34. Lovgren, K. 1988. The serum antibody response distributed in subclasses and isotypes after intranasal and subcutaneous immunization with influenza virus immunostimulating complexes. Scand. J. Immunol. 27 (2): 241-245.

35. de Vries, P., R. S. van Binnendijk, P. van der Marel, A. L. van Wezel, H. O. Voorma, B. Sundquist, F. G. Uytdehaag, and A. D. Osterhaus. 1988. Measles virus fusion protein presented in an immune-stimulating complex (iscom) induces haemolysis-inhibiting and fusion-inhibiting antibodies, virus-specific T cells and protection in mice. J. Gen. Virol. 69 (Pt 3): 549-559.

36. Fekadu, M., J. H. Shaddock, J. Ekstrom, A. Osterhaus, D. W. Sanderlin, B. Sundquist, and B. Morein. 1992. An immune stimulating complex (ISCOM) subunit rabies vaccine protects dogs and mice against street rabies challenge. Vaccine. 10 (3): 192-197.

37. Morgan, A. J., S. Finerty, K. Lovgren, F. T. Scullion, and B. Morein. 1988. Prevention of Epstein-Barr (EB) virus-induced lymphoma in cottontop tamarins by vaccination with the EB virus envelope glycoprotein gp340 incorporated into immune-stimulating complexes. J. Gen. Virol. 69 (Pt 8): 2093-2096.

38. Pyle, S. W., B. Morein, J. W. Bess, Jr., L. Akerblom, P. L. Nara, S. M. Nigida, Jr., N. W. Lerche, W. G. Robey, P. J. Fischinger, and L. O. Arthur. 1989. Immune response to immunostimulatory complexes (ISCOMs) prepared from human immunodeficiency virus type 1 (HIV-1) or the HIV-1 external envelope glycoprotein (gp120). Vaccine. 7 (5): 465-473.

39. Fossum, C., M. Bergstrom, K. Lovgren, D. L. Watson, and B. Morein. 1990. Effect of iscoms and their adjuvant moiety (matrix) on the initial proliferation and IL-2 responses: comparison of spleen cells from mice inoculated with iscoms and/or matrix. Cell. Immunol. 129 (2): 414-425.

40. Takahashi, H., T. Takeshita, B. Morein, S. Putney, R. N. Germain, and J. A. Berzofsky. 1990. Induction of $CD8^+$ cytotoxic T cells by immunization with purified HIV-1 envelope protein in ISCOMs [see comments]. Nature. 344 (6269): 873-875.

41. Jacobsen, N. E., W. J. Fairbrother, C. R. Kensil, A. Lim, D. A. Wheeler, and M. F. Powell. 1996. Structure of the saponin adjuvant QS-21 and its base-catalyzed isomerization product by 1H and natural abundance 13C NMR spectroscopy. Carbohydr. Res. 280 (1): 1-14.

42. Kensil, C. R. 1996. Saponins as vaccine adjuvants. Crit. Rev. Ther. Drug Carrier Syst. 13 (1-2): 1-55.

43. Sasaki, S., K. Sumino, K. Hamajima, J. Fukushima, N. Ishii, S. Kawamoto, H. Mohri, C. R. Kensil, and K. Okuda. 1998. Induction of systemic and mucosal immune responses to human immunodeficiency virus type 1 by a DNA vaccine formulated with QS-21 saponin adjuvant via intramuscular and intranasal routes. J. Virol. 72 (6): 4931-4939.

44. Boyaka, P. N., M. Marinaro, R. J. Jackson, F. W. van Ginkel, C. R. Kensil, and J. R. McGhee. 1999. The mucosal adjuvanticity of orally administered saponin QS-21 involves IL-12. Submitted.

45. Kensil, C. R., and R. Kammer. 1998. QS-21: a water-soluble triterpene glycoside adjuvant. Exp. Opin. Invest. Drugs 7: 1475-1482.

46. Mosmann, T. R., and R. L. Coffman. 1989. TH1 and TH2 cells: different patterns of lymphokine secretion lead to different functional properties. Annu. Rev. Immunol. 7: 145-173.

47. Morgan, D. A., F. W. Ruscetti, and R. Gallo. 1976. Selective *in vitro* growth of T lymphocytes from normal human bone marrows. Science. 193 (4257): 1007-1008.

48. Philip, P. A., and L. Flaherty. 1997. Treatment of malignant melanoma with interleukin-2. Semin Oncol. 24 (1 Suppl 4): S32-S38.

49. Bukowski, R. M. 1997. Natural history and therapy of metastatic renal cell carcinoma: the role of interleukin-2. Cancer. 80 (7): 1198-1220.

50. Goodman, M., L. Cabral, and P. Cassileth. 1998. Interleukin-2 and leukemia. Leukemia. 12 (11): 1671-1675.

51. De Paoli, P., S. Zanussi, C. Simonelli, M. T. Bortolin, M. D'Andrea, C. Crepaldi, R. Talamini, M. Comar, M. Giacca, and U. Tirelli. 1997. Effects of subcutaneous interleukin-2 therapy on CD4 subsets and *in vitro* cytokine production in HIV$^+$ subjects. J. Clin. Invest. 100 (11): 2737-2743.

52. Kemeny, M. M., G. I. Botchkina, M. Ochani, M. Bianchi, C. Urmacher, and K. J. Tracey. 1998. The tetravalent guanylhydrazone CNI-1493 blocks the toxic effects of interleukin-2 without diminishing antitumor efficacy. Proc. Natl. Acad. Sci. USA. 95 (8): 4561-4566.

53. Hank, J. A., J. Surfus, J. Gan, M. Albertini, M. Lindstrom, J. H. Schiller, K. M. Hotton, M. Khorsand, and P. M. Sondel. 1999. Distinct clinical and laboratory activity of two recombinant interleukin-2 preparations. Clin. Cancer Res. 5 (2): 281-289.

54. Trinchieri, G. 1995. Interleukin-12: a proinflammatory cytokine with immunoregulatory functions that bridge innate resistance and antigen-specific adaptive immunity. Annu. Rev. Immunol. 13: 251-276.

55. Hsieh, C. S., S. E. Macatonia, C. S. Tripp, S. F. Wolf, A. O'Garra, and K. M. Murphy. 1993. Development of Th1 CD4$^+$ T cells through IL-12 produced by *Listeria*- induced macrophages [see comments]. Science. 260 (5107): 547-549.

56. Boyaka, P. N., M. Marinaro, R. J. Jackson, S. Menon, H. Kiyono, E. Jirillo, and J. R. McGhee. 1999. IL-12 is an effective adjuvant for induction of mucosal immunity. J. Immunol. 162 (1): 122-128.

57. Clerici, M., D. R. Lucey, J. A. Berzofsky, L. A. Pinto, T. A. Wynn, S. P. Blatt, M. J. Dolan, C. W. Hendrix, S. F. Wolf, and G. M. Shearer. 1993. Restoration of HIV-specific cell-mediated immune responses by interleukin-12 *in vitro*. Science. 262 (5140): 1721-1724.

58. Fuchs, D., G. Weiss, and H. Wachter. 1993. Neopterin, biochemistry and clinical use as a marker for cellular immune reactions. Int. Arch. Allergy. Immunol. 101 (1): 1-6.

59. Atkins, M. B., M. J. Robertson, M. Gordon, M. T. Lotze, M. DeCoste, J. S. DuBois, J. Ritz, A. B. Sandler, H. D. Edington, P. D. Garzone, J. W. Mier, C. M. Canning, L. Battiato, H. Tahara, and M. L. Sherman. 1997. Phase I evaluation of intravenous recombinant human interleukin 12 in patients with advanced malignancies. Clin. Cancer Res. 3 (3): 409-417.

60. Leonard, J. P., M. L. Sherman, G. L. Fisher, L. J. Buchanan, G. Larsen, M. B. Atkins, J. A. Sosman, J. P. Dutcher, N. J. Vogelzang, and J. L. Ryan. 1997. Effects of single-dose interleukin-12 exposure on interleukin-12- associated toxicity and interferon-gamma production. Blood. 90 (7): 2541-2548.

61. Gupta, R. K., E. H. Relyveld, E. B. Lindblad, B. Bizzini, S. Ben-Efraim, and C. K. Gupta. 1993. Adjuvants - a balance between toxicity and adjuvanticity. Vaccine. 11 (3): 293-306.

62. Stewart, G. T. 1985. Whooping cough and pertussis vaccine: a comparison of risks and benefits in Britain during the period 1968-83. Dev. Biol. Stand. 61: 395-405.

63. Manclark, C. R., and J. L. Cowell. 1984. Pertussis. In: Bacterial Vaccines. R. Germanier, ed. Academic Press, New York. p. 69-106.

64. Munoz, J. J. 1985. Biological activities of pertussigen (pertussis toxin). In: Pertussis Toxin. R.D. Sekura, J. Moss and M. Vaughan, eds. Academic Press, Orlando. p. 1-18.

65. Roberts, M., A. Bacon, R. Rappuoli, M. Pizza, I. Cropley, G. Douce, G. Dougan, M. Marinaro, J. McGhee, and S. Chatfield. 1995. A mutant pertussis toxin molecule that lacks ADP-ribosyltransferase activity, PT-9K/129G, is an effective mucosal adjuvant for intranasally delivered proteins. Infect. Immun. 63 (6): 2100-2108.

66. Luderitx, O., C. Galanos, and E. T. Rietschel. 1982. Endotoxins of Gram-negative bacteria. Pharm. Ther. 15: 383.

67. Brade, H., L. Brade, and E. T. Rietschel. 1988. Structure-activity relationships of bacterial lipopolysaccharides (endotoxins). Current and future aspects. Zentralbl Bakteriol Mikrobiol Hyg [A]. 268 (2): 151-179.

68. Qureshi, N., and K. Takayama. 1990. Structure and function of lipid A. In: The Bacteria, Vol. XI. Molecular Basis of Bacterial Pathogenesis. B.H. Iglewski and V.L. Clark, eds. Academic Press, San Diego, CA. p. 318-39.

69. Ulrich, J. T., J. L. Cantrell, G. L. Gustafson, J. A. Rudbach, and J. R. Hiernaux. 1991. The adjuvant activity of monophosphoryl lipid A. In: Topics in Vaccine Adjuvant Research. D.R. Spriggs and W.C. Koff, eds. CRC Press, Boca Raton, FL. p. 133-43.

70. Rudbach, J. A., J. L. Cantrell, J. T. Ulrich, and M. S. Mitchell. 1990. Immunotherapy with bacterial endotoxin. In: Endotoxin. Plenum Press, New York. p. 665-76.

71. Ulrich, J. T., and K. R. Myers. 1995. Monophosphoryl lipid A as an adjuvant: past experiences and new directions. In: Vaccine Design: The Sub-unit and Adjuvant Approach. M.F Powell and M.J. Newman, eds. Plenum Press, New York. p. 495-524.

72. Munford, R. S., and C. L. Hall. 1986. Detoxification of bacterial lipopolysaccharides (endotoxins) by a human neutrophil enzyme. Science. 234 (4773): 203-205.

73. Parant, M. 1979. Biologic properties of a new synthetic adjuvant, muramyl dipeptide (MDP). Springer Seminars in Immunopathology. 2: 101-118.

74. Ott, G., G. Van Nest, and R. L. Burke. 1992. The use of muramyl peptides as vaccine adjuvants. In: Vaccine Research and Development. W.C. Koff and H.K. Six, eds. Marcel Dekker. p. 89-114.

75. Keitel, W., R. Couch, N. Bond, S. Adair, G. Van Nest, and C. Dekker. 1993. Pilot evaluation of influenza virus vaccine (IVV) combined with adjuvant. Vaccine. 11 (9): 909-913.

76. Bessler, W. G., and G. Jung. 1992. Synthetic lipopeptides as novel adjuvants. Res. Immunol. 143 (5): 548-553; discussion 579-80.

77. Spangler, B. D. 1992. Structure and function of cholera toxin and the related *Escherichia coli* heat-labile enterotoxin. Microbiol. Rev. 56 (4): 622-647.

78. Lycke, N., and J. Holmgren. 1986. Intestinal mucosal memory and presence of memory cells in lamina propria and Peyer's patches in mice 2 years after oral immunization with cholera toxin. Scand. J. Immunol. 23 (5): 611-616.

79. Vajdy, M., and N. Y. Lycke. 1992. Cholera toxin adjuvant promotes long-term immuno-logical memory in the gut mucosa to unrelated immunogens after oral immunization. Immunology. 75 (3): 488-492.

80. Hajishengallis, G., S. K. Hollingshead, T. Koga, and M. W. Russell. 1995. Mucosal im-munization with a bacterial protein antigen genetically coupled to cholera toxin A2/B subunits. J. Immunol. 154 (9): 4322-4332.

81. Agren, L. C., L. Ekman, B. Lowenadler, and N. Y. Lycke. 1997. Genetically engineered nontoxic vaccine adjuvant that combines B cell targeting with immunomodulation by cholera toxin A1 subunit. J. Immunol. 158 (8): 3936-3946.

82. Harford, S., C. W. Dykes, A. N. Hobden, M. J. Read, and I. J. Halliday. 1989. Inactiva-tion of the *Escherichia coli* heat-labile enterotoxin by *in vitro* mutagenesis of the A-subunit gene. Eur. J. Biochem. 183 (2): 311-316.

83. Tsuji, T., T. Inoue, A. Miyama, K. Okamoto, T. Honda, and T. Miwatani. 1990. A single amino acid substitution in the A subunit of *Escherichia coli* enterotoxin results in a loss of its toxic activity. J. Biol. Chem. 265 (36): 22520-22525.

84. Pizza, M., M. R. Fontana, M. M. Giuliani, M. Domenighini, C. Magagnoli, V. Giannelli, D. Nucci, W. Hol, R. Manetti, and R. Rappuoli. 1994. A genetically detoxified derivative of heat-labile *Escherichia coli* enterotoxin induces neutralizing antibodies against the A subunit. J. Exp. Med. 180 (6): 2147-53.

85. Pizza, M., M. Domenighini, W. Hol, V. Giannelli, M. R. Fontana, M. M. Giuliani, C. Magagnoli, S. Peppoloni, R. Manetti, and R. Rappuoli. 1994. Probing the structure-ac-tivity relationship of *Escherichia coli* LT-A by site-directed mutagenesis. Mol. Microbiol. 14 (1): 51-60.

86. Fontana, M. R., R. Manetti, V. Giannelli, C. Magagnoli, A. Marchini, R. Olivieri, M. Domenighini, R. Rappuoli, and M. Pizza. 1995. Construction of nontoxic derivatives of cholera toxin and characterization of the immunological response against the A subunit. Infect. Immun. 63 (6): 2356-2360.

87. Lycke, N., T. Tsuji, and J. Holmgren. 1992. The adjuvant effect of *Vibrio cholerae* and *Escherichia coli* heat-labile enterotoxins is linked to their ADP-ribosyltransferase activ-ity. Eur. J. Immunol. 22 (9): 2277-2281.

88. Douce, G., C. Turcotte, I. Cropley, M. Roberts, M. Pizza, M. Domenghini, R. Rappuoli, and G. Dougan. 1995. Mutants of *Escherichia coli* heat-labile toxin lacking ADP-ribosyltransferase activity act as nontoxic, mucosal adjuvants. Proc. Natl. Acad. Sci. USA. 92 (5): 1644-1648.

89. Di Tommaso, A., G. Saletti, M. Pizza, R. Rappuoli, G. Dougan, S. Abrignani, G. Douce, and M. T. De Magistris. 1996. Induction of antigen-specific antibodies in vaginal secre-tions by using a nontoxic mutant of heat-labile enterotoxin as a mucosal adjuvant. Infect. Immun. 64 (3): 974-979.

90. Giuliani, M. M., G. Del Giudice, V. Giannelli, G. Dougan, G. Douce, R. Rappuoli, and M. Pizza. 1998. Mucosal adjuvanticity and immunogenicity of LTR72, a novel mutant of *Escherichia coli* heat-labile enterotoxin with partial knockout of ADP-ribosyltransferase activity. J. Exp. Med. 187 (7): 1123-1132.

91. Douce, G., M. Fontana, M. Pizza, R. Rappuoli, and G. Dougan. 1997. Intranasal immunogenicity and adjuvanticity of site-directed mutant derivatives of cholera toxin. Infect. Immun. 65 (7): 2821-2828.

92. Dickinson, B. L., and J. D. Clements. 1995. Dissociation of *Escherichia coli* heat-labile enterotoxin adjuvanticity from ADP-ribosyltransferase activity. Infect. Immun. 63 (5): 1617-1623.

93. Chong, C., M. Friberg, and J. D. Clements. 1998. LT(R192G), a non-toxic mutant of the heat-labile enterotoxin of *Escherichia coli*, elicits enhanced humoral and cellular immune responses associated with protection against lethal oral challenge with *Salmonella* spp. Vaccine. 16 (7): 732-740.

94. O'Neal, C. M., J. D. Clements, M. K. Estes, and M. E. Conner. 1998. Rotavirus 2/6 viruslike particles administered intranasally with cholera toxin, *Escherichia coli* heat-labile toxin (LT), and LT-R192G induce protection from rotavirus challenge. J. Virol. 72 (4): 3390-3393.

95. de Haan, L., I. K. Feil, W. R. Verweij, M. Holtrop, W. G. Hol, E. Agsteribbe, and J. Wilschut. 1998. Mutational analysis of the role of ADP-ribosylation activity and GM1-binding activity in the adjuvant properties of the *Escherichia coli* heat-labile enterotoxin towards intranasally administered keyhole limpet hemocyanin. Eur. J. Immunol. 28 (4): 1243-1250.

96. Guidry, J. J., L. Cardenas, E. Cheng, and J. D. Clements. 1997. Role of receptor binding in toxicity, immunogenicity, and adjuvanticity of *Escherichia coli* heat-labile enterotoxin. Infect. Immun. 65 (12): 4943-4950.

97. Yamamoto, S., H. Kiyono, M. Yamamoto, K. Imaoka, K. Fujihashi, F. W. Van Ginkel, M. Noda, Y. Takeda, and J. R. McGhee. 1997. A nontoxic mutant of cholera toxin elicits Th2-type responses for enhanced mucosal immunity. Proc. Natl. Acad. Sci. USA. 94 (10): 5267-5272.

98. Yamamoto, S., Y. Takeda, M. Yamamoto, H. Kurazono, K. Imaoka, K. Fujihashi, M. Noda, H. Kiyono, and J. R. McGhee. 1997. Mutants in the ADP-ribosyltransferase cleft of cholera toxin lack diarrheagenicity but retain adjuvanticity. J. Exp. Med. 185 (7): 1203-1210.

99. Yamamoto, M., D. E. Briles, S. Yamamoto, M. Ohmura, H. Kiyono, and J. R. McGhee. 1998. A nontoxic adjuvant for mucosal immunity to pneumococcal surface protein A. J. Immunol. 161 (8): 4115-4121.

100. Yamamoto, M., H. Kiyono, S. Yamamoto, E. Batanero, M.-N. Kweon, S. Otake, M. Azuma, Y. Takeda, and J. R. McGhee. 1999. Direct effects on antigen-presenting cells and T lymphocytes explain the adjuvanticity of a nontoxic cholera toxin mutant. J. Immunol. 162: 7015-7021.

From: *Development of Novel Antimicrobial Agents: Emerging Strategies*
ISBN 1-898486-23-9 © 2001 Horizon Scientific Press, Wymondham, UK.

9

Attenuated Bacteria as Delivery Vehicles for *Helicobacter pylori* Vaccines

Thomas G. Blanchard and Steven J. Czinn

Abstract

Helicobacter pylori is the etiologic agent of gastritis and most peptic ulcers. Antibiotic therapy is complicated and prohibitively expensive in developing nations where the prevalence of infection is greater than 80%. Helicobacter vaccine development has shown great promise in rodent models but not in large animals or humans, largely owing to the lack of a suitable means of stimulating effective immunity by mucosal routes. Several live attenuated bacterial vaccines have been developed for use in humans and others are under development including *Vibrio cholerae*, *Mycobacterium bovis*, *Listeria monocytogenes*, *Shigella flexneri*, and *Salmonella typhi*. The success of these vaccines and the advantages conferred by using live bacterial vaccine vectors has generated an interest in exploiting this technology to deliver recombinant proteins to vaccinate against heterologous pathogens. Attenuated recombinant *Salmonella typhimurium* has now been employed for the development of a vaccine against *Helicobacter pylori* infection of mice. These initial studies indicate recombinant bacterial vaccine vectors may prove to be an effective solution to the lack of suitable mucosal adjuvants for Helicobacter immunization.

Introduction

The development of efficacious vaccines against enteric and urogenital pathogens has been hindered by lack of suitable delivery vehicles and appropriate adjuvants for the stimulation of immunity at mucosal tissues. Despite the identification of protective antigens for use in subunit vaccines and demonstrated success with prototype vaccines in animal models, testing in humans is often not possible because the mechanisms employed for stimulating mucosal immunity in animals are not safe for use in humans. This dilemma has now caught up with those involved in the development of a *Helicobacter pylori* (*H. pylori*) vaccine. However, several attenuated bacterial pathogens now exist with a proven ability to stimulate mucosal immunity in humans. These bacteria are safe and in some cases are already in use to vaccinate against potentially debilitating or lethal infectious diseases. The possible exploitation of these attenuated bacterial pathogens to deliver antigens from other pathogens is now being explored. Progress in Helicobacter vaccine development, the complications involved with existing mucosal adjuvants, and the recent application of live, recombinant, attenuated bacterial vectors to Helicobacter immunity to circumvent these problems is discussed below.

Helicobacter pylori **Vaccine Development**

H. pylori-related Disease and Therapy

The Gram-negative bacterium *H. pylori* colonizes the human stomach, residing in the mucus overlying the gastric epithelium and occasionally attaching itself to the epithelial cells. It is the predominant cause of gastritis and the etiologic agent of peptic ulcers (1,2). Additionally, long-term infection with *H. pylori* significantly increases the risk of developing gastric cancer in some populations (3). It is one of the most prevalent bacterial pathogens among all of mankind with rates of infection ranging from 20–90% depending upon the socioeconomic conditions of the population. The morbidity and mortality associated with these diseases combined with the economic burden placed on health care systems makes *H. pylori* a significant human pathogen.

Antimicrobial treatments for *H. pylori* infection have been developed and with proper compliance are over 90% successful. However, a pharmaceutical approach to *H. pylori* treatment has many limitations. The current protocols involve the administration of multiple agents several times a day for up to three weeks (4). These antibiotics are often disagreeable to patients after prolonged administration and symptoms such as nausea, diarrhea, abdominal pain, and pseudomembranous colitis can ensue (5,6). These adverse effects contribute to patient noncompliance and reduce the efficacy of the therapy. In countries where the prevalence of *H. pylori* infection is over 80% the cost of applying antimicrobial therapies is prohibitive. And since such a high percentage of certain populations are *H. pylori* positive, widespread treatment of *H. pylori* could also result in the development of antibiotic resistant strains. Such strains have already been reported in patients that have been treated with triple therapy who failed to cure infection (7,8). Additionally, eradication of *H. pylori* by pharmaceutical agents does not confer resistance to re-infection with *H. pylori*. Finally, pharmaceutical therapies do not address the particular need of those nonsymptomatic individuals who might go on to develop gastric adenocarcinoma, due in part to *H. pylori* infection (9).

Helicobacter Vaccination in Animal Models

Since its initial isolation by Warren and Marshall in 1983 (1) and the subsequent satisfaction of Koch's postulate linking *H. pylori* to gastric disease in 1985 (10,11), *H. pylori* vaccine research has progressed rapidly and met with many early successes in animal models. The development of animal models such as *H. pylori* infection of swine (12), *H. mustelae* infection of the domestic ferret (13), *H. felis* infection of mice (14), and finally *H. pylori* infection of mice (15–18) facilitated the testing of prototype *H. pylori* vaccines. These models suggest that the host can be protected against Helicobacter infection by immunization.

Relying primarily on murine models of Helicobacter infection and immunity many important observations were made in rapid succession. Successful vaccination was accomplished by oral inoculation of bacterial lysates in combination with small amounts of the mucosal adjuvant cholera toxin (CT) as described by Czinn and Nedrud (19). While initial reports of successful oral immunization were accomplished with whole cell bacterial lysates (20, 21), several groups described the use of purified or recombinant *H. pylori* proteins such as urease, catalase, heat shock protein, and a cytotoxin to obtain immunity (18, 22–25). These studies paved the way for the development of a well-characterized and safe subunit vaccine. Further reports documented the long-term efficacy of the oral prototype Helicobacter vaccination (26).

Therapeutic Vaccination for Helicobacter Infection in Animal Models

While an efficacious vaccine against *H. pylori* infection would provide a means of protecting pediatric populations, in many countries the majority of adults are already *H. pylori*-positive. A vaccine capable of inducing the host immune response to eradicate an established *H. pylori* infection would serve as a cost-effective solution to pharmaceutical treatment. During natural infection, *H. pylori* persists despite the induction of a local and systemic immune response. Yet immunization of mice could successfully prevent chronic infection of the gastric mucosa. The immune response induced by immunization has been shown to be quantitatively equivalent to that induced in infected mice (27, 28). The success of vaccination in mice could not be attributed to the chronology of the immune response in relation to infection since animals treated for infection by antimicrobial therapy can be re-infected despite the presence of the immune response generated to the initial infection (29, 30). This suggested that oral immunization somehow induces an immune response that is qualitatively different from the immune response induced by infection and therefore, such an immune response might be effective even if the vaccine was administered therapeutically.

Therapeutic immunization has now been demonstrated in several animal models. Up to 50% of mice experimentally infected with *H. felis* could be cured of the infection by oral immunization with Helicobacter antigens and CT (31, 32). Ferrets infected with the endogenous bacteria *H. mustelae* have also been treated by therapeutic immunization with *H. pylori* urease and CT with a 30% success rate (33). Ghiara *et al.* (34) have employed one of the recently described *H. pylori*/mouse models to test the effects of therapeutic vaccination against *H. pylori* in mice. When recombinant *H. pylori* VacA or CagA were administered orally with CT-related *E. coli* heat labile toxin (LT), 92% and 70% of the mice achieved eradication of the bacteria respectively. Seventy percent of these mice were protected against a subsequent challenge with *H. pylori*. The most recent application of therapeutic immunization was actually performed in a human phase II clinical trial. In a study which clearly illustrates the shortcomings of bacterial toxin mucosal adjuvants, over 60% of patients receiving *H. pylori* urease and LT as an oral therapeutic vaccine formulation experienced diarrhea without achieving eradication of *H. pylori* (35).

Mucosal Adjuvants

Toxicity of Mucosal Adjuvants in Humans

One of the central issues in the study of mucosal immunology and vaccine development has been the difficulty in stimulating mucosal immune responses. Induction of an immune response in the gut mucosa generally requires oral administration of large amounts of antigen. Although several bacterial products or viruses such as streptococcal M protein, *E. coli* pili proteins, and Reovirus have a distinct specificity for the intestinal mucosa and are immunogenic in small doses, most protein antigens, including Helicobacter proteins, are weak antigens by the oral route. Therefore, a major focus of research for oral vaccines has been the development of mucosal adjuvants and/or delivery vehicles to optimize the immunogenicity of prototype oral vaccines. Bacterial toxins such as *Vibrio cholerae* toxin (CT) and *E. coli* heat labile toxin (LT) are the most effective mucosal adjuvants described to date (see also chapter 8), and at least for laboratory animal studies, have greatly facilitated our understanding of mucosal immunology.

Most prophylactic and therapeutic Helicobacter immunization studies performed to date have employed either the oral or nasal route of immunization for stimulating gastric immunity and have employed the use of bacterial toxin adjuvants as originally described by Czinn and Nedrud (19). Several studies in pigs, ferrets, and mice have demonstrated that oral and nasal Helicobacter vaccination is unsuccessful in the absence of these adjuvants (21, 22, 36-39). Additionally, when recombinant *H. pylori* urease was administered orally to patients infected with *H. pylori* in a human phase I clinical trial without an adjuvant, the incidence of *H. pylori* infection was not reduced (40). Therefore, any successful oral or nasal Helicobacter vaccine will require the use of a mucosal adjuvant. However, both CT and LT possess potent toxicity that will preclude their use in humans. Even the small doses of these adjuvants that are suitable for use in mice have toxic effects in humans. As mentioned above, in the recent phase II clinical study examining the efficacy of administering a therapeutic *H. pylori* urease oral vaccine in combination with LT, 60% of the subjects experienced significant diarrhea (35). Consequently, several alternative approaches are being investigated to either make these bacterial toxins safe for use in humans or to find viable alternatives.

Strategies for Generating Safe Mucosal Adjuvants

Several distinct approaches have been employed to circumvent the toxic nature of using bacterial toxins as adjuvants. Each of these approaches attempts to reduce the participation of the A subunit of the toxin. CT and LT are composed of two protein subunits, a pentameric B subunit that forms a donut like structure, and a single A subunit that projects through the B subunit pentamer. The B subunit binds to target cells via G_{M1} ganglioside present on epithelial cells, facilitating entry of the A subunit into the cell cytoplasm. Cleavage of the A subunit into the enzymatically active A1 peptide and the A2 peptide anchor results in the transfer of ADP-ribose from NAD to a G protein which is part of the adenyl cyclase complex. As a result, adenyl cyclase is irreversibly activated resulting in an intracellular accumulation of cAMP that promotes the efflux of water and electrolytes from the cell.

One plausible mechanism that has been employed in other model systems but has yet to be described for Helicobacter antigens is to exploit the B subunit of CT or LT as a carrier molecule to deliver antigens to the gut epithelium. In one of the earliest reports on the effectiveness of CT as a mucosal antigen, oral administration of horseradish peroxidase conjugated to the B subunit of CT enhanced the peroxidase-specific mucosal IgA response (41). Although this approach has been used in numerous other systems, it has yet to be described for Helicobacter (reviewed in 42, 43).

A second approach is to substitute purified B subunit for CT. Although several studies have been reported describing the use of biochemically purified B subunit instead of CT holotoxin, these studies are difficult to interpret because commercially purified B subunit is typically contaminated with small amounts of holotoxin (44-47). Previous studies using Helicobacter mouse models have demonstrated that whereas commercially prepared B subunit possessed some mucosal adjuvant properties (37), holotoxin free recombinant B subunit, although itself highly immunogenic, does not possess adjuvant activity (48).

Another approach that is being aggressively pursued by a number of laboratories around the world is to conserve the structural integrity of the holotoxin but to reduce the enzymatic activity of the A subunit by introducing genetic mutations. Although many of these mutant CT and LT molecules have been described (49–53), most of these molecules have not been tested for adjuvanticity or have been tested with model proteins having no infectious challenge model. The oral adjuvanticity of these molecules is frequently diminished but some mutants retain adjuvanticity when administered intranasally. Since intranasal immunization has proven to be successful for the induction of Helicobacter immunity in mice, these mutants may ultimately be useful (38, 54, 55). Recent studies employing an LT mutant resistant to proteolytic cleavage of the A subunit have achieved both prophylactic and therapeutic protection against *H. pylori* in the mouse model (34, 56). Although such studies are encouraging, definitive testing in human clinical trials has not been performed, and previous results as to the adjuvanticity of prototype vaccines in mice have not been predictive for humans.

Live, Attenuated, Bacterial Pathogens as Vaccines and Antigen Delivery Vehicles

Live, attenuated, bacterial pathogens have been employed successfully for decades to vaccinate against infectious disease. The use of these attenuated bacteria as recombinant, multivalent, vaccine vectors have several distinct advantages. First, the nature of the infection of the bacteria employed as the vector lends itself to prolonged presence in the body and therefore results in prolonged antigen persistence. Ideally, this would lead to enhanced immunogenicity and increased vaccine efficacy. Second, multiple recombinant proteins can be expressed in a single bacterial vector to create multivalent vaccines with stronger efficacy or to immunize against multiple pathogens with one recombinant organism. Third, many attenuated bacterial pathogens can be administered by the oral route. This not only circumvents the number of intramuscular injections required to immunize children, but more importantly provides a means of inducing mucosal immunity against enteric and urogenital pathogens. Finally, attenuated bacterial pathogens can be used to induce specific arms of the immune response. Organisms such as *V. cholerae* infect the intestinal mucosa without invading the epithelium. Therefore, *V. cholerae* is effective at inducing protective secretory IgA responses at mucosal tissues. Alternatively, organisms such as *Mycobacterium bovis* (*M. bovis*), *Salmonella typhi* (*S. typhi*), *Listeria Monocytogenes* (*L. monocytogenes*), and *Shigella flexneri* (*S. flexneri*) can invade host cells and replicate intracellularly. Such organisms can be used to generate MHC Class I restricted CD8+ T cell responses as well as humoral immunity. Although several of these attenuated organisms may be problematic for pregnant or immunocompromised patients, advances in attenuation and susceptibility to specific antibiotics may circumvent these problems.

V. cholerae

V. cholerae induces debilitating diarrhea and is often fatal due to severe dehydration of the infected subject. The bacteria reside in the intestine and mediate disease without entering host tissue. The pathology associated with *V. cholerae* infections is due to the activity of the secreted cholera toxin that binds to the host intestinal epithelium (discussed above). Protection against *V. cholerae* is accomplished by the secretion of pathogen-specific secretory IgA into the lumen of the gut that neutralizes the toxin and also prevent adherence of the bacteria to the gut walls. Attenuated strains of *V. cholerae* expressing recombinant proteins therefore would be desirable when attempting to induce a sIgA response against the heterogenous pathogen. Attempts at attenuating *V. cholerae* involve the deletion of the A subunit or both the A and B subunit of the toxin. Although several early vaccine candidates were found to induce protective immunity in humans against infectious challenge, the vaccine strains themselves were not well tolerated (57, 58). However, the recently licensed strain CVD 103-HgR is well tolerated and has provided up to 100% protection against the classical strain and up to 67% protection against the El Tor strain of *V. cholerae* (59). The usefulness of expressing heterologous antigens in this strain has been demonstrated by expression of the Shiga-like toxin I of pathogenic *E. coli* (60). CVD 103-HgR expressing the B subunit of this toxin have been used to inoculate rabbits. Shiga-like toxin specific antibodies were induced and reduced intestinal fluid loss in response to challenge with shiga-like toxin was achieved. These experiments suggest the potential usefulness of employing recombinant *V. cholerae* bacteria to deliver heterologous antigens for the induction of protective immunity by intestinal sIgA.

Bacille Calmette-Guerin

Bacille Calmette-Guerin (BCG) is an attenuated strain of *Mycobacterium bovis* (*M. bovis*) that has been in use as a vaccine against tuberculosis in one form or another since 1921. Although there is some controversy as to the efficacy of the BCG vaccine there are several compelling arguments for continued research into its use as a vaccine vector for heterologous organisms. Mycobacterium are well recognized for their potent adjuvanticity properties and the BCG strain provides for long-term generation of antigen due to its long-term persistence following subcutaneous injection. It is also widely accepted as a safe and immunogenic vaccine that can be delivered shortly after birth. One prohibitive characteristic of Mycobacteria is the difficulty associated with performing genetic manipulation and consequently of introducing foreign genes into the vector. However, advances in mycobacterial plasmid vector development, transformation procedures, and homologous recombination now allow for the insertion of heterologous genes into the BCG chromosome. Many studies, predominantly in mice, employing antigens from a variety of pathogens have shown that recombinant BCG expressing heterologous antigens is effective at inducing an immune response against the recombinant antigen (61–65). The immune responses include cytotoxic T cells but also potent humoral antibody responses as well. It is of particular interest that recombinant BCG delivered by the appropriate route could induce secretory antibodies at mucosa with obvious implications for vaccinating against enteric pathogens and possibly venereal diseases as well.

Monocytogenes

L. monocytogenes invades systemic lymphoid tissue via the intestinal epithelium. Immune evasion is accomplished by escape from phagosomes and direct cell to cell spreading. Since the bacterium resides in the cytoplasm of the host cell, antigens are expressed in the context of MHC class I molecules and CD8+ cytotoxic T cells are activated. Therefore, an attenuated recombinant *L. monocytogenes* vector would be suited for the induction of cytotoxic T cells. Such a characteristic would be desirable to vaccinate against most viruses. Attenuations have now been accomplished that prevent the cell-to-cell spreading of the bacteria and animal models have been used to demonstrate the induction of CD8+ cytotoxic T cell responses against antigens from several different viruses (66–70). In one model, protective immunity was achieved in mice against lymphatic choriomeningitis virus (LCMV) challenge following immunization with *L. monocytogenes* expressing the LCMV antigens. However, it is important to note that such vaccines are poor inducers of humoral immunity and where antibodies, or cytotoxic T cells in combination with antibodies are required for optimal immunity, such vectors would not be appropriate.

Shigella

Shigella spp. invade intestinal tissue via transport across the M cells and subsequently disseminating among the intestinal epithelium by cell to cell spreading. Similar to *L. monocytogenes*, Shigella resides in the cell cytosol and as such is a prime candidate for generating MHC class I restricted CD8+ cytotoxic T cell responses. However, effective attenuation without loss of immunogenicity has been difficult to achieve. Strategies for attenuation have focused primarily on limiting its ability to survive within the cytosol using attenuations similar to those described below for Salmonella, or limiting its ability to spread between cells (71). Success in animal models employing attenuated strains of *S. flexneri* expressing heterogenous antigens have been limited. However, mice and guinea pigs have been used to induce the production of sIgA against enterotoxigenic *E. coli* antigens by immunization with recombinant *S. flexneri* (72). Information regarding the efficacy of such multivalent vaccines to generate MHC class I restricted cytotoxic T cells has not yet been reported.

Salmonella

Attenuated Salmonella vectors may be the most promising candidate for live multisubunit vaccines available. In addition to sharing many of the benefits associated with BCG including familiarity, safety, and the ability to induce both cellular and humoral immunity, Salmonella are easy to manipulate genetically and there are multiple animal models available. *S. typhi* Ty21a is a chemically mutagenized attenuated vaccine strain for vaccination of humans against typhoid fever. Although it requires multiple large doses to attain acceptable efficacy, its creation and use has paved the way for the development of improved Salmonella attenuations and vaccines. Two of the most common targeted mutations employed for the generation of attenuated strains include mutations that alter regulatory genes and mutations that create an auxotrophism for precursors in biosynthetic pathways.

Infection of mice with *S. typhimurium* has provided a useful animal model for development of effective attenuation and preliminary evaluation of the efficacy of expressing heterologous antigens. Since the initial description by Hoiseth and Stocker in 1981 of an effective *aroA* mutant for attenuation and vaccination against *S. typhimurium* in mice (73), numerous laboratories have employed similar strains to express viral, bacterial, and parasitic antigens including both proteins and LPS. The induction of an immune response to the heterologous antigen and/or induction of protective immunity against a second pathogen are too numerous to review here. The flexibility of the Salmonella model has only begun to be exploited. Foreign genes are now easily introduced into Salmonella on plasmids or by direct integration into the bacterial chromosome to insure better stability. The elucidation of various secretion systems in Salmonella (74) should now make it possible to design a vaccine that either directs the foreign antigen to the cytosol, to the outer membrane, or actually secretes it from the cell. Additionally, recombinant attenuated *S. typhimurium* strains have been engineered to secrete several different functional immunoregulatory cytokines including IL-1, IL-4, and IL-6 (75-77). Thus, multivalent Salmonella vectors could be designed to produce both the relevant foreign antigen(s) in combination with a cytokine that might increase the immunogenicity of vaccination or help direct the effector arm of the immune response.

Live, Attenuated *S. typhimurium* and *H. pylori* Vaccination

As described above, unlike the bacterial toxin adjuvants, attenuated recombinant bacterial strains such as *S. typhi*, BCG, and *V. cholerae* have already been demonstrated to be effective, safe delivery vehicles for the induction of local and systemic immunity by the oral route in humans. Thus, similar multivalent, recombinant, microorganisms expressing *H. pylori* antigens may provide a safe alternative to the problematic issue of developing a mucosal adjuvant. Two separate laboratory groups have recently reported the successful use of attenuated, recombinant *S. typhimurium* strains for vaccination of mice against *H. pylori* (55, 78). While these two studies are remarkably similar, the differences between the two are significant enough to warrant individual discussion.

Gomez-Duarte *et al.* (78) employed *S. typhimurium* 3261 (*aroA* mutation) expressing both the A and B subunits of *H. pylori* urease. The urease genes were carried on a plasmid and were under the control of a constitutive bacterial promotor. Protein expression was achieved at all stages of bacterial growth, including the stationary phase without any adverse effects on the *S. typhimurium*. Although the plasmid also carried the gene for beta-lactamase, selective pressure with ampicillin was not necessary to maintain plasmid stability either *in vitro* or *in vivo*. Thus, 100% of the *S. typhimurium* which were recovered from the spleens of mice inoculated with the recombinant *S. typhimurium* seven days previously, were ampicillin resistant. When mice were immunized with the recombinant *S. typhimurium,* urease-specific serum and intestinal antibody responses were induced as determined by ELISA. These results are consistent with previous studies performed in other model systems.

All mice receiving the recombinant *S. typhimurium* expressing *H. pylori* urease were protected from challenge with *H. pylori* as determined by failure of gastric antral biopsies to generate any urease activity *in vitro*. Conversely, gastric biopsies from *H. pylori*-challenged mice inoculated with PBS or *S. typhimurium* 3261 without plasmid generated a significant degree of urease activity due to *H. pylori* urease. These results are particularly interesting since only one inoculating dose of the recombinant *S. typhimurium* 3261 was required to achieve protective immunity. This is in contrast to the four or more doses of antigen and bacterial toxin typically administered to mice to achieve similar levels of protection against Helicobacter challenge. The authors subsequently employed a streptomycin-resistant strain of mouse-adapted *H. pylori* to challenge another group of immunized mice to facilitate the recovery of *H. pylori* from the gastric biopsies. The bacterial load correlated well with the presence or absence of urease activity. Although *H. pylori* could be cultured from some mice immunized with urease-expressing *S. typhimurium*, the overall bacterial load for this group was only 63 c.f.u./gastric biopsy compared to almost 3000 c.f.u. in PBS treated mice. Thus, mice could be protected, or the amount of *H. pylori* significantly decreased by immunization with recombinant attenuated *S. typhimurium*. Of extreme interest was the ability to achieve protection with only one dose of bacteria. If such an observation were consistently obtained, the problem of host immunity to the delivery vehicle which might proclude subsequent use for booster immunizations could be avoided.

Corthesy-Theulaz *et al.* (55) used the attenuated *S. typhimurium phoP^c* strain for recombinant expression of the *H. pylori* urease A and B subunits. The urease genes were carried on one of two plasmid systems. The first system employed a plasmid designed for high levels of constitutive protein expression. The second one was designed for phase variation of protein expression where only a subset of the *S. typhimurium* would be expressing the urease subunits at any given time. Although both systems resulted in expression of the urease subunits, the Salmonella designed for constitutive expression rapidly cured itself of the plasmid when selective antibiotic pressure was not maintained. This loss of plasmid occurred

Table 1. Comparison of live, attenuated, recombinant *S. typhimurium* subunit vaccines for Helicobacter immunization with the mucosal adjuvant LT.

	Oral Immunization		Intranasal Immunization	
Antigen	UreA, UreB[a]	UreA, UreB[b]	UreA, UreB[a]	UreA, UreB[c]
Adjuvant or delivery vehicle	LT	*S. typhimurium*	LT	*S. typhimurium*
Mouse strain	Outbred Swiss-Webster	Balb/c	Outbred Swiss-Webster	Balb/c
# of immunizations	4	1	4	2
Challenge strain	*H. pylori* X47-2AL	*H. pylori* P49 *H. pylori* P76	*H. pylori* X47-2AL	*H.pylori* P49
Antibody induction	↑ serum IgG ↑ secretory IgA	↑ serum IgG ↑ secretory IgA	↑ serum IgG ↑ secretory IgA	↑ serum IgG ↑ secretory IgA
Cytokine response	?[d]	?	?	IFN-γ, IL-10
Test for vaccine efficacy	Urease assay Colony counts	Urease assay Colony counts	Urease assay Colony counts	Urease assay Histologic sections
% of mice protected	100	100	100	60
Long lasting immunity	Yes[e]	?	?	?
Therapeutic immunity	Yes[f]	?	?	?
Post-immune gastritis	Yes	?	Yes	?
Protection in other animal models	?	?	?	?

[a]This data predominantly taken from Kleanthous *et al.* (54).
[b] Data from Gomez-Duarte *et al.* (78).
[c] Data from Corthesy-Theulaz *et al.* (55).
[d] Although there have been several reports detailing the dominant cytokines in immunized mice following challenge with Helicobacter, there are currently no reports detailing the cytokine responses induced by immunization prior to challenge.
[e] Long term protection studies have not been performed with *H. pylori*. However, the same laboratory has reported long-term protection against the *H. felis* (79).
[f] Therapeutic immunization has been applied successfully to a *H. pylori* mouse model using the cytotoxin and CagA proteins in combination with a detoxified LT mutant (34).

both *in vitro* and when administered to mice. This observation contrasts with those of Gomez-Duarte *et al.* (78) and helps to emphasize the importance of selecting appropriate vectors or of incorporating the heterologous gene into the Salmonella chromosome. Salmonella carrying the plasmid designed for phase variation expression maintained the plasmid as evidenced by recovery of plasmid positive Salmonella from mice four weeks after inoculation.

Both strains of recombinant *S. typhimurium phoP[c]* were used to immunize Balb/c mice by two intranasal inoculations given at two week intervals. Mice immunized with *S. typhimurium phoP[c]* constitutively expressing *H. pylori* urease failed to elicit a urease-specific antibody response and were not protected from challenge with a mouse-adapted strain of *H. pylori*. This is consistent with the rapid loss of the plasmid from the Salmonella. Mice immunized with *S. typhimurium phoP[c]* expressing urease under phase variation control responded with urease-specific serum antibody and CD4+ T cell responses as measured by proliferation and cytokine production. Spleen cells from these mice secreted both IFN-γ and IL-10. Almost two thirds of these mice were protected from challenge with *H. pylori*. Both the gastric biopsy urease test and direct enumeration of bacteria on histologic sections confirmed either the complete, or near absence of bacteria.

As outlined in Table 1, these two studies employing *S. typhimurium* vectors to vaccinate against *H. pylori* had similar designs but distinctly different outcomes. Both groups used Balb/c mice, the P49 challenge strain of *H. pylori*, and urease A and B antigens. Yet one group achieved 100% protection, while the other group achieved 60%. The wide disparity in results could be due to the route of administration (oral vs intranasal) or to the expression systems employed (constitutive *vs* phase variation). A definitive determination will require further studies but both studies are encouraging.

Table 1 also demonstrates that several important issues have not yet been addressed with the *S. typhimurium* experiments. First, while oral immunization induced 100% protection, the duration of immunity remains unknown. A practical Helicobacter vaccine will require long-term immunity since both adults and children are susceptible to infection. Long-term immunity against *H. pylori* infection in mice has already been demonstrated. Second, although the development of prophylactic immunity by immunization with multivalent attenuated *S. typhimurium* is an important observation, the potential for use in therapeutically immunizing infected mice is of equal if not greater importance given the high prevalence of people currently infected in developing nations. Additionally, gastric disease could not be studied in these mice since a low inflammation strain of mice, Balb/c, was used. Most laboratories have observed the induction of severe gastritis when immunized mice are challenged with *H. pylori*. Although the importance of this inflammation either for benefit or detriment remains unknown, proper characterization of this vaccine model will require histologic analysis.

Finally, given the failure of murine models of Helicobacter immunity to predict success in humans with LT adjuvant, these Salmonella vaccines should be tested in larger animal models. Despite the description of feline, canine, swine, and nonhuman primate models of *H. pylori* infection, a Helicobacter vaccine study demonstrating significant efficacy using CT or LT adjuvants has never been reported in any of these models. While the current successful use of attenuated *S. typhi* in humans indicates it might be possible to use Salmonella vectors to vaccinate against Helicobacter, success in other animal models would go a long way towards generating the rationale for human clinical trials.

Concluding Remarks

The development of a *H. pylori* vaccine has stagnated due to the lack of a safe and efficacious mucosal adjuvant for use in humans. Although prophylactic and therapeutic vaccine studies in mice suggest the possibility of vaccination against *H. pylori*, the adjuvants that make such vaccinations possible, CT and LT, are not safe in humans. The current use of live attenuated *S. typhi* bacteria to protect against typhoid fever in humans indicate that such bacteria might be effective in delivering *H. pylori* antigens to induce protective immunity against *H. pylori* at the gastric mucosa. Recent reports in which *H. pylori* urease A and B expressed in attenuated strains of *S. typhimurium* induce protective immunity in mice are encouraging, particularly in light of the small number of doses required to achieve protective immunity. These studies should be expanded to some of the larger *H. pylori* animal models to increase the predictive value of these studies for efficacy in humans.

References

1.   Warren, J.R. and Marshall, B.J. 1983. Unidentified curved bacilli on gastric epithelium in active chronic gastritis. Lancet i: 1273-1275.
2.   N.I.H. Consensus Conference, N.C. 1994. *Helicobacter pylori* in peptic ulcer disease. JAMA 272: 65-69.
3.   Eurogast Study Group. 1993. An international association between *Helicobacter pylori* infection and gastric cancer. Lancet 341: 1359-1362.
4.   Soll, A.H. 1996. Medical treatment of peptic ulcer disease: practice guidelines. JAMA 275: 622-629.
5.   Bell, G.D., Powell, K., Burridge, S.M., Pallecaros, A., Jones, P.H., Gant, P.W., Harrison, G. and Trowell, J.E. 1992. Experience with 'triple' anti-Helicobacter eradication therapy: side effects and the importance of testing the pre-treatment bacterial isolate for metronidazole resistance. Aliment. Pharmacol. Ther. 6: 427-435.
6.   Rauws, E.A.J. 1993. Reasons for failure of *Helicobacter pylori* treatment. Eur. J. Gastroenterol. Hepatol. 5 (suppl 2): S92-S95.
7.   Jorgensen, M., Daskalopoulos, G., Warburton, V., Mitchell, H.M. and Hazell, S.L. 1996. Multiple strain colonization and metronidazole resistance in *Helicobacter pylori*-infected patients: Identification from sequential and multiple biopsy specimens. J. Inf. Dis. 174: 631-635.
8.   Megraud, F. 1997. Resistance of *Helicobacter pylori* to antibiotics. Aliment. Pharmacol. Ther. 11 suppl 1: 43-53.
9.   Correa, P. 1992. Human gastric carcinogenesis: A multistep and multifactorial process- first american cancer society award lecture on cancer epidemiology and prevention. Cancer Res. 52: 6735-6740.
10. Marshall, B.J., Armstrong, J.A. and McGechie, D.B. 1985. Attempt to fulfill Koch's postulate for pyloric Campylobacter. Med. J. Austral. 142: 436-439.
11. Morris, A. and Nicholson, G. 1987. Ingestion of *Campylobacter pyloridis* causes gastritis and raised fasting gastric pH. Am. J. Gastroenterol. 82: 192-199.
12. Krakowka, S., Morgan, D.M., Kraft, W.G. and Leunk, R.D. 1987. Establishment of gastric *Campylobacter pylori* infection in the neonatal gnotobiotic piglet. Infect. Immun. 55: 2789-2796.
13. Fox, J.G., Correa, P., Taylor, N.S., Lee, A., Otto, G., Murphy, J.C. and Rose, R. 1990. *Helicobacter mustelae*-associated gastritis in ferrets: an animal model of *Helicobacter pylori* gastritis in humans. Gastroenterol. 99: 352-361.
14. Lee, A., Fox, J.G., Otto, G. and Murphy, J. 1990. A small animal model of human *Helicobacter pylori* active chronic gastritis. Gastroenterol. 99: 1315-1323.
15. Karita, M., Kouchiyama, T., Okita, L. and Nakazawa, T. 1991. New small animal model for human gastric *Helicobacter pylori* infection: success in both nude and euthymic mice. Am. J. Gastroenterol. 11: 1596-1603.
16. Karita, M., Li, Q., Cantero, D. and Okita, K. 1994. Establishment of a small animal model for human *Helicobacter pylori* infection using germ-free mouse. Am. J. Gastroenterol. 89: 208-213.
17. McColm, A.A., Bagshaw, J., Wallis, J. and McLaren, A. 1995. Screening of anti-Helicobacter therapies in mice colonized with *H. pylori*. Gut. 37: A92.
18. Marchetti, M., Arico, B., Burroni, D., Figura, N., Rappuoli, R. and Ghiara, P. 1995. Development of a mouse model of *Helicobacter pylori* infection that mimics human disease. Science. 267: 1655-1658.

19. Czinn, S.J. and Nedrud, J.G. 1991. Oral immunization against *Helicobacter pylori*. Infect. Immun. 59: 2359-2363.

20. Czinn, S.J., Cai, A. and Nedrud, J.G. 1993. Protection of germ-free mice from infection by *Helicobacter felis* after active oral or passive IgA immunization. Vaccine. 11: 637-642.

21. Chen, M., Lee, A., Hazell, S., Hu, P. and Li, Y. 1993. Immunisation against gastric infection with *Helicobacter* species: first step in the prophylaxis of gastric cancer? Zentralblatt für Bakteriol. 280: 155-165.

22. Lee, C.K., Weltzin, R., Thomas, W.D.J., Kleanthous, H., Ermak, T.H., Soman, G., Hill, J.E., Ackerman, S.K. and Monath, T.P. 1995. Oral Immunization with recombinant *Helicobacter pylori* urease induces secretory IgA antibodies and protects mice from challenge with *Helicobacter felis*. J. Inf. Dis. 172: 161-172.

23. Ferrero, R.L., J.-M. Thiberge, M. Huerre and A. Labigne. 1994. Recombinant antigens prepared from the urease subunits of *Helicobacter* spp.: Evidence of protection in a mouse model of gastric infection. Infect. Immun. 62: 4981-4989.

24. Ferrero, R.L., Thiberge, J.-M., Kansau, I., Wuscher, N., Huerre, M. and Labigne, A. 1995. The *groES* homolog of *Helicobacter pylori* confers protective immunity against mucosal infection in mice. Proc. Natl. Acad. Sci. USA. 92: 6499-6503.

25. Radcliff, F.J., Hazell, S.L., Kolesnikow, T., Doidge, C. and Lee, A. 1997. Catalase, a novel antigen for *Helicobacter pylori* vaccination. Infect. Immun. 65: 4668-4674.

26. Radcliff, F.J., Chen, M. and Lee, A. 1996. Protective immunization against Helicobacter stimulates long-term immunity. Vaccine. 14: 780-784.

27. Blanchard, T.G., Nedrud, J.G., Reardon, E. and Czinn, S.J. 1999. Qualitative and quantitative analysis of the local and systemic antibody responses in mice and humans with Helicobacter immunity and infection. J. Inf. Dis. 179: 725-728.

28. Sellman, S., Blanchard, T.G., Nedrud, J.G. and Czinn, S.J. 1995. Vaccine Strategies for prevention of *Helicobacter pylori* infections. Eur. J. Gastroenterol. Hepatol. 7 (suppl 1): S1-S6.

29. Batchelder, M., Fox, J., Hayward, A., Yan, L., Shames, B., Murphy, J. and Palley, L. 1996. Natural and experimental *Helicobacter mustelae* reinfection following successful antimicrobial eradication in ferrets. Helicobacter. 1: 34-42.

30. Czinn, S.J., Bierman, J.C., Diters, R.W., Blanchard, T.J. and Leunk, R.D. 1996. Characterization and therapy for experimental infection by *Helicobacter mustelae* in ferrets. Helicobacter. 1: 43-51.

31. Doidge, C., Gust, I., Lee, A., Buck, F., Hazell, S. and Manne, U. 1994. Therapeutic immunization against Helicobacter infection (Letter). Lancet. 343: 913-914.

32. Corthesy-Theulaz, I., Porta, N., Glauser, M., Saraga, E., Vaney, A.-C., Haas, R., Kraehenbuhl, J.P., Blum, A.L. and Michetti, P. 1995. Oral immunization with *Helicobacter pylori* urease B subunit as a treatment against Helicobacter infection in mice. Gastroenterol. 109: 115-121.

33. Cuenca, R., Blanchard, T.G., Czinn, S.J., Nedrud, J.G., Monath, T.P., Lee, C.K. and Redline, R.W. 1996. Therapeutic immunization against *Helicobacter mustelae* infection in naturally infected ferrets. Gastroenterol. 110: 1770-1775.

34. Ghiara, P., Rossi, M., Marchetti, M., Di Tommaso, A., Vindigni, C., Ciampolini, F., Covacci, A., Telford, J.L., De Magistris, M.T., Pizza, M., Rappuoli, R. and Del Giudice, G. 1997. Therapeutic intragastric vaccination against *Helicobacter pylori* in mice eradicates an otherwise chronic infection and confers protection against reinfection. Infect. Immun. 65: 4996-5002.

35. Michetti, P., Kreiss, C., Kotloff, K.L., Porta, N., Blanco, J.-L., Bachmann, D., Herranz, M., Saldinger, P.F., Corthesy-Theulaz, I., Losonsky, G., Nichols, R., Simon, J., Stolte, M., Ackerman, S., Monath, T.P. and Blum, A.L. 1999. Oral immunization with urease and *Escherichia coli* heat-labile enterotoxin is safe and immunogenic in Helicobacter pylori-infected adults. Gastroenterol. 116: 804-812.

36. Eaton, K.A. and Krakowka, S. 1992. Chronic active gastritis due to *Helicobacter pylori* in immunized Gnotobiotic piglets. Gastroenterol. 103: 1580-1586.

37. Lee, A. and Chen, M. 1994. Successful immunization against gastric infection with *Helicobacter* species: Use of a cholera toxin B-subunit-whole-cell vaccine. Infect. Immun. 62: 3594-3597.

38. Weltzin, R., Kleanthous, H., Guirdkhoo, F., Monath, T.P. and Lee, C.K. 1997. Novel intranasal immunization techniques for antibody induction and protection of mice against gastric *Helicobacter felis* infection. Vaccine. 15: 370-376.

39. Whary, M.T., Palley, L.S., Batchelder, M., Murphy, J.C., Yan, L., Taylor, N.S. and Fox, J.G. 1997. Promotion of ulcerative duodenitis in young ferrets by oral immunization with *Helicobacter mustelae* and muramyl dipeptide. Helicobacter. 2: 65-77.

40. Kreiss, C., Buclin, T., Cosma, M., Corthesy-Theulaz, I. and Michetti, P. 1996. Safety of oral immunisation with recombinant urease in patients with *Helicobacter pylori* infection (Letter). Lancet. 347: 1630-1631.

41. McKenzie, S.J. and Halsey, J.F. 1984. Cholera toxin B subunit as a carrier protein to stimulate a mucosal immune response. J. Immunol. 133: 1818-1824.

42. Elson, C.O. and Dertzbaugh, M.T. 1994. Mucosal adjuvants. In: Handbook of Mucosal Immunology. P. L. Ogra, J. Mestecky, L. M. E., W. Strober, J. R. McGhee and J. Bienenstock, eds. Academic Press, San Diego. p. 391-402.

43. Holmgren, J., Lycke, N. and Czerkinsky, C. 1993. Cholera toxin and cholera B subunit as oral-mucosal adjuvant and antigen vector systems. Vaccine. 11: 1179-1184.

44. Hashigucci, K., Ogawa, H., Ishidate, T., Yamashita, R., Kamiya, H., Watanabe, K., Hattori, N., Sato, T., Suzuki, Y., Nagamine, T., Aizawa, C., Tamura, S., Kurata, T. and Oya, A. 1996. Antibody responses in volunteers induced by nasal influenza vaccine combined with *Escherichia coli* heat-labile enterotoxin B subunit containing a trace amount of the holotoxin. Vaccine 14: 113-119.

45. Tamura, S.-i., Yamanaka, A., Shimohara, M., Tomita, T., Komase, K., Tsuda, Y., Suzuki, Y., Nagamine, T., Kawahara, K., Danbara, H., Aizawa, C., Oya, A. and Kurata, T. 1994. Synergistic action of cholera toxin B subunit (and *Escherichia coli* heat-labile toxin B subunit) and a trace amount of cholera whole toxin as an adjuvant for nasal influenza vaccine. Vaccine. 12: 419-426.

46. Wilson, A.D., Clarke, C.J. and Stokes, C.R. 1990. Whole cholera toxin and B subunit act synergistically as an adjuvant for the mucosal immune response of mice to keyhole limpet haemocyanin. Scand. J. Immunol. 31: 443-451.

47. Spiegel, S. 1990. Cautionary note on the use of the B subunit of cholera toxin as a ganglioside Gm1 probe: Detection of cholera toxin A subunit in B subunit preparations by a sensitive adenylate cyclase assay. J. Cell. Biochem. 42: 143-152.

48. Blanchard, T.G., Lycke, N., Czinn, S.J. and Nedrud, J.G. 1998. Recombinant cholera toxin B subunit is not an effective mucosal adjuvant for oral immunization of mice against *H. felis*. Immunol. 94: 22-27.

49. Burnette, W.N., Mar, V.L., Platler, B.W., Schlotterbeck, J.D., McGinley, M.D., Stoney, K.S., Rohde, M.F. and Kaslow, H.R. 1991. Site-specific mutagenesis of the catalytic subunit of cholera toxin: substituting lysine for arginine 7 causes loss of activity. Infect. Immun. 59: 4266-4270.

50. Cieplak, W.J., Mead, D.J., Messer, R.J. and Grant, C.C.R. 1995. Site-directed mutagenic alteration of potential active-site residues of the A subunit of *Escherichia coli* heat-labile enterotoxin. J. Biol. Chem. 270: 30545-30550.

51. Dickinson, B.L. and Clements, J.D. 1995. Dissociation of *Escherichia coli* heat-labile enterotoxin adjuvanticity from ADP-ribosyltransferase activity. Infect. Immun. 63: 1617-1623.

52. Douce, G., Trucotte, C., Cropley, I., Roberts, M., Pizza, M., Domenighini, M., Rappuoli, R. and Dougan, G. 1995. Mutants of *Escherichia coli* heat labile toxin lacking ADP-ribosyltransferase activity act as nontoxic, mucosal adjuvants. Proc. Natl. Acad. Sci. USA. 92: 1644-1648.

53. Yamamoto, S., Kiyono, H., Yamamoto, M., Imaoka, K., Fujihashi, K., Van Ginkel, F.W., Noda, M., Takeda, Y. and McGhee, J.R. 1997. A nontoxic mutant of cholera toxin elicits Th2-type responses for enhanced mucosal immunity. Proc. Natl. Acad. Sci. USA 94: 5267-5272.

54. Kleanthous, H., Myers, G.A., Georgakopoulos, K.M., Tibbitts, T.J., Ingrassia, J.W., Gray, H.L., Ding, R., Zhang, Z.Z., Lei, W., Nichols, R., Lee, C.K., Ermak, T.H. and Monath, T.P. 1998. Rectal and intranasal immunizations with recombinant urease induce distinct local and serum immune responses in mice and protect against *Helicobacter pylori* infection. Infect. Immun. 66: 2879-2886.

55. Corthesy-Theulaz, I.E., Hopkins, S., Bachmann, D., Saldinger, P.F., Porta, N., Haas, R., Zheng-Xin, Y., Meyer, T., Bouzourene, H., Blum, A.L. and Kraehenbuhl, J.P. 1998. Mice are protected from *Helicobacter pylori* infection by nasal immunization with attenuated *Salmonella typhimurium phoPc* expressing urease A and B subunits. Infect. Immun. 66: 581-586.

56. Marchetti, M., Rossi, M., Giannelli, V., Giuliani, M.M., Pizza, M., Censini, S., Covacci, A., Massari, P., Pagliaccia, C., Manetti, R., Telford, J.L., Douce, G., Dougan, G., Rappuoli, R. and Ghiara, P. 1998. Protection against *Helicobacter pylori* infection in mice by intragastric vaccination with *H. pylori* antigens is achieved using a non-toxic mutant of *E. coli* heat-labile enterotoxin (LT) as adjuvant. Vaccine. 16: 33-37.

57. Kaper, J.B., Lockman, H., Baldini, M.M. and Levine, M.M. 1984. Recombinant nontoxigenic *Vibrio cholerae* strains as attenuated cholera vaccine candidates. Nature. 308: 655-658.

58. Levine, M.M., Kaper, J.B., Herrington, D., Losonsky, G., Morris, J.G., Clements, M., Black, R.E., Tall, B. and Hall, R. 1988. Volunteer studies of deletion mutants of *Vibrio cholerae* 01 prepared by recombinant techniques. Infect. Immun. 56: 161-167.

59. Levine, M.M., Kaper, J.B., Herrington, D., Ketley, J., Losonsky, G., Tacket, C.O., Tall, B. and Cryz, R. 1988. Safety, immunogenicity and efficacy of recombinant live oral cholera vaccines, CVD 103 and CVD 103-HgR. Lancet. 2: 467-470.

60. Acheson, D.W.K., Levine, M.M., Kaper, J.B. and Keusch, G.T. 1996. Protective immunity to shiga-like toxin I following oral immunization with shiga-like toxin I B-subunit-producing *Vibrio cholerae* CVD 103-HgR. Infect. Immun. 64: 355-357.

61. Abdelhak, S., Louzir, H., Timm, J., Blel, L., Benlasfar, Z., Lagranderi, M., Georghie, M., Dellagi, K. and Gicquel, B. 1995. Recombinant BCG expressing the leishmania surface antigen Gp63 induces protective immunity against *Leishmania major* infection in Balb/c mice. Microbiol. 141: 1585-1592.

62. Aldovini, A. and Young, R.A. 1991. Humoral and cell mediated immune responses to live recombinant BCG-HIV vaccines. Nature. 351: 479-482.

63. Langermann, S., Palaszynski, S., Sadziene, A., Stover, C.K. and Koenig, S. 1994. Systemic and mucosal immunity induced by BCG vector expressing outer-surface protein A

of *Borrelia burgdorferi*. Nature. 372: 552-555.

64. Langermann, S., Palaszynski, S.R., Beurlein, J.E., Koenig, S., Hanson, M.S., Briles, D.E. and Stover, C.K. 1994. Protective humoral response against pneumococcal infection in mice elicited by recombinant bacille-Calmette-Guerin vaccines expressing pneumococcal surface protein A. J. Exp. Med. 180: 2277-2286.

65. Winter, N., Lagranderie, M., Gangloff, S., Leclerc, C., Gheorghiu, M. and Gicquel, B. 1995. Recombinant BCG strains expressing the SIV-mac251 *nef* gene induce proliferative and CTL responses againstr nef synthetic peptides in mice. Vaccine. 13: 471-478.

66. Frankel, F.R., Hegde, S., Lieberman, J. and Paterson, Y. 1995. Induction of cell-mediated immune responses to human immunodeficiency virus type I Gag protein by using *Listeria monocytogenes* as a live vaccine vector. J. Immunol. 155: 4775-4782.

67. Goossens, P.L., Milon, G., Cossart, P. and Saron, M.-F. 1995. Attenuated *Listeria monocytogenes* as a live vector for induction of CD8+ T cells in vivo: A study with the nucleoprotein of the lymphocytic choriomeningitis virus. Int. Immunol. 7: 797-805.

68. Ikonomidis, G., Paterson, Y., Kos, F.J. and Portnoy, D.A. 1994. Delivery of a viral antigen to the class I processing and presentation pathway by *Listeria monocytogenes*. J. Exp. Med. 180: 2209-2218.

69. Shen, H., Slifka, M.K., Matloubian, M., Jensen, E.R., Ahmed, R. and Miller, J.F. 1995. Recombinant *Listeria monocytogenes* as a live vaccine vehicle for the induction of protective anti-viral cell-mediated immunity. Proc. Natl. Acad. Sci. USA. 92: 3987-3991.

70. Slifka, M.K., Shen, H., Matloubian, M., Jensen, E.R., Miller, J.F. and Ahmed, R. 1996. Antiviral cytotoxic T-cell memory by vaccination with recombinant *Listeria monocytogenes*. J. Virol. 70: 2902-2910.

71. Noriega, F.R., Wang, J.Y., Losonsky, G., Maneval, D.R., Hone, D.M. and Levine, M.M. 1994. Construction and characterization of attenuated *aroA virG Shigella flexneri 2a* strain CVD 1203, a prototype live oral vaccine. Infect. Immun. 62: 5168-5172.

72. Noriega, F.R., Losonsky, G., Wang, J.Y., Formal, S.B. and Levine, M.M. 1996. Further characterization of ΔaroA ΔvirG *Shigella flexneri* 2a strain CVD1203 as a mucosal Shigella vaccine and as a live-vector vaccine for delivering antigens of enterotoxigenic *Escherichia coli*. Infect. Immun. 64: 23-27.

73. Hoiseth, S.K. and Stocker, B.A.. 1981. Aromatic dependent *Salmonella typhimurium* are non-virulent and effective as live vaccines. Nature. 291: 238.

74. Mecsas, J. and Strauss, E.J. 1996. Molecular mechanisms of bacterial virulence: Type III secretion and pathogenicity islands. Emerging Inf. Dis. 2: 271-288.

75. Carrier, M.J., Chatfield, S.N., Dougan, G., Nowicka, U.T., O'Callaghan, D., Beesley, J.E., Milano, S., Cillari, E. and Liew, F.Y. 1992. Expression of human IL-1 beta in *Salmonella typhimurium*. A model system for delivery of recombinant therapeutic proteins *in vivo*. J. Virol. 148: 1176-1181.

76. Denich, K., Borlin, P., O'Hanley, P.D., Howard, M. and Heath, A.W. 1993. Expression of the murine interleukin-4 gene in an attenuated *aroA* strain of *Salmonella typhimurium*: Persistence and immune response in BALB/c mice and susceptibility to macrophage killing. Infect. Immun. 61: 4818-4827.

77. Dunstan, S.J., Ramsay, A.J. and Strugnell, R.A. 1996. Studies of immunity and bacterial invasiveness in mice given a recombinant *Salmonella* vector encoding murine interleukin-6. Infect. Immun. 64: 2730-2736.

78. Gomez-Duarte, O.G., Lucas, B., Yan, Z.X., Panthel, K., Haas, R. and Meyer, T.F. 1998. Protection of mice against gastric colonization by *Helicobacter pylori* by single oral dose immunization with attenuated *Salmonella typhimurium* producing urease subunits A and B. Vaccine. 16: 460-471.

79. Ermak, T.H., Ding, R., Ekstein, B., Hill, J., Myers, G.A., Lee, C.K., Pappo, J., Kleanthous, H.K. and Monath, T.P. 1997. Gastritis in urease-immunized mice after *Helicobacter felis* challenge may be due to residual bacteria. Gastroenterol. 113: 1118-1128.

From: *Development of Novel Antimicrobial Agents: Emerging Strategies*
ISBN 1-898486-23-9 © 2001 Horizon Scientific Press, Wymondham, UK.

10

Development of an Adhesin Vaccine to Prevent Urinary Tract Infection

Matthew A. Mulvey, Scott J. Hultgren and Solomon Langermann

Abstract

Colonization of the mucosal epithelium is a critical step in the early stages of bacterial infection. Adherence to host cells usually involves a specific interaction between bacterial surface proteins called adhesins and surface structures, often glycoconjugates, on the target cell. Interactions between pathogenic bacteria and host cells may also induce signaling cascades, both in the bacteria as well as in the host cells, resulting in local inflammation, invasive disease and triggering of innate host defenses. Recent advances in our understanding of the biological basis for bacterial binding, coupled with the ability to purify adhesins away from whole bacteria, allow for new approaches to prophylaxis and therapy. By using purified adhesins as candidate vaccines, it has been possible to show that such proteins elicit protective antibody responses that prevent the interaction of the bacteria with the host and subsequent infection. This chapter will focus on studies with uropathogenic *Escherichia coli* (*E. coli*) and their pilus-associated FimH adhesin proteins. These studies clearly demonstrate both that adhesins play a crucial role in bacterial attachment and that an adhesin-based vaccine protects against colonization and infection. Furthermore, FimH is also critical for invasion of the bladder epithelium that may result in chronic infection, hence a FimH vaccine might also protect against recurrent infections. Development of a FimH vaccine to prevent urinary tract infections (UTIs) addresses an important medical disorder of significant morbidity especially in women. More broadly these studies point to the potential for developing adhesin-based vaccines that could target a wide range of pilus-associated adhesins ubiquitous among Gram-negative bacteria.

Escherichia coli Urinary Tract Infections: The Need for a Vaccine

Acute urinary tract infections (UTIs) are among the most common disorders prompting medical evaluation. A retrospective epidemiological study published in 1992 showed that forty percent of adult women in the United States experience at least one UTI sometime during their lifetime (1). This results in approximately seven million office visits each year with an estimated annual health care cost exceeding $1 billion. A more recent prospective study has shown that the incidence of UTIs is actually much higher. Among 796 women followed in either a university cohort or HMO cohort in Seattle, Washington, the annual incidence rate of UTIs was between 0.5 and 0.7% in individuals at highest risk (2). Thus urinary tract infections in young women constitute a significant problem in terms of morbidity and health care costs and new, cost-effective approaches toward preventing urinary tract infections are needed. *Escherichia coli* are the main causative agents of UTIs. They account for greater than 85% of cases of acute bladder infection (cystitis) and kidney infections (pyelonephritis) as well as greater than 60% of recurrent bouts of cystitis and at least 35% of recurrent pyelonephritis infections (3,4). They are also responsible for the majority of cases of asymptomatic bacteriuria. In most cases UTIs are limited to the bladder and do not result in pyelonephritis. One method of eliminating acute UTIs as well as preventing recurrent UTIs and ascending infections is the use of regular or intermittent antimicrobial prophylaxis. But concern about the emergence of antibiotic resistant bacterial strains limits the long-term feasibility of this approach. Recent studies have demonstrated a significant increase in the prevalence of resistance to several commonly used antimicrobial agents among a large group of isolates from women with uncomplicated cystitis and from children with UTIs (5,6). Given the high incidence, burden of discomfort and cost associated with *E. coli* UTIs, along with the emerging antibiotic resistance issues, efforts are now underway to develop a vaccine against UTIs that could prevent such infections altogether.

Rationale for an Adhesin-Based Vaccine

Most bacterial pathogens, including uropathogenic *E. coli*, must attach to host cells, and colonize mucosal sites such as the bladder, in order to initiate infection. In many cases bacterial adhesin-lectin proteins that mediate attachment to sugar moieties, such as glycolipids and glycoproteins, on the cell surface (7-9) mediate the initial attachment event. Adhesion properties of uropathogenic *E. coli* have been extensively studied, leading to the identification of a number of lectin-like adhesins. Most of these adhesins are associated with proteinaceous, filamentous polymeric organelles, known as pili or fimbriae, expressed on the surface of the bacteria (10). These include the FimH adhesin of type 1 pili that mediates binding to α-D mannosides (11-14), the PapG adhesin of P pili, that confers specificity for the α-D-Gal(1-4) α-D-gal moiety of blood group P antigens (15-19) and the S-pilus adhesin on S pili which mediates binding to sialyl-galactosides (20-21).

Pili are composed of single or multiple types of protein subunits, called pilins or fimbrins, which are typically arranged in a helical fashion. They are often expressed peritrichously around individual bacteria and range from a few fractions of a micrometer to greater than 20 μm in length and vary from less than 2 nm up to 11 nm in diameter. The primary function of pili is to act as scaffolding for the presentation of the associated adhesin proteins. High resolution electron microscopy of type 1, P and S pili of *Escherichia coli* has revealed that these structures are composite fibers, consisting of a thick pilus rod attached to a thin, short distally located tip fibrillum with the adhesin located at the distal end of the tip fibrillum (22-25). Their location at the distal tip of the pilus allows adhesins to mediate interactions of bacteria with each other, with inanimate surfaces, and with tissues and cells in susceptible host organisms. Such interactions facilitate the formation of bacterial communities such as biofilms and are often critical to the successful colonization of host organisms by both commensal and pathogenic bacteria (26).

Epidemiological and experimental evidence from studies on uropathogenic *E. coli* has suggested that type 1 pili direct attachment of uropathogenic *E. coli* to mannose-containing receptors in the vaginal tract and bladder and are important in the initial colonization of the lower urinary tract and establishment of UTIs (27-32). Recent studies using mutant strains of *E. coli* lacking the FimH adhesin have confirmed the critical role for FimH in colonization of the bladder by uropathogenic *E. coli* (33-34). Follow-up studies have shown that FimH binds specifically to mannosylated glycoproteins known as uroplakins that line the bladder mucosa (35). P pili on the other hand are important in ascending UTIs and pyelonephritis (36-40). S pili are thought to be important in invasive disease as well as in *E. coli* infections of the kidney particularly in children (41).

Given the critical role of type 1 pili and in particular FimH in the initial stages of bladder colonization, we reasoned that antibodies targeting FimH should prevent the initiation of urinary tract infections. Antibodies against the FimH adhesins would be expected to opsonize bacteria for clearance by the immune system and prevent microbial attachment and colonization of mucosal surfaces at the earliest stage of the disease process. Furthermore, by blocking attachment at the site of entry, the FimH vaccine might also be expected to protect against any subsequent bacterial invasion or ascending infection into the kidney. We have spent the past several years testing this molecule as a vaccine against cystitis in various animal models and in surrogate *in vitro* assays. We have focused our efforts on developing a subunit vaccine that contains the FimH adhesin along with its cognate chaperone, FimC, which is critical for stabilization of the full-length adhesin as will be explained below.

Assembly of Type 1 Pili and Choice of the FimCH Complex as a Vaccine

Type 1 pili are assembled via a highly conserved pathway referred to as the chaperone/usher pathway. This pathway is involved in the assembly of over 25 pilus organelles in Gram-negative bacteria (42-44). For type 1 pili, a chaperone protein called FimC, located in the bacterial periplasmic space forms a stable complex with each of the pilus subunits prior to their ordered assembly into a pilus organelle on the bacterial surface (45). FimC helps each of the pilus subunits including the adhesin to fold properly prior to assembly. In addition to directing the folding of pilus subunits, the chaperone also protects the subunits from degradation by proteases while shuttling them to the bacterial outer membrane for assembly. The chaperone-subunit complexes are targeted to a large channel protein in the bacterial outer membrane called the usher where the chaperone is released, exposing interactive surfaces on the subunits that facilitate their assembly into the pilus (46-47). Each of the pilus subunits are then incorporated into the pilus depending, in part, upon the kinetics with which they are partitioned to the usher in complex with the chaperone (48).

Earlier biochemical and genetic studies demonstrated a critical role for the pilus chaperones in maintaining the stability of subunit proteins. However the recent studies on the X-ray crystal structure of the FimH adhesin in a complex with its chaperone FimC (and PapK with its cognate PapD chaperone), shed new light on the nature of the chaperone subunit interaction (49-50). With regard to the FimCH complex the following was noted. The FimH adhesin has two domains: one interacts with the FimC chaperone (the pilin domain) while the other interacts with mannose (the lectin-binding domain). The pilin domain that interacts with FimC, as well as FimC itself, has an immunoglobulin-like fold made up of antiparallel β-strands. However, the seventh strand is missing from the FimH immunoglobulin fold, exposing a hydrophobic groove on the surface of the pilin domain. This groove would render the FimH unstable were it not for the fact that the FimC chaperone temporarily shares its seventh strand with the FimH subunit without parting with the shared strand, in a process called donor strand complementation. This complementation during formation of the FimCH complex stabilizes the FimH adhesin.

Table I. *In vivo* pre-clinical evaluation of FimCH vaccine

Experiment	Species (n)	Antigen/Formulation/Adjuvant	Dose (µg)	Route	Controls	Evaluations	Conclusions
Dose ranging/ Proof of principle	C3H/HeJ mice	FimCH 20mM HEPES buffer, pH 7.0 CFA/IFA	0.6 3.0 15.0 30.0	IP	CFA/IFA only FimC in 20mM HEPES pH 7.0 (30 µg)	Serum IgG titers to FimH (T3) IgG antibodies to FimH in urine and vaginal secretions Functional inhibitory titers (see *in vitro* assays) Protection against challenge with 10^8 cfu uropathogenic *E. coli* (multiple cystitis isolates)	Long lasting immune (> 1 yr.) response to FimH IgG antibodies to FimH transudate into urine and vaginal secretions Antibodies to FimH block attachment of uropathogenic *E. coli* to bladder cells. >99% reduction in colonization of the bladder mucosa *in vivo*. $P < 0.0001$ vs. FimC, naïve.
Passive protection with antibodies to FimCH	New Zealand White Rabbits	FimCH 20mM HEPES buffer, pH 7.0 CFA/IFA	$200.0, 1^0$ $50.0, 2^0$ $50.0, 3^0$	SC	FimC in 20mM HEPES pH 7.0 Pre-Immune	Confirmation that protection is antibody-mediated Protection against challenge with 10^8 cfu of diverse uropathogenic *E. coli* strains.	>99% reduction in colonization of the bladder mucosa *in vivo* with hyperimmune sera to FimCH. $P < 0.0001$ vs. FimC, naïve. Significant protection seen with 9/10 strains tested (primary and recurrent isolates) $P < 0.01$-0.0001. One strain showed ~ 3/4 log reduction ($P < 0.5$).
Test Adjuvant Suitable for Humans * Protection * Dose ranging * Immunization Regimen * Potency	C3H/HeJ mice	FimCH 20mM HEPES buffer, pH 7.0 or PBS pH 7.0 MF59	0.16 0.8 4 20	IM SC IP	MF59 only FimC in 20mM HEPES pH 7.0	Prime and boost at 4 weeks and 18 weeks vs. Prime and boost at 18 weeks only. Serum, urine and vaginal IgG titers to FimH (T3). Functional inhibitory titers Protection against challenge	4 week boost + 18 week boost enhances immune response (vs. no 4 week boost). IgG antibodies to FimH in serum; IgG to FimH transudates into urine and vaginal secretions. Potent inhibitory titers after each boost. >99% reduction in colonization of the bladder *in vivo*. $P < 0.0001$ vs. FimC, naïve.
Primate Studies * Potency * Protection * Effect of Vaccine on Normal Gut Flora	Cynomolgus monkeys (*Macaca fascicularis*, healthy adult females)	FimCH PBS pH 7.0 MF59	100.0	IM	MF59 only	a.) Immune response to FimCH: Functional Inhibitory titers and presence of FimH antibodies in secretions b.) Protection c.) Effect of FimH antibodies on normal gut flora	Vaccination with FimCH induces protective antibodies in monkeys; protects monkeys from cystitis (colonization with uropathogenic *E. coli*) and from inflammatory responses due to cystitis Vaccine has no observed effects on normal flora of GI tract

Table 2. *In vitro* pre-clinical evaluation of FimCH vaccine

Experiment	Cells/Substrate	Sampling	Controls	Evaluations	Conclusions
Inhibition of binding of type 1-piliated (FimH-expressing) *Escherichia coli* isolates	J-82 Human bladder cells	Antibodies to FimH: or sera to FimC alone Serial dilutions of 1/50 to 1/400 Mouse, rabbit and monkey serum antibodies. Undiluted vaginal wash samples from immunized monkeys.	Pre-immune sera Vaginal wash samples from control monkeys.	Inhibition of binding to bladder epithelial cells, after incubation of clinical isolates in the presence of the antibodies to FimH	94% of random urine isolates of *E. coli* were inhibited (49/52). 100% (24/24) of *E. coli* isolates from were inhibited. Murine, rabbit and monkey serum antibodies to FimCH inhibited binding. Serum inhibitory titers correlated with overall endpoint titers to FimH. Vaginal wash samples with FimH antibodies inhibited binding.
Inhibition of binding by type 1-piliated (FimH-expressing) *Escherichia coli* isolates	TriMannose (FimH has lectin-like binding activity;binds to mannose)	Antibodies to FimH: Serial dilutions of 1/50 to 1/400 Mouse, rabbit and monkey serum antibodies. Undiluted vaginal wash samples from immunized monkeys.	Pre-immune sera or sera to FimC alone Vaginal wash samples from control monkeys.	Inhibition of binding to bladder epithelial cells, after incubation of clinical isolates in the presence of the antibodies to FimCH	Results correlated with J-82 bladder assay above. Easier assay to perform on large number of samples compared to J-82 assay (higher throughput)
FimH sequence tested conservation in *Escherichia coli* strains from urinary tract infections	>50 random urinary tract infection isolates 24 isolates from women with recurrent infections	PCR amplification of *fimH* gene. Sequencing of *fimH* gene	NA	Comparison of deduced amino acid FimH sequences among isolates	>97% conservation of FimH among all strains

These novel findings based on the crystal structure explain in part why initial attempts at developing a subunit adhesin vaccine based on only FimH without its cognate chaperone protein FimC were unsuccessful. The use of hyper-expression systems to produce and purify large amounts of adhesin most likely failed because FimH expressed as an independent moiety, with an exposed hydrophobic groove, enters a nonproductive pathway in the bacterium and is rapidly degraded (51-52). In contrast, hyper-expression of recombinant FimH in combination with FimC as a FimCH complex allows for high level expression of the full-length adhesin. Since the FimCH complex is a stable compound, and importantly presents the lectin binding domain of FimH in the proper conformation, comparable to its the orientation at the pilus tip, the complex makes an ideal vaccine candidate for eliciting functional antibodies to type 1 piliated uropathogenic *E. coli.*

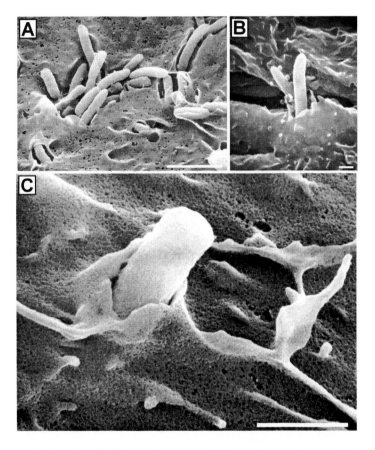

Figure 1. Internalization of type 1-piliated *E. coli* by bladder epithelial cells. (A and B) Scanning electron micrographs of the mouse bladders 2 hours after infection with a type 1-piliated clinical cystitis isolate, NU14, show the host epithelial cell membrane zippering around and enveloping attached bacteria. (C) Human bladder epithelial cells grown in culture also internalize adherent type 1-piliated bacteria. Biochemical and genetic analyses indicate that the internalization process is mediated by the type 1 pilus adhesin, FimH. Scale bars indicate 3 μm (A) and 1 μm (B and C).

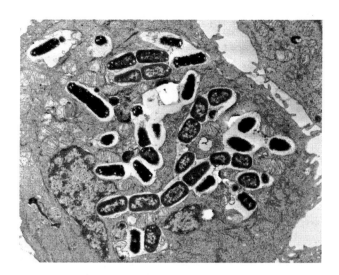

Figure 2. Transmission electron micrograph of cultured 5637 human bladder cells 24 hours after infection with the uropathogenic *E. coli* strain, NU14. The large numbers of intracellular bacteria are not present during early time points after infection and are indicative of intracellular bacterial replication. Various microscopic and biochemical assays have confirmed that many uropathogenic strains of *E. coli* can proliferate within bladder epithelial cells.

Pre-clinical Studies with the FimCH Vaccine

The FimCH vaccine under development for UTIs is made up of an ~ 59 kDa complex composed of FimC (29 kDa) and FimH (~30 kDa) in a 1:1 equimolar ratio. The recombinant form of the complex is produced in the *E. coli* periplasm and is purified from periplasmic extracts by standard chromatographic methods. The FimCH protein has been formulated in a number of different buffers compatible with its solubility profile and has been shown to be stable under long-term storage conditions either frozen or at 4°C. While some of the early pre-clinical studies in mice with FimCH were done using complete or incomplete Freund's adjuvant to enhance the immune response to the vaccine, most of the pre-clinical studies in mice and primates have been done with a 1:1 formulation of the FimCH vaccine mixed with a proprietary adjuvant called MF59. MF59 is surfactant-stabilized emulsion adjuvant MF59 (53) (Chiron Corporation, Emeryville, CA), which has been shown to be a potent and safe adjuvant in human subjects with several vaccines, is an oil-in-water formulation that contains surfactants (Tween 20 and Span 85) along with metabolizable oil (squalene) emulsified under high-pressure conditions. In experimental animals and humans, MF59 augments serum antibody responses to a variety of different antigens by 3 to >100-fold over alum.

The results of many pre-clinical studies performed in both mice and primates are summarized in Tables 1 and 2. The data demonstrate that immunization with the FimCH complex elicited long-lasting immune responses to the FimH adhesin, at doses as low as 0.3- 0.8 micrograms, depending upon the adjuvant tested. More importantly, the vaccine induced functional antibodies as measured by the ability of the antibodies to block attachment of type 1-piliated uropathogenic *E. coli* to free mannose or mannsoylated receptors on bladder epithelial cells. This was determined using two different *in vitro* surrogate assays for protective antibody responses: 1.) Interrupting attachment to immobilized mannose (the receptor for FimH) in a 96-well assay format or 2.) Blocking attachment of uropathogenic *E. coli* to J-82 bladder cells.

Sera from mice vaccinated with the FimCH vaccine was shown to block attachment of 49/52 (94%) clinical *E. coli* UTI isolates induced to express type 1 pili. The anti-FimH antisera was also able to block 100% (24/24) of *E. coli* strains obtained from women with recurrent urinary tract infections including multiple isolates from a single individual obtained from sequential infections. In addition, the antisera blocked attachment of pyelonephritic isolates of *E. coli* as well as isolates obtained from women with asymptomatic bacteriuria.

Immunization with FimCH has been shown to reduce *in vivo* colonization of the bladder mucosa by greater than 99% in the murine cystitis model (34). This has been confirmed using many different uropathogenic *E. coli* strains, including strains from women with primary cystitis or recurrent infections. Furthermore the FimH vaccine protected against colonization and disease by uropathogenic strains of *E. coli* capable of expressing multiple adhesins such as P pili and S pili, in addition to type 1 pili. This is an important finding given that bacterial adhesion is a multifaceted process, and in some cases may entail multiple mechanism of adherence and more than a single adhesin. Immunoglobulin G specific for FimH was detected in the urine of protected mice, consistent with our original hypothesis that antibodies directed against an adhesin protein might protect along the mucosal surface.

The murine model for cystitis has proven to be useful in screening the FimCH vaccines and assessing cross-protection against diverse clinical UTI isolates *in vivo*. However we were interested in validating the concept of a FimCH vaccine in a non-human primate model of UTI prior to initiating clinical trials of the FimH vaccine in humans. Monkeys and humans have similar bladder, ureteral, and renal physiology and both exhibit natural occurrence of bladder infections and pyelonephritis, and vesicoureteral reflux facilitating ascending UTIs (54-55). Vaccination and challenge studies in the primate model corroborated the original murine studies (56). Monkeys vaccinated with purified recombinant FimCH chaperone-adhesin complex with the MF59 adjuvant developed long-lasting serum IgG antibodies to FimH as well as functional inhibitory titers as measured by the ability of the antibodies to block *in vitro* binding of type 1-piliated *E. coli* to mannose. The vaccine protected the monkeys from

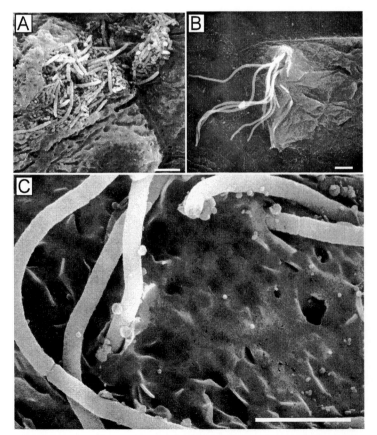

Figure 3. Scanning electron micrographs of mouse bladders 6 hours after infection with uropathogenic *E. coli*. (A) Numerous bacteria apparently in the process of exiting their host bladder cells can be observed at this time point. (B and C) The bacteria are often filamentous and partially septated. Similar filamentous forms of bacteria have been observed in *in vitro* assays using cultured bladder cells. Scale bars indicate 5 μm (A and B) and 3 μm (C).

bladder infection by uropathogenic *E. coli* and from an inflammatory response typically associated with cystitis. Furthermore, in the monkey studies we were able to demonstrate a direct correlation between the presence of inhibitory antibodies in local (vaginal wash) secretions and protection against colonization and infection.

Taken together the pre-clinical data suggest that a FimCH vaccine may have utility in preventing cystitis in humans, by blocking colonization when infectious *E. coli* are first introduced into the bladder. What about blocking recurrent infections that affect so many women after a primary bout of UTI? This issue will in all likelihood be addressed and may be answered during upcoming human clinical trials with the FimCH vaccine. However, in the meantime, as we gain greater insight into the mechanism of recurrent infections at the molecular level, and develop laboratory models for such infections, these issues are also being addressed experimentally. The next section will explore the role of FimH in invasion of the bladder epithelium and the implication this has for recurrent UTIs as well as for FimH adhesin-based vaccines to protect against such infections.

Adhesin-mediated Invasion of Host Cells; Another Reason for a FimH Vaccine

Research over the past several years has demonstrated that interactions between bacterial adhesins and host receptor molecules can act as trigger mechanisms, activating signal transduction cascades and altering gene expression in both bacterial and host cells (57-58). The activation of host signaling pathways following bacterial attachment can result in dramatic rearrangements of the eukaryotic cytoskeleton that can lead to the internalization of adherent bacteria. In such cases, the host cell membrane can zipper around and envelope adherent bacteria in response to direct, sequential interactions between host cell membrane receptors and specific bacterial adhesins, which are sometimes referred to as invasins. In other cases bacteria introduce effector molecules into their target host cells, triggering intense ruffling of the host cell membrane that results in bacterial uptake (58).

Several factors with invasin-like properties have identified in a variety of bacterial pathogens other than uropathogenic *E. coli*. The two best characterized invasins, the prototypical invasin protein encoded by *Yersinia* and internalin expressed by *Listeria*, directly mediate bacterial internalization into host cells via a zipper-like mechanism by interacting with β_1-integrin and E-cadherin, respectively (59-60). Recent work has indicated that the type 1 pilus adhesin, FimH, can also function as an invasin, mediating the entry of uropathogenic *E. coli* into host bladder epithelial cells (61). This study demonstrated that both uropathogenic and laboratory K12 strains expressing type 1 pili were taken up by bladder epithelial cells, while non-piliated strains and piliated mutants lacking the FimH adhesin were non-invasive (61). Furthermore, bladder epithelial cells in culture efficiently internalized FimH-coated latex beads in the absence of any other potential virulence factors. In contrast, BSA-coated beads, which occasionally adhere non-specifically to bladder epithelial cells, were rarely internalized.

The uptake of uropathogenic *E. coli* by bladder epithelial cells was noted in electron microscopic (EM) studies more than twenty years ago (62-63). Bacteria were observed within membrane bound vesicles and free within the cytoplasm of the superficial facet cells that line the lumenal surface of the bladder. It was proposed at the time that the bladder cells internalized bacteria as part of a host defense mechanism. A more recent study, however, suggested that bacterial internalization into bladder epithelial cells could benefit the pathogen (64). Using a murine cystitis model, it was shown that a subpopulation of type 1-piliated *E. coli*

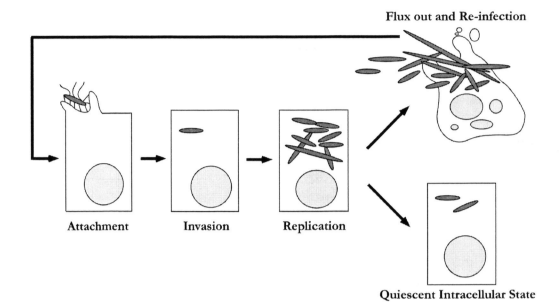

Figure 4. Flux model of urinary tract infections. Upon entering the bladder, type 1-piliated *E. coli* can attach to and invade the superficial facet cells lining the lumen of the bladder. These events are mediated by the FimH adhesin. Once internalized, many uropathogenic strains of *E. coli* can proliferate, forming discrete foci containing multiple bacteria within the host bladder cells. During the infection process, as-of-yet-undefined factor(s) can trigger the activation of a suicide response by infected bladder cells. These cells will eventually exfoliate and be cleared from the body with the flow of urine, carrying any associated bacteria with them. To avoid clearance by this mechanism, some uropathogens appear able to flux out of their dying host cells and re-infect surrounding tissue. Rather than moving in and out of host bladder cells, it is also possible that some internalized uropathogens can enter a quiescent state within their host cells. See text for additional details.

inoculated into the bladder could enter the epithelium and that these intracellular bacteria appeared to have a survival advantage over their extracellular counterparts. Scanning EM of mouse bladders shortly after infection with type 1-piliated *E. coli* showed occasional bacteria being enveloped by host bladder epithelial cells (Figure 1A and B). The envelopment and uptake of type 1-piliated bacteria by human bladder epithelial cells grown in culture has also been observed (Figure 1C) and appears to mimic bacterial internalization by murine bladder cells *in vivo* (61). *In vitro* studies indicate that the internalization process requires localized rearrangement of the host cytoskeleton and the activation of phosphoinositide 3-kinase and as-of-yet undefined host protein tyrosine kinases.

Assays designed to measure intracellular bacterial replication following invasion demonstrated that many uropathogenic strains of *E. coli*, but not internalized K12 strains, could proliferate within host bladder cells grown in culture (61). Transmission EM (Figure 2) and confocal microscopy of bladder cells 1 to 3 days after infection with, a clinical cystitis isolate, showed discrete pockets containing large numbers of bacteria within host cells. Massive foci of intracellular bacteria also developed *in vivo* within superficial bladder cells several hours after the inoculation of mouse bladders with other clinical isolates. Current data suggest that uropathogens, after replicating intracellularly, are able to exit their host cells and infect surrounding tissue (65). Histological examination and scanning EM of mouse bladders 6 hours after infection with uropathogenic *E. coli* show signs of this process occurring *in vivo* (Figure 3A-C). The bacteria apparently exiting the host cells are often multinucleate and filamentous (Figure 3B and C), a situation that could arise due to rapid bacterial replication without complete septation of daughter cells during an intracellular growth phase. These filamentous forms could potentially allow bacteria to spread between host cells while maintaining contact with the bladder epithelium, thus reducing the risk to the pathogen of being rinsed away with the flow of urine.

These and other observations indicate that the pathogenesis of UTIs may be more dynamic and complicated than previously assumed. The capacity of uropathogens to move in and out of bladder epithelial cells may facilitate bacterial dissemination within the urinary tract and may provide a means for uropathogens to evade both innate and adaptive host defenses. By entering bladder epithelial cells, uropathogenic *E. coli* may escape the cleansing flow of urine through the bladder lumen and avoid contact with anti-microbial substances and immune cells. The intracellular environment of bladder epithelial cells may also provide a better source of nutrients for invading microbes than available at host cell surfaces and within the urine. The sanctuary provided by bladder epithelial cells, however, is limited. In response to infection by type 1-piliated *E. coli* strains, but not *fimH* mutants, bladder epithelial cells can exfoliate (62-64). Exfoliation occurs via an apoptosis-like mechanism involving the activation of proteolytic enzymes known as caspases and host DNA fragmentation (64). The urine of patients with UTIs often contains exfoliated bladder epithelial cells with associated bacteria (29) and it has been demonstrated that the exfoliation and clearance of infected bladder cells in response to infection can act as a host defense mechanism (64).

The ability of uropathogenic *E. coli* to replicate intracellularly may allow invading uropathogens to multiply within a protected niche and subsequently escape from their host bladder cells before the host cells can complete the exfoliation process (Figure 4). However, the exfoliation of bladder epithelial cells in response to infection may not always be entirely beneficial to the host and may leave lower layers of the bladder epithelium more susceptible to infection. The interplay between host cells and uropathogens is further complicated by recent studies suggesting that uropathogens can also enter a quiescent state within some host bladder epithelial cells (65). These bacteria could potentially serve as a source for recurrent infections. The central role of the type 1 pilus adhesin, FimH, in mediating both bacterial attachment and internalization into host bladder epithelial cells makes it an ideal target for vaccine development. Antibodies directed at FimH could not only act prophylactically, preventing the initial colonization of the bladder epithelium by type 1-piliated *E. coli*, but may also disrupt the ability of uropathogens to persist and disseminate within the urinary tract by fluxing in and out of the bladder epithelium.

Concluding Remarks: FimH as a Paradigm for Other Adhesin-based Vaccines

The need for effective prophylactic vaccines against a wide range of bacterial infections becomes critical as the prevalence of antibiotic resistant pathogens increases. Blocking the primary stages of infection, bacterial attachment to host cell receptors and colonization of the mucosal surface, may be the most effective strategy to prevent infections. Bacterial adhesion typically requires an interaction between a bacterial adhesin and the host cell receptor. Recent pre clinical studies in our laboratory with the FimH adhesin, have confirmed that antibodies targeting a bacterial adhesin can impede colonization, block infection and prevent disease. Evidence from the FimH vaccine studies supports the paradigm that prophylactic vaccination with adhesins should be an effective means to block bacterial infections that initiate at mucosal surfaces. With recent advances in the identification, characterization and isolation of other adhesins, similar approaches are being explored for a variety of other diseases beyond urinary tract infections.

References

1. Kunin, C.M. 1994. Urinary tract infections in females. Clin. Infect. Dis. 18:1-12.
2. Hooton, T.M., Scholes, D., Hughes, J.P., Winter, C., Roberts, P.L., Stapleton, A.R., Stergachis, A. and Stamm, W.E. 1996. A prospective study of risk factors for symptomatic urinary tract infection in young women. New Engl. J. Med. 335: 468-474.
3. Muhldorfer, I. and Hacker, J. 1994. Genetic aspects of *Esherichia coli* virulence. Microb. Pathog. 16:171-181.
4. Barnett, B.J. and Stephens, D.S. 1997. Urinary tract infection: an overview. Am J. med. Sci. 314:245-249.
5. Gupta K., Scholes, D. and Stamm, W.E. 1999. Increasing prevalence of antimicrobial resistance among uropathogens causing acute uncomplicated cystitis in women. JAMA. 281:736-738.
6. Allen, U.D., MacDonald, N., Futie, L., Chan, F. and Stephens, D. 1999. Risk factors for resistance to "first-line" antimicrobials among urinary tract isolates of *Escherichia coli* in children. CMAJ. 160:1436-1440.
7. Karlsson, K.-A. 1989. Animal glycosphingolipids as membrane attachment sites for bacteria. Ann. Rev. Biochem. 58:309-350.
8. Sharon, N. 1996. Carbohydrate-lectin interactions in infectious disease. Adv Exp Med Biol. 408:1-8
9. Lingwood, C.A. 1998. Oligosaccharide receptors for bacteria: a view to kill. Curr. Op. Chem. Biol. 2:695-700.
10. Soto, G.E. and Hultgren, S.J. 1999. Bacterial adhesins: common themes and variations in architecture and assembly. J. Bacteriol. 181: 1059-1071.
11. Ofek, I., Mirelman, D. and Sharon, S. 1977. Adherence of *Escherichia coli* to human mucosal cells mediated by mannose receptors. Nature. 265:623-625.
12. Hanson, M.S. and Brinton, C.C. 1988. Identification and characterization of *E. coli* type-1 pilus tip adhesion protein. Nature. 332: 265-268.
13. Maurer, L. and Orndorff, P.E. 1987. Identification and characterization of genes determining receptor binding and pilus length of *Escherichia coli* type 1 pili. J. Bacteriol. 169:640-645.
14. Abraham, S.N., Goguen, J.D., Sun, D., Klemm, P. and Beachey, E.H. 1987. Identification of two ancillary subunits of *Escherichia coli* type 1 fimbriae by using antibodies against synthetic oligopeptides of *fim* gene products. J. Bacteriol. 169:5530-5536.
15. Leffler, H. and Svanborg-Eden, C. 1980. Chemical identification of a glycosphingolipid receptor from *Escherichia coli* attaching to human urinary tract epithelial cells and agglutinating human erythrocytes. FEMS Microbiol. Lett. 8:127-134.
16. Lindberg, F.P., Lund, B. and Normark, S. 1984. Genes of pyelonephritic *E. coli* required for digalactoside specific agglutination of human cells. EMBO J. 3:1167-1173.
17. Hull, R.A. and Hull, S.I. 1984. Frequency of gene sequences necessary for pyelonephritis-associated pili expression among isolates of *Enterobactereiaceae* from human extraintestinal infections. Infect. Immun. 43:1064-1067.
18. Hull R.A. and Hull S.I. 1994. Adherence mechanisms in urinary tract infections. In: Molecular Genetics of Bacterial Pathogenesis. V.L. Miller, J.B. Kaper, D.A. Portnoy, eds. ASM Press, Washington, D.C. p. 79-90.
19. Jones, C.H., Dodson, K. and Hultgren, S.J. 1996. Structure, function and assembly of adhesive P pili. In: Urinary Tract Infections: Molecular Pathogenesis and Clinical Management. H.L.T. Mobley and J.W. Warren eds. ASM Press, Washington, D.C. p. 175-219.

20. Moch, T., Hoschutzsky, H., Hacker, J., Kroncke, D. and Jann, K. 1987. Isolation and characterization of the α-sialyl-β-2,3-galactosyl-specific adhesin from fimbriated *Escherichia coli*. Proc. Natl. Acad. Sci. USA. 84:3462-3466.

21. Schmoll, T., Morschhauser, J., Ott, M., Ludwig, B., van Die, I. and Hacker, J. 1990. Complete genetic organization and functional aspects of the *Escherichia coli* S fimbrial adhesin determinant: nucleotide sequence of the genes *sfa* B, C, D, E,F. Microb. Pathog. 9:331-343.

22. Hultgren, S.J., Abraham, S., Caparon, M., Falk, P., St, Geme III, J.W. and Normark, S. 1993. Pilus and nonpilus bacterial adhesins: assembly and function in cell recognition. Cell. 73:887-901.

23. Jacob-Dubuisson, F., Kuehn, M. and Hultgren, S.J. 1993 A novel secretion apparatus for the assembly of adhesive bacterial pili. Trends in Microbiol. 1:50-55.

24. Kuehn, M.J., Heuser, J., Normark, S. and Hultgren, S.J. 1992. P Pil in uropathogenic *E. coli* are composite fibres with distinct fibrillar adhesive tips. Nature. 356:252-255.

25. Jones, C.H., Pinkner, J.S., Roth, R., Heuser, J., Nicholes, A.V., Abraham, S.N. and Hultgren, S.J. 1995. FimH adhesin of type 1 pili is assembled into a fibrillar tip structure in the *Enterobacteriacae*. Proc. Natl. Acad. Sci. USA. 92:2081-2085.

26. Costerton, J.W., Stewart, P.S. and Greenberg, E.P. 1999. Bacterial biofilms: a common cause of persistent infections. Science. 284: 1318-1322.

27. Ofek, I. Mosek, A. and Sharon, N. 1981. Mannose-specific adherence of *Escherichia coli* freshly excreted in the urine of patients with acute urinary tract infections and of isolates subcultured from the infected urine. Infect. Imm. 34: 708-714.

28. Abraham S.N., Babu J.P., Giampapa C.S., Hasty D.L., Simpson W.A. and Beachey E. H. Protection against *Escherichia coli*-induced urinary tract infections with hybridoma antibodies directed against type 1 fimbriae or complementary D-mannose receptors. Infect. Immun. 1985. 48:625-628.

29. Kisielius, P.V., Schwan, W.R., Amundsen, S.K., Duncan, J.L. and Schaeffer, A.J. 1989. *In vivo* Expression and Variation of *Escherichia coli* Type 1 and P pili in the urine of adults with acute urinary tract infections. Infect. Immun. 57: 1656-1662.

30. Pere, A., Nowicki, B., Saxen, H., Siitonen, A. and Korhonen, T.K. 1987. Expression of P, type-1 and type 1C Fimbriae of *Escherichia coli* in the urine of patients with acute urinary tract infection. J. Infect. Dis. 156:567-574.

31. Schaeffer, A.J., Jones, J.M. and Dunn, J.K. 1981. Association of *in vitro Escherichia coli* adherence to vaginal and buccal epithelial cells with susceptibility to women with recurrent urinary tract infections. N. Engl. J. Med. 304:1062-1066.

32. Gaffney, R.A., Venegas, M.F., Kanerva, C., Navas, E.L. anderson, B.E., Duncan, J.L. and Schaeffer, A.J. 1995. Effect of vaginal fluid on adherence of type 1 piliated *Escherichia coli* to epithelial cells. J. Inf. Dis. 172:1528-1535.

33. Connell, H., Agace, W., Klemm, P., Schenbri, M., Marild, S. and Svanborg, C. 1996. Type 1 fimbrial expression enhances *Escherichia coli* virulence for the urinary tract. Proc. Natl. Acad. Sci. USA 93:9827-9832.

34. Langermann S., Palaszynski S., Barnhart M., Auguste, G., Pinkner, J.S., Burlein, J., Barren, P., Koenig, S., Leath, S., Jones, C.H. and Hultgren, S.J. 1997. Prevention of mucosal *Escherichia coli* infection by FimH-adhesin-based systemic vaccination. Science 276:607-611.

35. Wu, X.-R., Sun, T.-T. and Medina, J.J. 1996. *In vitro* binding of type 1-fimbriated *Escherichia coli* to uroplakins Ia and Ib: relation to urinary tract infections. Proc. Natl. Acad. Sci. USA. 93:9630-9635.

36. Vaisanen, V., Elo, J., Tallgren, L.G.,Siitogen, A., Makela, P.H., Svanborg-Eden, C., Kallenius, G., Svenson, S.B., Hultberg, H. and Korhonen, T. 1981. Mannose resistant hemagglutination and P antigen recognition are characteristics of *Escherichia coli* causing primary pyelonephritis. Lancet. 2: 1366-1369.

37. Kallenius, G., Svenson, S.B., Hultberg, H., Mollby, R., Helin, I., Cedergen, B. and Windberg, J. 1981. Occurrence of P-fimbriated *Escherichia coli* in urinary tract infection. Lancet. II: 1369-1372.

38. Elo, J., Tallgren, L.G., Vaisanen, V., Korhonen, T.K., Svenson, S.B. and Makela, P.H. 1985. Association of P and other fimbriae with clinical pyelonephritis in children. Scand. J. Urol. Nephrol.19:281-4.

39. O'Hanley, P., Lark, D., Falkow, S. and Schoolnik, G. 1985. Molecular basis of *Escherichia coli* colonization of the upper urinary tract in BALB/c mice. Gal-Gal pili immunization prevents *Escherichia coli* pyelonephritis in the BALB/c mouse model of human pyelonephritis. J. Clin. Invest. 75:347-60.

40. Roberts, J.A., Marklund, B.I., Ilver, D., Haslam, D., Kaack, M.B., Baskin, G., Louis, M., Mollby, R., Winberg, J. and Normark, S. 1994. The Gal(alpha 1-4)Gal-specific tip adhesin of *Escherichia coli* P-fimbriae is needed for pyelonephritis to occur in the normal urinary tract. Proc. Natl. Acad. Sci. USA. 91:11889-93.

41. Siitonen, A., Takala, A., Ratiner, Y.A., Pere, A. and Makela, P.H. 1993. Invasive *Escherichia coli* infections in children: bacterial characteristics in different age groups and clinical entities. Pediatr. Infect. Dis. J. 12:606-12.

42. Hung, D.L., Knight, S.D., Woods, R.M., Pinkner, J.S. and Hultgren, S.J. 1996. Molecular basis of two subfanilies of immunoglobulin-like chaperones. EMBO J. 15:379-3905.

43. Hung, D.L. and Hultgren, S.J. 1998. Pilus biogenesis via the chaperone/usher pathway: An integration of structure and function. J. Struct. Biol. 124:201-220.

44. Thanassi, D.G., Saulino, E.T. and Hultgren, S.J. 1998. The chaperone/usher pathway: a major terminal branch of the general secretory pathway. Curr. Opin Microbiol. 1: 223-31

45. Jones, C.H., Pinkner, J.S., Nicholes, A.V., Slonim, L.N., Abraham, S.N. and Hultgren, S.J. 1993. FimC is a periplasmic PapD-like chaperone that directs assembly of type 1 pili in bacteria. Proc. Natl. Acad. Sci. USA. 90:8397-8401.

46. Dodson, K., Jacob-Dubuisson, F., Striker, R.T. and Hultgren, S.J. 1996. Outer-membrane PapC molecular usher discriminately recognizes periplasmic chaperone-pilus subunit complexes. Proc. Natl. Acad. Sci. USA. 90:3670-3674.

47. Thanassi, D.G., Saulino, E.T., Lombardo, M.J., Roth, R., Heuser, J. and Hultgren, S.J. 1998. The PapC usher forms an oligomeric channel: implications for pilus biogenesis across the outer membrane. Proc. Natl. Acad. Sci. USA. 95:3146-3151.

48. Saulino, E.T., Thanassi, D.G., Pinkner, J. and Hultgren, S.J. 1998. Ramifications of kinetic partitioning on usher-mediated pilus biogenesis. EMBO J. 17:2177-2185.

49. Choudhury, D., Thompson, A., Stojanoff, V., Langermann, S., Pinkner, J., Hultgren,S.J. and Knight, S.D. 1999. X-ray Structure of the FimC-FimH Chaperone-Adhesin Complex from Uropathogenic *Escherichia coli*. Science. 285:1061-1066.

50. Sauer, F.G., Fütterer, K., Pinkner, J.S., Dodson, K.W., Hultgren, S.J. and Waksman, G. 1999. Structural Basis of Chaperone Function and Pilus Biogenesis. Science. 285:1058-1061.

51. Hultgren, S.J, Normark S. and Abraham, S.N. 1991. Chaperone-assisted assembly and molecular architecture of adhesive pili. Annu Rev Microbiol. 45:383-415

52. Kuehn, M.J., Normark, S. and Hultgren, S.J. 1991. Immunoglobulin-like PapD chaperone caps and uncaps interactive surfaces of nascently translocated pilus subunits. Proc. Natl. Acad. Sc.i USA. 88:10586-90.

53. Ott G., Barchfeld G.L., Chernoff D., *et al.* 1996. MF59: Design and evaluation of a safe and potent adjuvant for human vaccines. In: Vaccine Design: The Subunit and Adjuvant Approach. Powell, M.F., ed. New York: Plenum Press. p. 277-296.

54. Roberts, J.A. 1992. Contribution of experimental pathology to the understanding of human pyelonephritis. J. Urol. 148:1721-1725.

55. Roberts, J.A. 1992. Vesicoureteral reflux and pyelonephritis in the monkey: a review. Urol. 148:1721-1725.

56. Langermann, S., Mollby, R., Burlein, J.E., Palaszynski, S.R., Auguste, C.G., DeFusco, A., Strouse, R., Schenerman, M.A., Hultgren, S.J., Pinkner, J.S., Winberg, J., Guldevall, L., Soderhall, M., Ishikawa, K., Normark, S. and Koenig, S. 2000. Vaccination with FimH adhesin protects cynomolgus monkeys from colonization and infection by uropathogenic *Escherichia coli.* J. Infect. Dis. 181:774-778.

57. Finlay, B.B. and Falkow, S. 1997. Common themes in microbial pathogenicity revisited. Microbiol. Mol. Biol. Rev. 53: 210-230.

58. Finlay, B.B. and Cossart, P. 1997. Exploitation of mammalian host cell function by bacterial pathogens. Science. 276: 718-725.

59. Isberg, R.R. 1990. Pathways for the penetration of enteropathogenic *Yersinia* into mammalian cells. Mol. Biol. Med. 7:73-82.

60. Parida, S.K., Domann, E., Rohde, M., Muller, S., Darji, A., Hain, T., Wehland, J. and Chakraborty, T. 1998. Internalin B is essential for adhesion and mediates the invasion of Listeria monocytogenes into human endothelial cells. Mol. Microbiol. 28:81-93.

61. Martinez, J.J., Mulvey, M.A., Schilling, J.D., Pinkner, J.S. and Hultgren, S.J. 2000. Type 1 pilus mediated invasion of bladder epithelial cells. EMBO J. 15: 2803-2812.

62. Fukushi, Y., Orikasa, S. and Kagayama, M. 1979. An electron microscopic study of the interaction between vesical epithelium and *E. coli.* Invest. Urol. 17: 61-68.

63. McTaggart, L.A., Rigby, R.C. and Elliot, T.S.J. 1990. The pathogenesis of urinary tract infections associated with *Escherichia coli, Staphylococcus saprophyticus* and *S. epidermidis.* J. Med. Microbiol. 32: 135-141.

64. Mulvey, M.A., Lopez-Boado, Y.S., Wilson, C.L., Roth, R., Parks, W.C., Heuser, J. and Hultgren, S.J. 1998. Induction and evasion of host defenses by type 1-piliated uropathogenic *Escherichia coli.* Science. 282: 1494-1497.

65. Mulvey, M.A., Schilling, J.D. and Hultgren, S.J. unpublished observations.

From: *Development of Novel Antimicrobial Agents: Emerging Strategies*
ISBN 1-898486-23-9 © 2001 Horizon Scientific Press, Wymondham, UK.

11

Antimicrobial Peptides in Innate Immunity

Tomas Ganz and Robert I. Lehrer

Abstract

Antibiotic peptides encoded by genes are increasingly recognized as effector molecules of host defense in plants and animals. In higher animals, the peptides are particularly abundant on epithelial surfaces and in the storage granules of phagocytic cells. In these settings, antimicrobial peptides are positioned to act during the earliest stages of the interaction between the host and its microbial invaders. The most abundant antimicrobial peptides of mammals include defensins, characterized by six conserved disulfide-bridged cysteines and cathelicidins, structurally heterogenous peptides that share a conserved precursor domain, cathelin. A rich array of additional antimicrobial peptides are also expressed by vertebrates and invertebrates. Studies of antimicrobial peptides are providing new insights into the dynamic interactions between microbes and their hosts, and generating new paradigms for the pathogenesis and treatment of diseases.

Innate Host Defense

Multicellular organisms must continually defend themselves against penetration and parasitization by microbes. In higher animals, the initial encounter between microbes and their potential hosts commonly takes place on epithelial surfaces (skin, the moist surfaces of the eyes, nose, airways and the lungs, mouth and the digestive tract, and the urinary and reproductive systems). The predominant host resistance mechanisms operative during early phases of initial infections do not require antibody-mediated specific antigen recognition, a relatively slow process that becomes optimal only after days of clonal expansion of antigen-specific lymphocytes. Instead, the initial recognition mechanisms take advantage of shared structural or functional characteristics of microbes. Whatever the recognition mechanisms deployed, successful host defense ends by the destruction, sequestration or metabolic inhibition of invading microbes, abetted by a large number of host-derived antimicrobial substances. These range in size from small inorganic molecules such as hydrogen peroxide, to large protein complexes such as are generated by the activation of the complement cascade. Antimicrobial peptides are conventionally defined as polypeptide antimicrobial substances that are encoded by genes and synthesized by ribosomes, and that contain fewer than 100 amino acid residues. This definition distinguishes them from most (but not all) peptide antibiotics of bacteria and fungi, which are synthesized by specialized metabolic pathways

Table I. Antimicrobial peptides: Structures, distribution and activity.

Structure	Representative Peptides	Species and Tissue	Antimicrobial Activity (reported)
4-disulfide α–helix+ β-sheet	plant defensins drosomycin	plants arthropod hemolymph	fungi
3-disulfide β-sheet-rich	α-defensins β-defensins	vertebrate neutrophils mammalian epithelia	bacteria, fungi, enveloped viruses
3-disulfide α-helix+ β-sheet	insect defensins	arthropod hemolymph molluscs	Gram+ bacteria
3-disulfide 2α-helices+ β-sheet	γ-thionins (crambin)	plants	bacteria, fungi, mammalian cells
2-disulfide β-sheet	protegrins	pig neutrophils	bacteria, fungi, enveloped viruses
	tachyplesins polyphemusins	horseshoe crab hemocytes	
1-disulfide cyclic	bactenecin-1 cyclic dodecapeptide	ruminant leukocytes	bacteria
	ranalexin brevinin	amphibian skin	
α-helix	cecropins magainin, PGLa LL-37	insect hemolymph amphibian skin mammalian leukocytes	bacteria
linear with repeating motifs	bactenecins 5 and 7 PR-39 indolicidin diptericin, apidaecin	mammalian leukocytes insect hemolymph	bacteria

and often incorporate exotic amino acids. Antimicrobial peptides are structurally diverse (Table I). Various regulatory systems and modes of delivery provide for tailoring specific tissue responses to microbial challenges. Certain antimicrobial peptides are present constitutively; the local synthesis or release of others is provoked by invading microbes; and yet other antimicrobial peptides can be brought into the area of invasion by mobile cells.

Distribution of Antimicrobial Peptides

Antimicrobial peptides are widely distributed in plants and animals. Epithelial surfaces of vertebrates secrete antimicrobial peptides from both barrier epithelia and glandular structures (1-4). Other epithelia, such as the tongue and buccal mucosa of pigs, contain antimicrobial concentrations of defensin peptides within the cells of the keratinized superficial layers (5), where these peptides could form an antimicrobial barrier. Exemplifying an inducible system, human β-defensin-2 accumulates within the keratinized layer in response to inflammatory stimuli (6, 7). Like in vertebrates, insect epithelia, most prominently the gut, secrete tissue-specific antimicrobial peptides (8, 9), a response which is likely to be important in insect resistance to intestinal parasites. In plants, antimicrobial peptides are found in seeds (especially during the vulnerable period of germination), leaves and other structures (10, 11).

Phagocytic cells contain several types of storage organelles (granules) for microbicidal substances and digestive enzymes (12, 13). In the process of phagocytosis, some of these granules empty their contents onto ingested microbes, generating a killing vacuole with very high concentrations of microbicidal and digestive substances. Other granules are secreted into the extracellular fluid where their contents kill microbes or inhibit their multiplication. Both types of granules have proven to be a rich source of antimicrobial peptides (14-16). In invertebrates, the fluid portion of blood as well as the granules of phagocytic cells (hemocytes) contain antimicrobial peptides (17, 18).

Structures and Mechanism of Action

Almost all antimicrobial peptides are cationic and amphipathic. The simplest antimicrobial peptide structures whose mechanism of action has been investigated are either α-helices or β-hairpins. Both types of peptides can form transmembrane channels. The length of a simple α-helix is approximately 1.5 Å per amino acid residue whereas that of a β-hairpin is roughly 3.5 Å per two residues. Since the hydrocarbon core of the phospholipid membrane is roughly 30 Å across it takes about twenty amino acids to span the membrane by either an α-helical or β-hairpin peptide. The simplest antimicrobial peptides of these two classes are the frog skin peptide magainin (23 amino acids) (19, 20) and the pig leukocyte peptide protegrin (16-18 amino acids) (21, 22). There are two major hypotheses about how the disruption of membrane integrity kills the target microbes. The loss of microbial viability may be due to the cumulative effects of energy drain due to the equilibration of intracellular and extracellular ion concentrations through the disrupted membrane. Alternatively, antimicrobial peptides may enter the target cell through the disrupted membrane, bind to as yet unknown intracellular molecules and interfere with their metabolic function. Either way, repair processes may limit or reverse these lesions when peptide concentrations are low or limited in time. Prolonged exposure to higher concentrations of antimicrobial peptides overwhelms the repair capacity of the microbe and the damage becomes irreversible.

The assembly of membrane pores by magainins (23) and tachyplesins (β-hairpin peptides from horseshoe crab hemocytes) (24) is favored by membranes that are rich in anionic phospholipids, a characteristic property of bacterial membranes. Conversely, the cell membranes of animals are rich in neutral phospholipids and cholesterol, substances that inhibit the incorporation of these peptides into membranes and the formation of pores. This mechanism explains why the concentrations necessary to kill eukaryotic cells are much higher than those required for killing most bacteria. Current evidence favors similar mechanisms of action for other peptides commonly found in the animal and plant kingdoms (25).

Defensins (26) are particularly abundant and widely distributed antimicrobial peptides characterized by a cationic β-sheet rich amphipathic structure stabilized by a conserved three-disulfide motif. They range in size from 29 to 47 amino acids, and are abundant in many vertebrate granulocytes, Paneth cells (specialized granule-rich intestinal host defense cells), and on epithelial surfaces. Like the structurally simpler magainins and protegrins, defensins also form pores in target membranes. There is evidence that the permeabilization of target cells is nonlethal unless followed by defensin entry into the cell and additional intracellular damage (27).

Regulation of Synthesis and Release

In invertebrates and plants, organisms that lack adaptive immunity, antimicrobial peptides constitute a major component of host defense (10, 28, 29). Many peptides of vertebrate animals resemble structurally and functionally their counterparts from plants and invertebrates (e.g. insect defensins and plant defensins) but a comprehensive evolutionary lineage has not yet been established. Both plants and invertebrates induce the synthesis of antimicrobial peptides in response to infection, employing similar transcriptional regulators, most prominently the rel/NF-κB family. In vertebrates, antimicrobial peptide synthesis is either constitutive or inducible by microbial macromolecules and/or cytokines. An epithelial β-defensin of the bovine trachea, the tracheal antimicrobial peptide (TAP), is synthesized in the airway epithelia when these are exposed to inhaled bacteria or lipopolysaccharide (30). This response is initiated by lipopolysaccharide receptors that ultimately signal to transcriptional regulators including the NF-κB complex, acting on NF-κB binding motifs in the promoter of the TAP gene. A similar mechanism may regulate the production of one of the human epithelial defensins, human β-defensin-2.

The signaling pathways that ultimately control the transcription of antimicrobial peptide genes are beginning to be defined. In *Drosophila*, at least two distinct pathways participate in the induction response. The antifungal response is induced by *toll*, a receptor that is used during embryonic development for dorsoventral patterning. The acute phase response in mammals, which involves the cytokine interleukin-1 (IL-1) (31), and the mammalian response to lipopolysaccharide and other bacterial substances utilize a similar pathway. Recently, several mammalian *toll*-like receptors have been implicated in the response to Gram-negative and Gram-positive bacteria (32, 33). The *Drosophila* antibacterial response involves a less extensively characterized *imd* (immune deficiency gene) system whose relationship to the *toll* family has not yet been elucidated.

In addition to transcriptional regulation of synthesis, stimulus-dependent degranulation provides an additional level of responsiveness and specificity. Thus the granulocytes of many vertebrates contain antimicrobial defensin peptides in their phagocytic granules and another class of antimicrobial peptides, cathelicidins, in granules destined for extracellular secretion (34, 35). Intestinal Paneth cells, positioned at the bottom of narrow crypts in the small intestine, release their defensin-rich granules (2) upon stimulation by cholinergic or bacterial stimuli, both of which are associated with food ingestion (36).

All known antimicrobial peptides are synthesized as larger precursors, containing one or multiple copies of the active peptide segment which are released by proteolytic processing. In the simplest cases the co-translational removal of an N-terminal signal peptide frees the active moiety but more commonly one or more anionic propieces are also removed during processing (1, 37, 38). An intriguing and as yet unexplained processing pattern is seen with cathelicidins, a group of peptides with a conserved 100 amino acid domain that is frequently proteolytically cleaved from the highly variable C-terminal antimicrobial domain (39). In phagocytes, the cathelicidins are commonly stored as inactive precursors in secretory granules. In many cases, the processing enzyme is neutrophil elastase contained in a separate set of storage granules. During phagocytosis, this binary system combines to generate active antimicrobial peptides. The function of the highly conserved cathelin domain is not yet known.

Spectrum of Activity

Many antimicrobial peptides display activity against Gram-positive and Gram-negative bacteria, yeasts and fungi, and even certain enveloped viruses and protozoa. Other peptides are more restricted in their spectrum. Even minor variations in peptide structure can influence activity, and a systematic understanding of the relationship between peptide structure and activity is an important area for future investigations. Evidence is accumulating that many peptides act synergistically with larger polypeptides whose antimicrobial activity is enzymatic (e.g. lysozyme) or is dependent on specific recognition of bacterial macromolecules (e.g. the bactericidal permeability-inducing protein, BPI) (40). Synergistic interactions between two antimicrobial peptides in the frog skin, magainin 2 and PGLa, have also been reported (41). In addition to their action on microbes, some antimicrobial peptides can function as regulatory molecules in the host. For example, *in vitro* studies suggest that defensins can attract phagocytes and lymphocytes to sites of infection, inhibit the release of cortisol from adrenal cells, induce the proliferation of fibroblasts and modify ionic fluxes in epithelial cells (26).

Biological Role and Consequences of Defects in the Function of Antimicrobial Peptides

Especially convincing evidence for the role of antimicrobial peptides has been obtained in plants, where genetic transplantation of antimicrobial peptides from one plant species to another can confer resistance to selected plant pathogens (10). In animals, antimicrobial peptides from insects were among the first to be studied by modern biochemical methods (17) and the application of *Drosophila* genetics to the study of antimicrobial peptides (28) has had a profound effect on the field. In insects, injury or infection elicits production of antimicrobial peptides in the "fat body" of insects (its equivalent of the vertebrate liver) that within a few hours renders the insect's hemolymph (its equivalent of blood) antimicrobial. Genetic disruption of the pathways that induce antimicrobial peptides renders the flies susceptible to fungal or bacterial infections. Linking specific insect antimicrobial peptides to their susceptibility to infection is complicated by the potentially broad effects of their induction pathways that may also activate non-peptide host defense systems.

Specific evidence for the biological role of antimicrobial peptides in mammals is also beginning to emerge. Inhibition of cathelicidin processing by a synthetic inhibitor of neutrophil elastase diminished the concentration of mature protegrins in pig wound fluid, decreased the microbicidal activity of the fluid and retarded the clearance of bacteria from the wound but a similar inhibitor of the neutrophil protease cathepsin G, an enzyme not involved in cathelicidin processing, had no effect (J. Shi and T. Ganz, submitted for publication).

Gene ablation techniques are emerging as a powerful tool for the study of the role of antimicrobial peptides in mammals, but as yet are routinely employed only in mice. Interestingly, mice are naturally deficient in neutrophil defensins, probably due to a mutation in the gene that is similar to mammalian neutrophil defensin genes (42). Since rats, hamsters, rabbits and several other mammals have abundant neutrophil defensins, it is likely that evolutionary compensation by another antimicrobial mechanism has developed in the mouse. On the other hand, mice express a large number (more than twenty) α-defensin genes (cryptdins) in the Paneth cells of the small intestine, as well as several epithelial β-defensin genes, both constitutive and inducible (43-48). Several laboratories are undertaking the phenotypic analysis of mice with disrupted defensin genes or disrupted genes that encode enzymes required for the processing and activation of defensins.

In a rare human disease, specific granule deficiency, the content of defensins (and probably several other antimicrobial peptides and proteins as well) in neutrophil granulocytes is severely reduced. The patients develop recurrent and severe bacterial infections. However, the interpretation and attribution of this defect is made complex by the multiple proteins affected (49). Cystic fibrosis is a common genetic disorder of epithelial host defense, characterized by persistent colonization of airway epithelia with *Pseudomonas aeruginosa* and other bacteria. Although the bacteria do not invade or spread to other organs, they induce a vigorous but ineffective inflammatory response that causes progressive lung dysfunction (50). The primary genetic defect in cystic fibrosis is the absence or malfunction of an epithelial chloride channel, "the cystic fibrosis transmembrane regulator". Recently, it has been proposed that the malfunction of the chloride channel raises the concentration of salt on the epithelial surface, and the high concentrations of salt inactivate antimicrobial peptides (51, 52). If confirmed, this scheme could explain the connection between the genetic defect and the clinical presentation in cystic fibrosis. More importantly, the new paradigm could lead to the development of more effective therapies for this devastating disease.

Concluding Remarks

Antimicrobial peptides are found in animals in circumstances that strongly suggest an important role in innate host defense. Concentrations of antimicrobial peptides in phagocytes and some epithelia are high enough to exert direct antimicrobial activity. Lower concentrations away from the primary sources of production may exert chemotactic or other regulatory effects that influence inflammation and adaptive immunity. Certain antimicrobial peptides have useful antimicrobial activity profiles and low costs of production that make them suitable for pharmaceutical development.

References

1. Zasloff, M. 1987. Magainins, a class of antimicrobial peptides from Xenopus skin: isolation, characterization of two active forms, and partial cDNA sequence of a precursor. Proc. Natl. Acad. Sci. USA. 84: 5449-5453.
2. Ouellette, A.J. and Selsted, M.E. 1996. Paneth cell defensins: endogenous peptide components of intestinal host defense. FASEB J. 10: 1280-1289.
3. Jones, D.E. and Bevins, C.L. 1992. Paneth cells of the human small intestine express an antimicrobial peptide gene. J. Biol. Chem. 267: 23216-23225.
4. Diamond, G., Zasloff, M., Eck, H., Brasseur, M., Maloy, W.L. and Bevins, C.L. 1991. Tracheal antimicrobial peptide, a cysteine-rich peptide from mammalian tracheal mucosa: peptide isolation and cloning of a cDNA. Proc. Natl. Acad. Sci. USA. 88: 3952-3956.
5. Shi, J., Zhang, G., Wu, H., Ross, C., Blecha, F. and Ganz, T. 1999. Porcine epithelial beta-defensin 1 is expressed in the dorsal tongue at antimicrobial concentrations. Infect Immun. 67: 3121-3127.
6. Harder, J., Bartels, J., Christophers, E. and Schroeder, J.-M. 1997. A peptide antibiotic from human skin. Nature. 387: 861-862.
7. Liu, L., Wang, L., Jia, H.P., Zhao, C., Heng, H.H.Q., Schutte, B.C., McCray, P.B.J. and Ganz, T. 1998. Structure and mapping of the human β-defensin HBD-2 gene and its expression at sites of inflammation. Gene. 222: 237-244.

8. Richman, A. and Kafatos, F.C. 1996. Immunity to eukaryotic parasites in vector insects. Curr. Opin. Immunol. 8: 14-19.
9. Richman, A.M., Dimopoulos, G., Seeley, D. and Kafatos, F.C. 1997. Plasmodium activates the innate immune response of *Anopheles gambiae* mosquitoes. EMBO J. 16: 6114-6119.
10. Fritig, B., Heitz, T. and Legrand, M. 1998. Antimicrobial proteins in induced plant defense. Curr. Opin. Immunol. 10: 16-22.
11. Broekaert, W.F., Terras, F.R., Cammue, B.P. and Osborn, R.W. 1995. Plant defensins: novel antimicrobial peptides as components of the host defense system. Plant Physiol. 108: 1353-1358.
12. Levy, O. 1996. Antibiotic proteins of polymorphonuclear leukocytes. Eur. J. Haematol. 56: 263-277.
13. Ganz, T. and Lehrer, R.I. 1997. Antimicrobial peptides of leukocytes. Curr. Opin. Hematol. 4: 53-58.
14. Ganz, T., Selsted, M.E., Szklarek, D., Harwig, S.S., Daher, K., Bainton, D.F. and Lehrer, R.I. 1985. Defensins. Natural peptide antibiotics of human neutrophils. J. Clin. Invest. 76: 1427-1435.
15. Selsted, M.E., Szklarek, D. and Lehrer, R.I. 1984. Purification and antibacterial activity of antimicrobial peptides of rabbit granulocytes. Infect. Immun. 45: 150-154.
16. Cowland, J.B., Johnsen, A.H. and Borregaard, N. 1995. hCAP-18, a cathelin/pro-bactenecin-like protein of human neutrophil specific granules. FEBS Lett. 368: 173-176.
17. Boman, H.G., Faye, I., Gudmundsson, G.H., Lee, J.Y. and Lidholm, D.A. 1991. Cell-free immunity in Cecropia. A model system for antibacterial proteins. Eur. J. Biochem. 201: 23-31.
18. Iwanaga, S., Muta, T., Shigenaga, T., Miura, Y., Seki, N., Saito, T. and Kawabata, S. 1994. Role of hemocyte-derived granular components in invertebrate defense. Ann. N. Y. Acad. Sci. 712: 102-116.
19. Bechinger, B., Zasloff, M. and Opella, S.J. 1993. Structure and orientation of the antibiotic peptide magainin in membranes by solid-state nuclear magnetic resonance spectroscopy. Protein Sci. 2: 2077-2084.
20. Ludtke, S.J., He, K., Heller, W.T., Harroun, T.A., Yang, L. and Huang, H.W. 1996. Membrane pores induced by magainin. Biochemistry. 35: 13723-13728.
21. Aumelas, A., Mangoni, M., Roumestand, C., Chiche, L., Despaux, E., Grassy, G., Calas, B. and Chavanieu, A. 1996. Synthesis and solution structure of the antimicrobial peptide protegrin-1. Eur. J. Biochem. 237: 575-583.
22. Fahrner, R.l., Dieckmann, T., Harwig, S.S., Lehrer, R.I., Eisenberg, D. and Feigon, J. 1996. Solution structure of protegrin-1, a broad-spectrum antimicrobial peptide from porcine leukocytes. Chem. Biol. 3: 543-550.
23. Matsuzaki, K. 1998. Magainins as paradigm for the mode of action of pore forming polypeptides. Biochim. Biophys. Acta. 1376: 391-400.
24. Matsuzaki, K., Fukui, M., Fujii, N. and Miyajima, K. 1991. Interactions of an antimicrobial peptide, tachyplesin I, with lipid membranes. Biochim. Biophys. Acta. 1070: 259-264.
25. Lohner, K., Latal, A., Lehrer, R.I. and Ganz, T. 1997. Differential scanning microcalorimetry indicates that human defensin, HNP-2, interacts specifically with biomembrane mimetic systems. Biochemistry. 36: 1525-1531.
26. Ganz, T. and Lehrer, R.I. 1995. Defensins. Pharmacol. Ther. 66: 191-205.

27. Lichtenstein, A. 1991. Mechanism of mammalian cell lysis mediated by peptide defensins. Evidence for an initial alteration of the plasma membrane. J. Clin. Invest. 88: 93-100.

28. Meister, M., Lemaitre, B. and Hoffmann, J.A. 1997. Antimicrobial peptide defense in Drosophila. Bioessays. 19: 1019-1026.

29. Hoffmann, J.A. 1995. Innate immunity of insects. Curr. Opin. Immunol. 7: 4-10.

30. Diamond, G., Russell, J.P. and Bevins, C.L. 1996. Inducible expression of an antibiotic peptide gene in lipopolysaccharide-challenged tracheal epithelial cells. Proc. Natl. Acad. Sci. USA. 93: 5156-5160.

31. O'Neill, L.A. and Greene, C. 1998. Signal transduction pathways activated by the IL-1 receptor family: ancient signaling machinery in mammals, insects and plants. J. Leukoc. Biol. 63: 650-657.

32. Poltorak, A., He, X., Smirnova, I., Liu, M.Y., Huffel, C.V., Du, X., Birdwell, D., Alejos, E., Silva, M., Galanos, C., Freudenberg, M., Ricciardi-Castagnoli, P., Layton, B. and Beutler, B. 1998. Defective LPS signaling in C3H/HeJ and C57BL/10ScCr mice: mutations in Tlr4 gene. Science. 282: 2085-2088.

33. Schwandner, R., Dziarski, R., Wesche, H., Rothe, M. and Kirschning, C.J. 1999. Peptidoglycan- and lipoteichoic acid-induced cell activation is mediated by toll-like receptor 2. J. Biol. Chem. 274: 17406-17409.

34. Rice, W.G., Ganz, T., Kinkade, J.M.Jr., Selsted, M.E., Lehrer, R.I. and Parmley, R.T. 1987. Defensin-rich dense granules of human neutrophils. Blood. 70: 757-765.

35. Sorensen, O., Arnljots, K., Cowland, J.B., Bainton, D.F. and Borregaard, N. 1997. The human antibacterial cathelicidin, hCAP-18, is synthesized in myelocytes and metamyelocytes and localized to specific granules in neutrophils. Blood. 90: 2796-2803.

36. Qu, X.D., Lloyd, K.C., Walsh, J.H. and Lehrer, R.I. 1996. Secretion of type II phospholipase A2 and cryptdin by rat small intestinal Paneth cells. Infect. Immun. 64: 5161-5165.

37. Valore, E.V., Ganz, T. 1992. Posttranslational processing of defensins in immature human myeloid cells. Blood. 79: 1538-1544.

38. Terry, A.S., Poulter, L., Williams, D.H., Nutkins, J.C., Giovannini, M.G., Moore, C.H. and Gibson, B.W. 1988. The cDNA sequence coding for prepro-PGS (prepro-magainins) and aspects of the processing of this prepro-polypeptide. J. Biol. Chem. 263: 5745-5751.

39. Zanetti, M., Gennaro, R. and Romeo, D. 1995. Cathelicidins: a novel protein family with a common proregion and a variable C-terminal antimicrobial domain. FEBS Lett. 374: 1-5.

40. Levy, O., Ooi, C.E., Weiss, J., Lehrer, R.I. and Elsbach, P. 1994. Individual and synergistic effects of rabbit granulocyte proteins on *Escherichia coli*. J. Clin. Invest. 94: 672-682.

41. Westerhoff, H.V., Zasloff, M., Rosner, J.L., Hendler, R.W., De Waal, A., Vaz Gomes, A., Jongsma, P.M., Riethorst, A. and Juretic, D. 1995. Functional synergism of the magainins PGLa and magainin-2 in *Escherichia coli*, tumor cells and liposomes. Eur. J. Biochem. 228: 257-264.

42. Eisenhauer, P.B. and Lehrer, R.I. 1992. Mouse neutrophils lack defensins. Infect. Immun. 60: 3446-3447.

43. Bals, R., Goldman, M.J. and Wilson, J.M. 1998. Mouse beta-defensin 1 is a salt-sensitive antimicrobial peptide present in epithelia of the lung and urogenital tract. Infect. Immun. 66: 1225-1232.

44. Grandjean, V., Vincent, S., Martin, L., Rassoulzadegan, M. and Cuzin, F. 1997. Antimicrobial protection of the mouse testis: synthesis of defensins of the cryptdin family. Biol. Reprod. 57: 1115-1122.

45. Huttner, K.M., Kozak, C.A. and Bevins, C.L. 1997. The mouse genome encodes a single homolog of the antimicrobial peptide human beta-defensin 1. FEBS Lett. 413: 45-49.

46. Morrison, G.M., Davidson, D.J., Kilanowski, F.M., Borthwick, D.W., Crook, K., Maxwell, A.I., Govan, J.R. and Dorin, J.R. 1998. Mouse beta defensin-1 is a functional homolog of human beta defensin-1. Mamm. Genome. 9: 453-457.

47. Morrison, G.M., Davidson, D.J. and Dorin, J.R. 1999. A novel mouse beta defensin, Defb2, which is upregulated in the airways by lipopolysaccharide. FEBS Lett. 442: 112-116.

48. Ouellette, A.J., Hsieh, M.M., Nosek, M.T., Cano-Gauci, D.F., Huttner, K.M., Buick, R.N. and Selsted, M.E. 1994. Mouse Paneth cell defensins: primary structures and anti-bacterial activities of numerous cryptdin isoforms. Infect. Immun. 62: 5040-5047.

49. Ganz, T., Metcalf, J.A., Gallin, J.I., Boxer, L.A. and Lehrer, R.I. 1988. Microbicidal/cytotoxic proteins of neutrophils are deficient in two disorders: Chediak-Higashi syndrome and "specific" granule deficiency. J. Clin. Invest. 82: 552-556.

50. Tummler, B. and Kiewitz, C. 1999. Cystic fibrosis: an inherited susceptibilityto bacterial respiratory infections. Mol. Med. Today. 5: 351-358.

51. Smith, J.J., Travis, S.M., Greenberg, E.P. and Welsh, M.J. 1996. Cystic fibrosis airway epithelia fail to kill bacteria because of abnormal airway surface fluid. Cell. 85: 229-236.

52. Goldman, M. anderson, G., Stolzenberg, E.D., Kari, U.P., Zasloff, M. and Wilson, J.M. 1997. Human beta-defensin-1 is a salt-sensitive antibiotic in lung that is inactivated in cystic fibrosis. Cell. 88: 553-560.

From: *Development of Novel Antimicrobial Agents: Emerging Strategies*
ISBN 1-898486-23-9 © 2001 Horizon Scientific Press, Wymondham, UK.

12

The Role of Membrane Lipid Composition in Cell Targeting of Antimicrobial Peptides

Karl Lohner

Abstract

Host defense peptides, which evolved in nature, show high specificity in their action towards their target cells, e.g. some exhibit toxicity towards bacteria. Most of these antimicrobial peptides kill bacteria by permeation or destruction of their cytoplasmic membrane, although alternative mechanism such as binding to DNA after translocation across the membrane may exist. In order to understand the molecular basis of the specific interaction of antimicrobial peptides with bacterial membranes, it is important to consider the membrane architecture of eukaryotic and prokaryotic cell membranes, which differ markedly in their complexity and lipid composition. Research in this field showed that antimicrobial peptides can discriminate between the major lipid classes of mammalian and bacterial cell membranes. Besides of electrostatic interactions between the cationic, amphipathic peptides and the negatively charged bacterial lipids, intrinsic membrane properties, like membrane curvature strain, modulate the mode of action and efficiency of these peptides. Understanding the mutual dependence of these lipid-peptide interaction is a key for the rational design of novel peptide antibiotics.

Introduction

As outlined in the first chapter of this book we face a dramatic increase in the number of bacteria which are resistant to conventional antibiotics. This situation has also become one of the dominant health issues of the World Health Organization in the last few years. In particular, in the field of hospital infection control, there is increasing concern regarding both the rapid emergence of resistant strains with high epidemic potential and the fact that bacteria show already reduced susceptibility to vancomycin, a drug of last resort. There are reports of about a dozen new antibiotic drugs that show promising antibacterial activity by interfering with protein synthesis, cell wall formation or DNA replication, but it is assumed that these antibiotics will not overcome resistance in the long term (1) and therefore, novel antibiotic agents have to be urgently developed.

One emerging strategy is based on host defense peptides which have evolved in nature to contend with invaders as an active system of defense (2, Ganz and Lehrer, this monography). These host defense peptides usually exhibit a high specificity towards their target cell, i.e. some of these peptides show toxicity towards humans, while others have toxicity which is

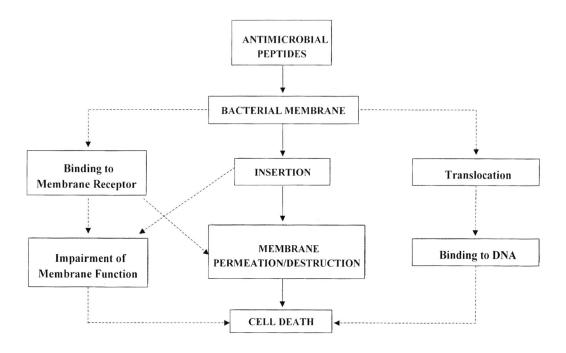

Figure 1. Possible mechanisms of antimicrobial peptides for killing of bacteria indicating that all pathways involve interaction with the bacterial membrane (see also text).

restricted to microorganisms. These latter antimicrobial peptides are of major interest as a novel source for antibiotics. The biological activity of antimicrobial peptides is thought to stem primarily from their ability to permeabilize the cell membrane of bacteria, although alternative mechanism(s) of bacterial killing may exist (Figure 1). It is generally accepted yet that the biological activity of most antimicrobial peptides is not related to binding to a specific membrane protein receptor site but that the site of action is the lipid bilayer matrix of the cell membrane. This has been concluded from studies on synthetic magainin, cecropins and melittin with all the amino acids in the D-configuration which exhibit antibacterial and hemolytic activity identical to that of the naturally occuring L forms (3,4). However, some peptides such as the proline-arginine-rich peptide, PR-39, or cecropin P were found to exhibit different antimicrobial activity in their L- and D-form which, in addition, appeared to depend strongly on the bacteria species (5). In respect of membrane permeation and destruction, two distinct mechanisms are particularly discussed. One model proposes that these peptides act by perturbing the barrier function of membranes by transmembrane pore formation (see Matsuzaki, this monography), while the other model suggests membrane destruction via a "carpet" mechanism (see Shai, this mongraphy). It has also been discussed that these mechanisms may represent gradual steps of membrane perturbation depending on the nature of the peptides, lipids and environmental conditions (6,7). Furthermore, there is also evidence that antimicrobial peptides may lead to membrane damage by changing the intrinsic membrane curvature strain which in some cases may result in the formation of non-bilayer lipid structures (7,8). It was also shown that some antimicrobial peptides have specificity for particular membrane lipid components, e.g. the hemolytic peptide cinnamycin appears to have a specific affinity for phosphatidylethanolamine (9) and sapecin for diphosphatidylglycerol (10). Moreover, recent permeabilization experiments with rabbit neutrophil defensin also indicated a preference for diphosphatidylglycerol (11). Finally, nisin Z was shown to form a well-defined complex with Lipid II resulting in pore formation (12, Pag and Sahl, this monography).

Furthermore, it has been suggested that, as a secondary mechanism of action, antimicrobial peptides may inhibit DNA biosynthesis which causes the ultimate cell death (e.g. 13-15). Recent experiments on the interaction of different classes of antimicrobial peptides with planar bilayers and with the inner membrane of *E. coli* showed that there was no clear correlation between membrane damage and the minimal inhibitory concentration of the peptides (16). Therefore, the authors also proposed that the known abilities of antimicrobial peptides to act on lipid membranes rather reflect their ability to cross the membrane and that their target(s), most likely DNA, is located in the intracellular space. Nevertheless, in order to gain access to these targets the peptides have to overcome the membrane barrier which again emphasizes a possible role of the membrane lipids regarding the specificitiy and/or efficiency of antimicrobial peptides. As will be outlined below it is also conceivable that antimicrobial peptides affect the packing properties of membranes in a way that may lead to conformational changes of membrane proteins which in turn could cause impairment of membrane function.

Thus, independent on the mechanism of killing bacterial membranes are an important partner in the modulation of the biological activity of antimicrobial peptides (Figure 1) and therefore, this chapter will mainly focuss on the membrane architecture of bacterial membranes and the role of their lipid composition in the target specificity of antimicrobial peptides. Understanding the mutual dependence of the lipid-peptide interaction is a key for the rational design of novel peptide antibiotics. This is likely to be accomplished only if the molecular basis of the action of these peptides is known.

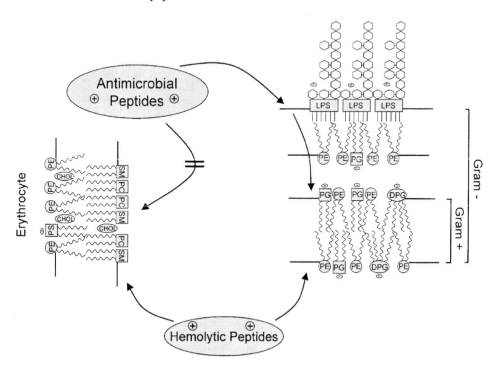

Figure 2. This scheme illustrates the principal membrane architecture of Gram-positive and Gram-negative bacteria as well as of erythrocytes. Latter can be considered as an archetype of mammalian plasma membranes. This figure also indicates the preferential interaction of antimicrobial peptides with bacterial membranes, while hemolytic peptides exhibit unspecific activity. Legend for lipids: PC, phosphatidylcholine; PE, phosphatidylethanolamine; PG, phosphatidylglycerol; DPG, diphosphatidylglycerol; PS, phosphatidylserine; SM, sphingomyelin; LPS, lipopolysaccharides; chol, cholesterol.

Membrane Architecture and Lipid Composition

An enormous variety of lipid classes are found in nature which exhibit a great diversity in their molecular structure and in their polymorphism when dispersed in aqueous environment. However the fundamental structural unit of biological membranes is a highly dynamic, liquid-crystalline phospholipid bilayer (17) which acts as a fundamental permeability barrier. While the bilayer forms the common matrix of cell membranes, considerable variations in the structure, complexity and lipid composition of membranes exist between prokaryotic and eukaryotic cells (Figure 2).

Prokaryotic Cell Membranes

The cell envelope of Gram-negative bacteria is a complex structure consisting of an inner membrane, an unique outer membrane layer and an intervening layer of peptidoglycan in the periplasmic space (e.g. 18). The outer membrane has a distinctive composition which is highly asymmetric. Lipopolysaccharides (LPS) are located exclusively in the outer layer and phospholipids, to a large extent phosphatidylethanolamine (PE), are confined to the inner layer of the outer membrane. The inner, cytoplasmic membrane is essentially a bilayer of lipids, which hosts the respiratory enzymes. There exists a wealth of information concerning the phospholipid composition of individual genera and species of Gram-negative bacteria

Table 1. Membrane phospholipid composition as percentages of the total of representative bacteria species (19, 23).

Bacteria Species	Cell Membrane	Phospholipid [1]				
		PE	PG	DPG	LysylPG	Others
Gram-negative						
Escherichia	OM [2]	91	3	6	0	0
coli	CM [3]	82	6	12	0	0
Salmonella	OM	81	17	2	0	0
typhimurium	CM	60	33	7	0	0
Pseudomonas	OM	87	13	0	0	0
cepacia	CM	82	18	0	0	0
Gram-positive						
Staphylococcus aureus	CM	0	57	5	38	Trace
Bacillus megaterium	CM	40	40	5	15	0
Bacillus subtilis	CM	10	29	47	7	6 [4]
Micrococcus luteus	CM	0	26	67	0	7 [5]

[1] PE, phosphatidylethanolamine; PG, phosphatidylglycerol; DPG, diphosphatidylglycerol;
[2] OM, outer membrane;   [3] CM, cytoplasmic membrane or inner membrane for Gram-negative bacteria
[4] including phosphatidic acid and glycolipids;   [5] almost exclusively phosphatidylinositol.

(19). The concept of a characteristic lipid composition of cell membranes is well accepted, although changes in lipid composition (mostly in quantitative than in qualitative terms) may occur depending on environmental conditions. The phospholipid composition of the outer and inner membranes of representative Gram-negative bacteria, such as *Escherichia coli, Salmonella typhimurium* and *Pseudomonas cepacia* is given in Table 1. Thereby, it is evident that phosphatidylethanolamine represents the major phospholipid class in the outer and inner membrane. There is also a considerable amount of negatively charged phosphatidylglycerol (PG) and diphosphatidylglycerol (DPG or cardiolipin) incorporated into the membrane. On the other hand, phosphatidic acid (PA) is only found in traces. This is most likely due to the rapid turnover of PA as it serves as a precursor for all other phospholipids (PE, PG and DPG) (20,21). Only limited insight exists regarding the distribution of phospholipids between the outer and inner leaflet of the cytoplasmic membrane of Gram-negative and -positive bacteria (for a recent review see 21). However, early studies on *Erwinia carotovora* indicated an asymmetric distribution of PE (22).

In Gram-positive bacteria we face a rather primitive situation dealing basically with a simple lipid bilayer membrane. This probably reflects an early evolutionary stage in which the genetic imperative for lipids is primarily the formation of a cell membrane (23). The phospholipids constitute up to 80% of the total cellular lipids and consist to a large extent of phosphatidylglycerol and derivatives of it, mostly diphosphatidylglycerol and aminoacyl derivatives (Table 1). The phospholipid composition from one group to another may vary to a larger extent as compared to Gram-negative bacteria. Especially, the genera of *Bacillus* contains quite a diversity of species which is also reflected in a large richness in lipid chemistry. For example, *Bacillus megaterium* exhibits a rather high content of PE as compared to other genera of Gram-positive bacteria, while *Micrococcus luteus* possesses an extremely high amount of DPG (Table 1). Equimolar mixtures of PG/DPG are reported for *Streptococcus pneumoniae*. In any case, a common feature is the presence of large amounts of negatively charged phospholipids. Many Gram-positive species are characterized by a high content of branched-fatty acids which are less common in Gram-negative bacteria. The proportion of such fatty acids can account for up to 90% as in *Bacillus subtilis*. Further, the most abundant fatty acids in *Staphylococcus aureus* or *Micrococcus luteus* are anteiso-branched ones. No polyunsaturated fatty acids, sterols or complex lipid structures like polysaccharides were reported (23).

Eukaryotic Cell Membranes

Cytotoxicity of antimicrobial peptides is usually tested by the hemolytic activity on human erythrocytes. Thus this part is restricted to the description of the plasma membrane of erythrocytes which serves as a model for eukaryotic cell membranes. This membrane is comprised to a large amount (about 60%) of phospholipids and to about 25 % of cholesterol (e.g. 24). In contrast to bacterial cytoplasmic membranes it is well established that the phospholipid classes are asymmetrically distributed between the outer and inner lipid leaflet of mammalian erythrocyte membranes (e.g. 25). The choline phosphatides, phosphatidylcholine (PC) and sphingomyelin (SM), occur predominantly in the external leaflet, whereas the aminophosphatides, phosphatidylserine (PS) and -ethanolamine (PE), are found almost exclusively in the inner leaflet of the bilayer. Latter is comprised to about 1/3 by the alkenyl analog, ethanolamine plasmalogen, which was shown to strongly promote the formation of non-bilayer structures (26,27). This asymmetric phospholipid distribution causes that uncharged, zwitterionic phospholipids are exposed to the outside of the cell which represents the first site of interaction with antimicrobial peptides. Therefore, in a first approach a model membrane consisting of PC, SM and cholesterol would be sufficient to mimic the human erythrocyte membrane. In many biological systems, these two choline phospholipids appear to occupy similar cellular compartments and their content is tightly regulated. Interestingly, the ratio of these choline phospholipids in erythrocyte membranes depends on the organism which in turn may at least partly affect the hemolytic activity of a given antimicrobial peptide. For example, it is conceivable that a lower hemolytic activity will be found for erythrocytes that have a higher content of SM owing to the rigidity of sphingomyelin membranes.

In order to gain insight into the specificity towards particular lipid components exhibited by antimicrobial peptides it is of interest to study their interaction with phospholipid model membranes consisting of mixtures of the various lipid components which mimic bacterial and mammalian cell membranes. However, the importance of such affinity specificity toward particular lipids on the peptide lytic specificity is not yet totally understood, and the occurrence of binding preferences to certain lipid types could only be somehow related to the peptide specificity toward given microorganisms (28).

The Possible Role of Bacterial Cell Walls in the Interaction with Antimicrobial Peptides

Gram-positive bacteria are not protected by an outer membrane like Gram-negative bacteria but instead have a cell wall formed by a peptidoglycan-teichoic acid network confering a negative charge to the cell surface of these bacteria. Teichoic acids exhibit a high variability in their chemical structure and can be covalently linked to the peptidoglycan or to membrane glycolipids. Recent experiments with staphylococcal mutants, which exhibit different degrees of D-alanine esterified teichoic acids, indicated a role of the surface charge in protection of the bacteria against cationic antimicrobial peptides (29). Thereby, mutants that lacked D-alanine were more sensitive to a number of antimicrobial peptides such as defensins, protegrins or magainin 2, most likely owing to their increased net negative surface charge as compared to the wild-type. The impact of surface charge for resistance mechanism(s) has also been shown for Gram-negative bacteria (see below).

As described in the previous chapter, Gram-negative bacteria have an outer membrane containing only lipopolysaccharides in its outer membrane leaflet resulting again in a negatively charged cell membrane surface. This peculiar lipid is composed of a polysaccharide moiety which is covalently linked to lipid A. The polysaccharide chain itself is divided into the O-antigen and the core region, respectively. The O-antigen region is characterized by a high variability regarding its chemical nature, while lipid A represents the conserved region of LPS. Lipid A is also the predominant mediator of endotoxic shock or sepsis (30), a serious complication in Gram-negative bacterial infections with a very high mortality rate of especially elder and immune-suppressed patients. Therefore, molecules that bind to LPS and neutralize its biological effects or enhance clearance are of interest for clinical applications. Two related mammalian proteins, bactericidal/permeability-increasing protein (BPI) and LPS-binding protein (LBP), have opposite effects on the LPS activities. While the latter is not toxic to Gram-negative bacteria and enhances the bioactivity of LPS, both BPI and its amino-terminal half exert a selective and potent antibacterial activity against Gram-negative bacteria and neutralize the endotoxic activity of LPS (31). These observations have prompted clinical studies of recombinant amino-terminal fragments of BPI (e.g. 32). Interestingly, BPI and LBP share a completely conserved region of 102 amino acid residues which lies at the interfaces between the tertiary structural elements creating two apolar lipid-binding pockets. Overall differences in charge between these two proteins suggest that the bactericidal activity of BPI is related to the high positive charge clustering at the tip of the NH_2-terminal domain which coincides with residues known to affect binding to LPS (33). Limulus anti-LPS factor (LALF) also blocks the endotoxic activity of LPS. Although LALF does not share sequence homology to the mammalian proteins, the LPS binding site, composed of an extended amphipathic loop with striking charge distribution, provides further evidence that these proteins posses a solvent-exposed, interchangeable LPS binding motif (34). High affinity LPS-binding and neutralizing peptides were synthesized based on the crystal structure of recombinant LALF demonstrating that cyclic 10-mer peptides exhibit higher binding affinity than the linear analog presumably by mimicking the three-dimensional characteristics of the exposed hairpin loop (35). The lipid A binding affinity of the cyclic peptide was compareable to the high LPS binding affinity of the cationic peptide antibiotic polymyxin B.

LPS may also exhibit an inhibitory activity to protect Gram-negative bacteria from the attack of antimicrobial or hemolytic peptides. Besides of representing a steric barrier, electrostatic interactions may result in the accumulation of these cationic peptides at the LPS layer, which extent to the extracellular space. In fact, as has been observed for polymyxin B, complex formation was also described between bacterial LPS and other antimicrobial peptides, such as tachyplesin (36) and cecropin A (37). Furthermore, it was shown that magainin alter the thermotropic phase behavior of the outer membrane-peptidoglycan complexes from wild-type *Salmonella typhimurium* and a series of LPS mutant strains (38) and more recently that incorporation of lipid A into PC liposomes enhances the interaction of magainin 2 (39). However, LPS can be modified in different environments which can result in resistance to antimicrobial peptides. Thus polymyxin B-resistant strains of *S. typhimurium* and *E. coli* showed extensive esterification of the lipid A monophosphate by arabinose as well as of diphosphates in LPS by aminoethanol (40-42). Moreover, modifications in both the fatty acid profile and degree of acylation of lipid A were detected (42,43). Reduced or blocked peptide binding to LPS may have implications on the peptide transfer to the periplasmic space in order to interact with the inner, cytoplasmic membrane which is assumed to be the site of action of most antimicrobial peptides. Hancock and coworkers proposed that the peptide transfer across the outer membrane is accomplished by the so-called self-promoted uptake system which involves direct binding of the cationic peptides to LPS (reviewed in 44,45). Studies with the antimicrobial peptides magainin (e.g. 38,46) and defensin (13) demonstrated that these peptides can permeabilize the outer membrane and are able to gain access to the inner, cytoplasmic membrane. Electron micrographs of *E. coli* cells exposed to magainin showed bleb formation on the surface of the bacterial cell. This process is accompanied by permeabilization of the outer membrane, but the lethal event appears to be depolarization of the inner membrane (47). Similarly, binding of BPI to LPS causes immediate growth arrest, but actual killing of the bacteria coincides with damage of the inner membrane (32).

Phospholipid Polymorphism and its Impact on the Interaction with Antimicrobial Peptides

Antimicrobial peptides can interact with phospholipid membranes in various ways. Thereby one has to distinguish between specific lipid effects caused by binding of the peptide to specific groups of the lipid and effects dependent on changes in the bulk physical properties of the membrane which are not necessarily independent of each other.

Preferential Interaction with Negatively Charged Phospholipids

A number of antimicrobial peptides were shown to interact preferentially with negatively charged phospholipids as expected because of their cationic nature. For example, the β-sheet peptides such as tachyplesin (36), human neutrophil peptide (48,49) or protegrin-1 from porcine leukocytes (50) discriminate between choline phospholipids and anionic phospholipids showing preferential interaction with the latter. The same behavior was found for the α-helical frog skin peptides, PGLa (51) and magainin, which killed more effectively bacteria containing inner membrane with higher amounts of phosphatidylglycerol (47). The effects of the cyclic antimicrobial peptide, gramicidin S, on the thermotropic phase behavior of liposomes composed of zwitterionic or anionic lipids also varied markedly with the structure and the charge of the lipid polar headgroup (52). Finally, monolayer penetration experiments demonstrated that nisin Z interacts preferentially with DPG and PG (53). Moreover, its biologically less active [Glu-32] mutant, obtained by introducing a negative charge in the cationic C-terminus of this peptide, inserted less efficiently (54), clearly emphasizing the role of electrostatic forces for the interaction between antimicrobial peptides and membrane lipids.

However, experiments with magainin indicated that besides of electrostatic interactions membrane curvature is an important parameter to determine the kind and efficiency of interaction with lipid membranes (55). Magainin effectively formed pores in PG model membranes well below lipid-to-peptide molar ratios of 100:1. However, pore formation in PS, DPG or PA liposomes was only observed at much higher ratios of 50:1 to 10:1 which was accompanied by some morphological changes of the liposomes. The different behaviour of these acidic phospholipids was attributed to their ability to form non-bilayer structures under certain conditions, i.e. reduced headgroup repulsion. Support for this assumption comes from the observation that incorporation of PE, known to induce negative curvature strain, inhibited magainin induced pore formation in negatively charged lipids. Similarly, protegrin affected differently the phase behaviour of the negatively charged liposomes composed of PG, PA and PS, respectively (unpublished data). These observations emphasize that the headgroup structure and composition are important in defining the two-dimensional organization in membranes and affected both the mode and efficiency of interaction.

Phase Separation

Data from differential scanning calorimetry and X-ray diffraction suggest that PGLa induces lipid segregation in PG liposomes, most likely resulting in peptide-depleted and peptide-rich domains (51). This can be deduced from the second phase transition observed at temperatures above the chain-melting transition of PG (Figure 3). Recent work in our laboratory showed that a second phase transition is also detected in such liposomes in the presence of protegrin-1. However, in contrast to PGLa this transition occurs below the chain-melting transition of PG (Figure 3). The nature of this difference has yet to be elucidated. There is also evidence that other biologically active amphipathic molecules, such as cardiotoxin (56) and synthetic peptides (8 and references therein) interact with model membranes to induce lateral separation of phospholipids into co-existing microenvironments. Again, one of which is associated with the interacting molecule. Re-ordering of the native membranes in this manner may contribute to the biological function of antimicrobial peptides and other membrane-active molecules. Exclusion of certain lipids from areas of the cell membrane may lead to local disruption of membrane order, resulting in increased permeability and cell lysis.

PGLa also induced phase separation in mixed PE/PG liposomes, the major phospholipid components of bacterial cytoplasmic membranes (7). Such liposomes represent a simple, but

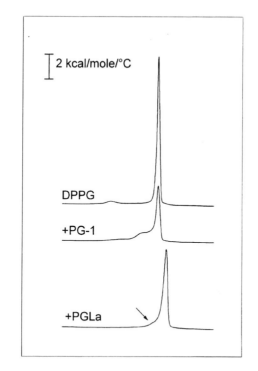

EXCESS HEAT CAPACITY

TEMPERATURE [°C]

Figure 3. Thermograms of DPPG liposomes in PBS buffer, pH 7.4, in the absence (top) and presence of protegrin-1 (PG-1, middle) and PGLa (bottom), respectively. Lipid-to-peptide molar ratio was adjusted to 25 and the scan rate was 0.25°C/min. The pretransition of the pure lipid around 33°C vanishes in the presence of the antimicrobial peptides but additional transitions are detected below (PG-1) and above (PGLa) the main transition of DPPG around 40°C which is hardly detected in the presence of PGLa (see arrow).

adequate model system to mimic bacterial cell membranes. Interestingly, studies on the mixing properties of these lipids revealed a gel phase immiscibility in the composition range between 5 and 40 mol% of PE (57). The partial immiscibility of the lipids was explained by the different molecular properties of the individual lipid components as determined by the nature of their headgroups and their molecular geometry. In the case of PG and also of the mammalian plasma membrane lipids, PC and SM, the cross sectional area of the polar lipid headgroup matches the cross sectional area of the hydrocarbon side chains, while PE is characterized by a smaller headgroup cross section (Figure 4). Hence, the molecular geometry of PG can be described by a cylindrical shape and of PE by a truncated cone. The former will preferentially pack in flat bilayers, whereas the latter will rather adopt highly curved lipid structures. It can be assumed that the molar ratio of PG and PE may be important in determining microscopic differences in their lateral organization, packing and/or mobility, which can be amplified by the interaction with other membrane constituents and in particular by interaction with membrane-active solutes in the environment. This packing properties may also facilitate membrane insertion of antimicrobial peptides. Evidence for this comes from studies on the insertion of alamethicin into model membranes composed of PC and PE with identical acyl chains, which depended strongly on the molar ratio of these lipids, emphasizing again the importance of the lipid packing properties (58). These authors also concluded that the specific physical characteristics of the phospholipids, i.e. the cross sectional area of the lipid headgroup and the hydrocarbon side chains, are important factors in determining whether and how the lipid bilayer will be affected by membrane-active peptides.

FLAT BILAYER   **TENDENCY TO CURVE**

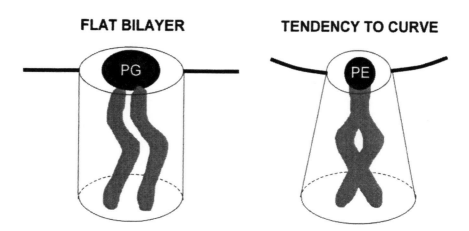

Figure 4. Schematic representation of the molecular shape of the major phospholipids of bacterial cell membranes, phosphatidylglycerol (PG) adopting the shape of a cylinder and phosphatidylethanolamine (PE) of a truncated cone. PG rather packs in a flat bilayer arrangement, while PE will favor highly curved supramolecular lipid structures such as the inverse hexagonal (H_{II}) phase.

Membrane Curvature

Phosphatidylethanolamine constitutes to a large extent the total lipid content in many bacteria (Table 1). As described above PE is characterized by a cone-shaped molecular geometry, (59,60). This property makes PE prone to form non-bilayer or nonlamellar structures. Diphosphatidylglycerol, another significant bacterial membrane lipid component, also readily forms nonlamellar phases when the negative headgroup charges are neutralised by binding of Ca^{2+} (61). Also binding of melittin, a cationic hemolytic peptide, to DPG induced the formation of an inverted hexagonal (H_{II}) phase (62). A prerequisite for the formation of such a phase or of bicontinuous cubic phases is that the opposing monolayers of the bilayer wish to bend towards the water region. This property leads to less stable bilayers. In the lamellar bilayer state this desire for monolayer curvature is physically frustrated, because energetically expensive voids would be formed in the core of the bilayer. Another way of viewing lipid curvature stress is to describe it as a change of the lateral pressure profile of the bilayer (63, 64). If the lipids exhibit a cylindrical shape they will have uniform lateral pressure through the bilayer with little negative curvature strain and little free volume. However, if the lipids have a cone-shape, then they will have an increased lateral pressure towards the centre of the bilayer resulting in a negative curvature stress and a larger free volume.

Many biological membranes have a lipid composition in a narrow window close to a lamellar-nonlamellar phase boundary e.g. several microorganisms e.g. *Acholeplasma laidlawii* or *Escherichia coli* have been shown to precisely regulate the amounts of bilayer and nonbilayer lipids (65, 66). The importance of the proper balance between these type of lipids has been widely discussed (66,67). The presence of nonlamellar phase forming lipids significantly increases membrane monolayer curvature stress, thereby conferring upon cell membranes a degree of nonlamellar-forming propensity, which is believed to be essential for normal membrane function such as membrane fusion (e.g. 68-70). It has been also suggested that this frustrated curvature stress may affect the conformational state and hence the activity of membrane proteins, e.g. membrane integration and functioning of important transport proteins were severely impaired in *E.coli* mutants which lack phosphatidylethanolamine (64).

It was suggested that owing to the molecular geometry of PE a high lateral packing force arises in the hydrophobic core of the membrane which presses against the surface of integral membrane proteins and thereby keeps them in the conformational state being important for their characteristic biological activities. In accordance with this assumption are observations that non-bilayer lipids are often required for functional reconstitution of membrane proteins (e.g. 64). One may speculate that upon membrane insertion of antimicrobial peptides these packing properties will be affected by changing the lateral packing force and consequently the conformation of integral membrane proteins. Such secondary effects may then be responsible for impairment of membrane function.

Several antimicrobial peptides were shown to affect the lamellar-nonlamellar phase transition (reviewed by 7,8). For example, alamethicin, when incorporated in small amounts in dielaidoyl PE liposomes, induces the formation of a cubic phase (71). It was suggested that alamethic promotes such lipid structures by changing the thickness and/or flexibility of the lipid bilayer. This is supported by the observation that adsorption of alamethicin onto diphytanoyl PC bilayer causes bilayer thinning, thereby inducing chain disorder over a large area (72,73). Furthermore, it was shown that magainin behaves in many aspects similar to alamethicin causing membrane thinning in PC/PS bilayers below the critical concentration for peptide insertion (74), which roughly correlates with the concentration required for cytolytic activity (75). A decrease of the thickness of diphytanoyl PC bilayers was also reported in the presence of the β-sheet antimicrobial peptide protegrin-1 (76). Huang and coworkers

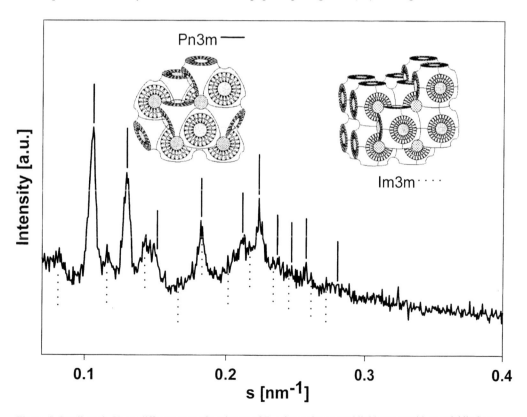

Figure 5. Small-angle X-ray diffractogram of a mixture of *E. coli* membrane total lipid extract with gramicidin S at a lipid-to-peptide molar ratio of 25:1. Data recorded at 25°C at the Austrian SAXS-beamline at the Synchrotrone ELETTRA, Trieste, Italy. Position of hkl-reflections corresponding to a bicontinuous cubic phase of space group Pn3m (solid line) and of space group Im3m (broken line), respectively, are indicated in the diffractogram. A schematic representation of the respective lipid structures is also shown in the figure.

proposed that this decrease is compensated by an increase of the hydrophobic cross sectional area of the lipid acyl chains. In case of PE this lateral expansion will further enhance the mismatch between the cross-sectional areas of the headgroup and hydrocarbon side chains, inducing the lipid monolayer to curl. Although there are different molecular mechanisms that may lead to formation of nonlamellar phases by amphipathic peptides, a significant increase in monolayer curvature stress is likely to be of major importance (e.g. 7,8) and may well be key to their membrane-disruptive properties, as suggested recently for the cyclic peptide gramicidin S of *Bacillus brevis* (77). The X-ray study showed that this antimicrobial peptide has considerable potential for disrupting the structural integrity of lipid bilayer membranes by markedly decreasing the energetic barriers against the formation of nonlamellar lipid phases. It was proposed that the formation of bicontinuous cubic lipid phases (Figure 5) is due to the limited flexibility of the β–turn of gramicidin S as well as to the clustered location of the ornithine side chains, which might facilitate an accommodation of the peptide in the lipid membrane that favors formation of such phases. That the membrane-disrupting capacity of gramicidin S may actually be a function of the nonlamellar phase-forming propensity of the target cell membrane has some interesting implications for the mechanism of action of anti-microbial peptide.

It can be envisaged that molecules like antimicrobial peptides interacting with such bilayers do not only induce a local curvature, but cause the liberation of longer-range forces that were already existing before the local interaction. Two intrinsically curved bilayers can be locked into a planar structure by the equilibrium of the two opposing monolayer curvatures. The monolayer forces are "frustrated" in the bilayer. This frustration can be relaxed very effectively into a curved bilayer by a local interaction with a membrane active molecule which releases tension at one surface and consequently the tension at the other side of the bilayer becomes dominant. This can amplify the local event into a gross morphological change of the bilayer structure leading finally to membrane destruction and cell death

.

Concluding Remarks

Most antimicrobial peptides kill bacteria by membrane permeation or destruction exhibiting a high specificity/activity towards bacterial membranes. In order to elucidate their molecular mechanism of action, it is important to consider the architecture of eukaryotic and prokaryotic cell membranes which differ markedly in their complexity and lipid composition. Research in this field has clearly shown that antimicrobial peptides discriminate between the lipids typical for mammalian (choline phospholipids) and bacterial (PE, PG and its derivatives) cell membranes. Thus, the mode of lipid-peptide interactions and consequently of membrane perturbation by antimicrobial peptides is at least partly due to the variations in the packing properties that exist between the lipid matrices of mammalian and bacterial cell membranes. An understanding of this mutual dependence at a molecular level will support the rational design of antimicrobial peptides as novel antibiotics.

Acknowledgement

Research in our laboratory on antimicrobial peptides is supported by grants of the Austrian Ministry of Science and Transportation as well as by the Österreichische Nationalbank (No. 7190).

References

1. Rouhi, M. 1995. Steps urged to combat drug-resistant strains. Chemical Eng. News. May 22: 7-8.
2. Boman, H.G. 1991. Antibacterial peptides: key components needed in immunity. Cell. 65: 205-207.
3. Wade, D., Boman, A., Wahlin, B., Drain, C.M., Andreu, D., Boman, H.G. and Merrifield, R.B. 1990. All-D amino acid-containing channel-forming antibiotic peptides. Proc. Natl. Acad. Sci. USA. 87: 4761-4765.
4. Oren, Z., Hong, J. and Shai, Y. 1997. A repertoire of novel antibacterial diastereomeric peptides with selective cytolytic activity. J. Biol. Chem. 272: 14643-14649.
5. Vunnam, S., Juvvadi, P. and Merrifield, R.B. 1997. Synthesis and antibacterial action of cecropin and proline-arginine-rich peptides from pig intestine. J. Pept. Res. 49: 59-66.
6. Lohner, K. and Epand, R. 1997. Membrane Interactions of Hemolytic and Antibacterial Peptides. In: Advances in Biophysical Chemistry. C.A. Bush, ed. JAI Press Inc., Greenwhich, Connecticut. Vol. 6: p. 53-66.
7. Lohner, K. and Prenner, E.J. 1999. Differential scanning calorimetry and X-ray diffraction studies of the specificity of the interaction of antimicrobial peptides with membrane-mimetic systems. Biochim. Biophys. Acta. 1462: 141-156.
8. Epand, R.M. 1998. Lipid polymorphism and protein-lipid interactions. Biochim. Biophys. Acta. 1376: 353-368.
9. Choung, S.Y., Kobayashi, T., Takemoto, K., Ishitsuka, H. and Inoue, K. 1988. Interaction of a cyclic peptide, Ro09-0198, with phosphatidylethanolamine in liposomal membranes. Biochim. Biophys. Acta. 940: 180-187.
10. Matsuyama, K. and Natori, S. 1990. Mode of action of sapecin, a novel antibacterial protein of *Sarcophaga peregrina* (flesh fly). J. Biochemistry 108: 128-132.
11. Hristova, K., Selsted, M.E. and White, S.H. 1997. Critical role of lipid composition in membrane permeabilization by rabbit neutrophil defensins. J. Biol. Chem. 272: 24224-24233.
12. Breukink, E., Wiedemann, I., van Kraaij, C., Kuipers, O.P., Sahl, H. and de Kruijff, B. 1999. Use of the cell wall precursor lipid II by a pore-forming peptide antibiotic. Science. 286: 2361-2364.
13. Lehrer, R.I., Barton, A., Daher, K.A., Harwig, S.S.L., Ganz, T. and Selsted, M.E. 1989. Interaction of human defensins with *Escherichia coli*. Mechanism of bactericidal activity. J. Clin. Invest. 84: 553-561.
14. Park, C.B., Kim, H.S. and Kim, S.C. 1998. Mechanism of action of the antimicrobial peptide buforin II: buforin II kills microorganisms by penetrating the cell membrane and inhibiting cellular functions. Biochem. Biophys. Res. Commun. 244: 253-257.
15. Sharma, S., Verma, I. and Khuller, G.K. 1999. Biochemical interaction of human neutrophil peptide-1 with *Mycobacterium tuberculosis* H37Ra. Arch. Microbiol. 171: 338-342.
16. Wu, M., Maier, E., Benz, R. and Hancock, R.E. 1999. Mechanism of interaction of different classes of cationic antimicrobial peptides with planar bilayers and with the cytoplasmic membrane of *Escherichia coli*. Biochemistry. 38: 7235-7242.
17. Bloom, M., Evans, E. and Mouritsen, O. G. 1991. Physical properties of the fluid lipid-bilayer component of cell membranes: a perspective. Quart. Rev. Biophys. 24: 293–367.
18. Hammond, S.M., Lambert, P.A. and Rycroft, A.N. 1984. The Bacterial Cell Surface. Croom Helm, London.

19. Wilkinson, S.G. 1988. Gram-negative bacteria. In: Microbial Lipids. C. Ratledge, and S.G. Wilkinson, eds. Academic Press, London. Vol. 1: p. 299-488.

20. Sparrow, C.P. and Raetz, CR. 1985. Purification and properties of the membrane-bound CDP-diglyceride synthetase from *Escherichia coli*. J. Biol. Chem. 260(22): 12084-12091.

21. Huijbregts, R.P., de Kroon, A.I. and de Kruijff, B. 2000. Topology and transport of membrane lipids in bacteria. Biochim Biophys Acta. 1469(1): 43-61.

22. Shukla, S.D., Green, C. and Turner, J.M. 1980. Phosphatidylethanolamine distribution and fluidity in outer and inner membranes of the gram-negative bacterium *Erwinia carotovora*. Biochem. J. 188(1): 131-135.

23. O'Leary, W.M. and Wilkinson, S. G. 1988. Gram-positive bacteria. In: Microbial Lipids. C. Ratledge, and S.G. Wilkinson, eds. Academic Press, London. Vol. 1, p. 117-201.

24. Yorek, M.A. 1993. Biological Distribution. In: Phospholipids Handbook. G. Cevc, ed. Marcel Dekker, Inc., New York. p. 745-775.

25. Rothman, J.E. and Leonard, J. 1977. Membrane asymmetry. Science. 195:743-753.

26. Lohner, K., Hermetter, A. and Paltauf, F. 1984. Phase behavior of ethanolamine plasmalogen. Chem. Phys. Lipids. 34: 163-170.

27. Lohner, K., Balgavy, P., Hermetter, A., Paltauf, F. and Laggner, P. 1991. Stabilization of non-bilayer structures by the etherlipid ethanolamine plasmalogen. Biochim. Biophys. Acta. 1061: 132-140.

28. Blondelle, S.E. and Lohner, K. 2000. Combinatorial libraries: A tool to design antimicrobial and antifungal peptide analogs having lytic specificities for structure-activity-relationship studies. Biopolymer. 55: 74-87.

29. Peschel, A., Otto, M., Jack, R.W., Kalbacher, H., Jung, G. and Gotz, F. 1999. Inactivation of the *dlt* operon in *Staphylococcus aureus* confers sensitivity to defensins, protegrins, and other antimicrobial peptides. J. Biol. Chem. 274: 8405-8410.

30. Takada, H. and Kotani, S. 1992. Molecular biochemistry and cellular Biology. In: Bacterial Endotoxic Lipopolysaccharides. D.C. Morrison, and J.L. Ryan, eds. CRC Press, Boca Raton, FL. Vol. 1, p. 107-130.

31. Iovine, N.M., Elsbach, P. and Weiss, J. 1997. An opsonic function of the neutrophil bactericidal/permeability-increasing protein depends on both its N- and C-terminal domains. Proc. Natl. Acad. Sci. USA. 94: 10973-10978.

32. Elsbach, P. 1998. The bactericidal/permeability-increasing protein (BPI) in antibacterial host defense. J. Leukoc. Biol. 64: 14-18.

33. Beamer, L.J., Carroll, S.F. and Eisenberg, D. 1998. The BPI/LBP family of proteins: a structural analysis of conserved regions. Protein Sci. 7: 906-914.

34. Schumann, R.R., Lamping, N. and Hoess, A. 1997. Interchangeable endotoxin-binding domains in proteins with opposite lipopolysaccharide-dependent activities. J. Immunol. 159: 5599-5605.

35. Ried, C., Wahl, C., Miethke, T., Wellnhofer, G., Landgraf, C., Schneider-Mergener, J. and Hoess, A. 1996. High affinity endotoxin-binding and neutralizing peptides based on the crystal structure of recombinant Limulus anti-lipopolysaccharide factor. J. Biol. Chem. 271: 28120-28127.

36. Nakamura, T., Furunaka, H., Miyata, T., Tokunaga, F., Muta, T., Iwanaga, S., Niwa, M., Takao, T. and Shimonishi, Y. 1988. Tachyplesin, a class of antimicrobial peptide from the hemocytes of the horseshoe crab (*Tachypleus tridentatus*). Isolation and chemical structure. J. Biol. Chem. 263: 16709-16713.

37. DeLucca, A.J., Jacks, T.J. and Brogden, K.A. 1995. Binding between lipopolysaccharide and cecropin A. Mol. Cell. Biochem. 151: 141-148.

38. Rana, F.R., Macias, E.A., Sultany, C.M., Modzrakowski, M.C. and Blazyk, K.J. 1991.

Interactions between magainin 2 and *Salmonella typhimurium* outer membranes: effect of lipopolysaccharide structure. Biochemistry. 30: 5858-5866.

39. Matsuzaki, K., Sugishita, K. and Miyajima, K. 1999. Interaction of an antimicrobial peptide, magainin 2, with lipopolysaccharide-containing liposomes as a model for outer membranes of Gram-negative bacteria. FEBS Lett. 449: 221-224.

40. Helander, I.M., Kilpelainin, I. and Vaara, M. 1997. Phosphate groups in lipopolysaccharides of *Salmonella typhimurium* rfaP mutants. FEBS Lett. 409: 457-460.

41. Numilla, K., Kilpelainin, I., Zahringer, U., Vaara, M. and Helander, I.M. 1995. Lipopolysaccharides of polymyxin B-resistant mutants of *Escherichia coli* are extensively substituted by 2-aminoethyl pyrophosphate and contain aminoarabinose in lipid A. Mol. Microbiol. 16: 271-278.

42. Guo, L., Lim, K.B., Gunn, J.S., Bainbridge, B., Darveau, R.P., Hackett, M. and Miller, S.I. 1997. Regulation of Lipid A Modifications by *Salmonella typhimurium* Virulence Genes *phoP-phoQ*. Science. 276: 250-253.

43. Guo, L., Lim, K.B., Poduje, C.M., Daniel, M., Gunn, J.S., Hackett, M. and Miller, S.I. 1998. Lipid A acylation and bacterial resistance against vertebrate antimicrobial peptides. Cell. 95: 189-198.

44. Hancock, R.E. 1997. The bacterial outer membrane as a drug barrier. Trends Microbiol. 5: 37-42.

45. Hancock, R.E. 1997. Peptide antibiotics. Lancet. 349: 418-22.

46. Rana, F.R. and Blazyk, K.J. 1991. Interactions between the antimicrobial peptide, magainin 2, and *Salmonella typhimurium* lipopolysaccharides. FEBS Lett. 293: 11-15.

47. Matsuzaki, K., Sugishita, K., Harada, M., Fujii, N. and Miyajima, K. 1997. Interactions of an antimicrobial peptide, magainin 2, with outer and inner membranes of Gram-negative bacteria. Biochim. Biophys. Acta. 1327: 119-130.

48. White, S.H., Wimley, W.C. and Selsted, M.E. 1995. Structure, function, and membrane integration of defensin. Curr. Opinion Struct. Biol. 5: 521-527.

49. Lohner, K., Latal, A., Lehrer, R.I. and Ganz, T. 1997. Differential scanning microcalorimetry indicates that human defensin, HNP-2, interacts specifically with biomembrane mimetic systems. Biochemistry. 36: 1525-1531.

50. Latal, A., Lehrer, R.I., Harwig, S.S.L. and Lohner, K. 1996. Interaction of enantiomeric protegrins with liposomes. Prog. Biophys. Mol. Biol. 65:121.

51. Latal, A., Degovics, G., Epand, R.F., Epand, R.M. and Lohner, K. 1997. Structural aspects of the interaction of peptidyl-glycylleucine-carboxyamide, a highly potent antimicrobial peptide from frog skin, with lipids. Eur. J. Biochem. 248: 938-946.

52. Prenner, E.J., Lewis, R.N.A.H., Kondejewski, L.H., Hodges, R.S. and McElhaney, R.N. 1999. Differential scanning calorimetric study of the effect of the antimicrobial peptide gramicidin S on the thermotropic phase behavior of phosphatidylcholine, phosphatidylethanolamine and phosphatidylglycerol lipid bilayer membranes. Biochim. Biophys. Acta. 1417: 211-223.

53. Demel, A.R., Peelen, T., Siezen, R., de Kruijff, B. and Kuipers, O. 1996. Nisin Z, mutant nisin Z and lacticin 481 interactions with anionic lipids correlate with antimicrobial activity. A monolayer study. Europ. J. Biochem. 235: 267-274.

54. Breukink, E., van Kraaij, C., Demel, A.R., Peelen, T., Siezen, R., de Kruijff, B., and Kuipers, O. 1997. The C-terminal region of nisin is responsible for the initial interaction of nisin with the target membrane. Biochemistry. 36: 6968-6976.

55. Matsuzaki, K., Sugishita, K., Ishibe, N., Ueha, M., Nakata, S., Miyajima, K. and Epand, R.M. 1998. Relationship of membrane curvature to the formation of pores by magainin 2. Biochemistry. 37: 11856-11863.

56. Carbone, M.A. and MacDonald, P.M. 1996. Cardiotoxin II segregates phosphatidylglycerol from mixtures with phosphatidylcholine: (31)P and (2)H NMR spectroscopic evidence. Biochemistry. 35: 3368-3378.

57. Latal, A., Degovics, G. and Lohner, K. 1998. Phase separation of enriched phosphatidylglycerol domains in mixtures with phosphatidylethanolamine. Chem. Phys. Lipids. 94: 161.

58. Heller, W.T., He, K., Ludtke, S.J., Harroun, T.A. and Huang, H.W. 1997. Effect of changing the size of lipid headgroup on peptide insertion into membranes. Biophys. J. 73: 239-244.

59. Cullis, P.R. and de Kruijff, B. 1979. Lipid polymorphism and the functional role of lipids in biological membranes. Biochim. Biophys. Acta. 559: 399-420.

60. Israelachvili, J.N., Horn, R.G. and Marcelja, S. 1980. Physical principles of membrane organization. Q. Rev. Biophys. 13: 121-200.

61. Vasilenko, I., de Kruijff, B. and Verkleij, A.J. 1982. Polymorphic phase behaviour of cardiolipin from bovine heart and from *Bacillus subtilis* as detected by ^{31}P-NMR and freeze-fracture techniques. Effects of Ca^{2+}, Mg^{2+}, Ba^{2+} and temperature. Biochim. Biophys. Acta. 684: 282-286.

62. Batenburg, A.M., Hibbeln, J.C., Verkleij, A.J. and de Kruijff, B. 1987. Melittin induces H_{II} phase formation in cardiolipin model membranes. Biochim. Biophys. Acta. 903: 142-54.

63. Cantor, R.S. 1999. Lipid composition and the lateral pressure profile in bilayers. Biophys. J. 76: 2625-2639.

64. De Kruijff, B. 1999. Biomembranes. Lipids beyond the bilayer. Nature 386: 129-130.

65. Morein, S., Andersson, A.-S., Rilfors, L. and Lindblom, G. 1996. Wild-type *Escherichia coli* cells regulate the membrane lipid composition in a ''window'' between gel and non-lamellar structures. J. Biol. Chem. 271: 6801–6809.

66. Rilfors, L., Wieslander, A. and Lindblom G. 1993. Regulation and physicochemical properties of the polar lipids in *Acholeplasma laidlawii*. In: Subcellular Biochemistry. S. Rottem, and I. Kahane, eds. Plenum Press, New York. Vol.20: p. 109-166.

67. McElhaney, R.N. 1992. Mycoplasma: Molecular Biology and Pathogenesis. In: J. Maniloff, R.N.McElhaney, L.R. Finch, and J.B. Baseman, eds. American Society for Microbiology, Washington DC. Chapter 8: p. 113-155.

68. Gruner, S.M. 1985. Intrinsic curvature hypothesis for biomembrane lipid composition: a role for nonbilayer lipids. Proc. Natl. Acad. Sci. USA. 82: 3665-3669.

69. Hui, S.W. and Sen, A. 1989. Effects of lipid packing on polymorphic phase behavior and membrane properties. Proc. Natl. Acad. Sci. USA. 86: 5825-5829.

70. Lohner, K. 1996. Is the high propensity of ethanolamine plasmalogens to form non-lamellar lipid structures manifested in the properties of biomembranes? Chem. Phys. Lipids. 81: 167-184.

71. Keller, S.L., Gruner, S.M. and Gawrisch, K. 1996. Small concentrations of alamethicin induce a cubic phase in bulk phosphatidylethanolamine mixtures: Biochim. Biophys. Acta. 1278: 241–246.

72. Wu, Y., He, K., Ludtke, S.J. and Huang, H.W. 1995. X-ray diffraction study of lipid bilayer membranes interacting with amphiphilic helical peptides: diphytanoyl phosphatidylcholine with alamethicin at low concentrations. Biophys. J. 68: 2361–2369.

73. He, K., Ludtke, S.J., Heller, W.T. and Huang, H.W. 1996. Mechanism of alamethicin insertion into lipid bilayers. Biophys. J. 71: 2669–2879.

74. Ludtke, S.J., He, K. and Huang, H.W. 1995. Membrane thinning caused by magainin 2. Biochemistry. 34: 16764–16769.

75. Ludtke, S.J., He, K., Wu, Y. and Huang, H.W. 1994. Cooperative membrane insertion of magainin correlated with its cytolytic activity. Biochim. Biophys. Acta. 1190: 181–184.
76. Heller, W.T., Waring, A.J., Lehrer, R.I., Harroun, T.A., Weiss, T. M., Yang, L. and Huang, H.W. 2000. Membrane thinning effect of the β-sheet antimicrobial protegrin. Biochemistry. 39: 139–145.
77. Staudegger, E, Prenner, E.J., Kriechbaum, M., Degovics, G., Lewis, R.N.A.H., McElhaney, R.N. and Lohner, K. 2000. X-ray studies on the interaction of gramicidin S with microbial lipid extracts: Evidence for cubic phase formation. Biochim. Biophys. Acta. 1468: 213-230.

From: *Development of Novel Antimicrobial Agents: Emerging Strategies*
ISBN 1-898486-23-9 © 2001 Horizon Scientific Press, Wymondham, UK.

13

Molecular Mechanisms of Membrane Perturbation by Antimicrobial Peptides

Katsumi Matsuzaki

Abstract

Recently, a large number of antimicrobial peptides have been discovered from animals as well as plants. These peptides are recognized as important components of innate defense mechanisms. Many of these molecules form cationic amphipathic secondary structures that can interact with anionic bacterial membranes. Peptide-induced membrane permeabilization is an effective mechanism of antimicrobial action, which enables rapid and broad-spectrum bacterial killing. In this chapter, the various modes of membrane perturbation reported to date will be critically reviewed, emphasizing that the mode is strongly dependent on the physicochemical properties not only of the peptide but also of the target membrane.

Introduction

During the last decade, a large number of antimicrobial peptides have been discovered from animals as well as plants (1-4 and see also chapter 11 in this monograph. These molecules are recognized as important components of innate defense mechanisms. These peptides are typically composed of 12–45 amino acid residues and have common physicochemical properties; they are highly basic (cationic) due to the presence of multiple Lys and Arg residues and form amphipathic secondary structures (α-helix and β-sheet), although there are also classes of peptides with unique structures such as the Pro-rich peptides. Table 1 summarizes several representative peptides.

Many, but not all, of these peptides are considered to kill bacteria by permeabilizing and/or disrupting bacterial membranes. For example, the addition of magainin 2 to *Escherichia coli* cells induces blebs on the cell surface and permeabilizes the inner membranes, leading to cell lysis (5). Electrostatic interactions between the cationic peptides and bacterial anionic components (lipopolysaccharides, acidic phospholipids etc.) play a crucial role in the binding process. This topic is described in detail by Lohner in the previous chapter.

The fundamental architecture of biomembranes is the lipid bilayer, in which membrane proteins are embedded in a mosaic fashion. The membrane-acting antimicrobial peptides appear to target the lipid matrix rather than the proteins because enantiomeric peptides composed of D-amino acids exhibit the same potency as naturally occurring L peptides, indicating that chiral molecules are not involved in the antimicrobial action (6, 7). Indeed,

Table 1. Representative antimicrobial peptides.

Peptide	Source	Primary Structure [a]
α-helical		
magainin 2	*Xenopus laevis*	H₂N-GIGKFLHSAKKFGKAFVGEIMNS-COOH
PGLa	*Xenopus laevis*	H₂N-GMASKAGAIAGKIAKVALKAL-NH₂
cecropin A	*Hyalophora cecropia*	H₂N-KWKLFKKIEKVGQNIRDGIIKAGPAVAVVGQATQIAK-NH₂
alamethicin	*Trichoderma viride*	Ac-UPUAUAQUVUGLUPVUUEF-ol
β-sheet		
tachyplesin I	*Tachypleus tridentatus*	H₂N-KWCFRVCYRGICYRRCR-NH₂
protegrin-1	Porcine leukocytes	H₂N-RGGRLCYCRRRFCVCVGR-NH₂

[a] U, α-aminoisobutyric acid; F-ol, phenylananinol. Acidic and basic residues are underlined and in bold face, respectively. The alamethicin helix may contain a 3_{10}-helix.

the peptides induce permeabilization of artificial lipid bilayers (8-11). The observation that magainin 2 shows higher antibacterial activity against bacteria with membranes containing larger amounts of acidic lipids also supports this hypothesis (5). In this chapter, the proposed mechanisms of membrane perturbation induced by antimicrobial peptides will be critically reviewed.

Dynamic Peptide–membrane Interactions

The lipid bilayer is composed of a hydrophilic polar head group region and a hydrophobic core, although recent studies have emphasized a more complicated nature of the interfacial region (12). Amphipathic peptides can be accommodated into the heterogeneous bilayer structure mainly through hydrophobic and electrostatic interactions. The incorporation of "aliens" causes stress in the lipid bilayer, leading to membrane perturbation. When the unfavorable energy reaches a threshold, the membrane barrier property becomes lost, which is the basis of the antimicrobial action of these peptides. Several modes of action have been proposed for peptide-mediated membrane permeabilization, as described in the following section.

Before going into details of the detailed molecular mechanisms of action, five points regarding the general aspects of the effects of these peptides should be emphasized. First, membrane perturbation is a consequence of peptide–lipid *interactions*. This implies that the mode of action is determined not only by the peptide species but also by the physicochemical properties of the lipid bilayer. In other words, a certain peptide can exhibit multiple modes of action depending on the lipid composition of the target membrane. Second, peptide–lipid interactions are very *dynamic*. The peptide first attacks the membrane from the external aqueous phase, then induces physicochemical perturbation of the membrane. The "active state" disappears with time. The investigator should always keep it in mind that the peptide–lipid interactions are time-dependent and therefore different states may be observed. Third, the above kinetic events proceed in seconds to minutes. The antimicrobial action should be much faster than bacterial doubling time, e.g. 20 min in the case of *E. coli*. Fourth, relatively high peptide/lipid molar ratios (P/L) of approx. 1/300 to 1/10 are needed to destroy the

A) zero   B) positive   C) negative

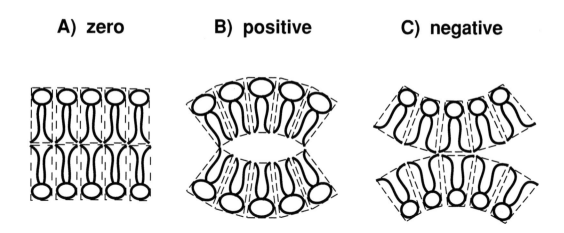

Figure 1. Membrane curvature. (A) Lipids with a cylindrical shape form a flat monolayer (zero curvature). Two monolayers of this class can construct a stable, strain-free bilayer. Reversed cone-shaped (B) and cone-shaped (C) lipids form convex (positive curvature) and concave (negative curvature) monolayers, respectively. If either monolayer forms a symmetrical bilayer, it will be less stable because of strain.

membrane barrier because membrane perturbation is based on physicochemical interactions between the peptide and the lipid. This is important from the viewpoint of selective toxicity. If a small degree of peptide accumulation was sufficient to permeabilize the membrane, the peptide would also be toxic to host cells for which the affinity of the peptide is low, but not zero. Finally, the physicochemical mechanisms based on general peptide–lipid interactions also provide broad antimicrobial spectra and may limit development of resistance.

Modes of Interactions

Lipid Bilayers

Model lipid membranes are usually used for biophysical investigation of peptide–lipid interactions. Hydration of most lipid films of various compositions produces closed, onionlike multilamellar vesicles (MLVs) (diameter ≈ 1 µm), which can be further sized by extrusion (filtration) or sonication into large unilamellar vesicles (LUVs) or small unilamellar vesicles (SUVs), respectively (13). MLVs are often employed for solid-state NMR experiments. Optically clear SUVs (~ 30 nm diameter) are suitable for spectroscopic measurements, especially for CD spectroscopy. However, high degrees of curvature impose strain on lipid packing. Therefore, strain-free LUVs of 100 nm are commonly used for most studies. It should be noted that in these closed systems, any stress, e.g. membrane expansion, asymmetrically imposed on one monolayer can be transmitted to the opposing lipid leaflet because of the constraint that the two monolayers should constitute a single closed bilayer.

Macroscopically oriented planar bilayers are useful to determine the orientations of both the peptide and lipid (14-22). This system might not mimic actual biomembranes in that the interlamellar space is limited even at maximum hydration and the ionic strength is usually low.

Physicochemical Properties of Lipid Bilayers

The physicochemical properties of lipid bilayers that influence peptide–lipid interactions include charge, fluidity, and curvature strain. Cationic antimicrobial peptides selectively bind to membranes containing acidic phospholipids such as PG or PS facilitated by electrostatic interactions (8, 9, 23-27). The enhanced binding due to the presence of negatively charged lipids can be explained by the Gouy-Chapman theory (28). Fluid bilayers are generally more susceptible to the peptides. For example, magainins more effectively permeabilize lipid bilayers in the liquid-crystalline phase than those in the gel phase (25). The incorporation of cholesterol into fluid state membranes suppresses magainin-induced membrane permeabilization (26).

The important regulatory role of curvature strain in the functions of membrane associated peptides and proteins has attracted a great deal of attention (29). If the head group of a lipid possesses a size comparable to the cross-sectional area of the acyl chains, i.e. the lipid has a cylindrical shape, the lipid forms a flat monolayer (zero curvature). Two monolayers of this class can construct a stable, strain-free bilayer (Figure 1A). A relatively larger polar head results in a convex monolayer (positive curvature), whereas a smaller hydrophilic group causes concave bending of the monolayer (negative curvature). If either monolayer forms a symmetrical bilayer, it will be less stable because of strain (Figure 1 B and C). As will be discussed in detail later, the association of a peptide to lipid bilayers modulates curvature strain, therefore stabilizing or destabilizing the membrane.

It should be noted that the three physicochemical parameters described above are not completely independent. For example, a reduction in pH neutralizes the net negative charge of PS, reducing interlipid repulsion and therefore decreasing the effective cross-sectional area of the head group. This imposes negative curvature strain on the membrane, which converts the bilayer phase to the hexagonal II phase (30-32). The introduction of a *cis* double bond in the acyl chain, which fluidizes the bilayer, simultaneously imposes more negative curvature because of the bulkiness of the unsaturated chain. The gel to liquid crystalline

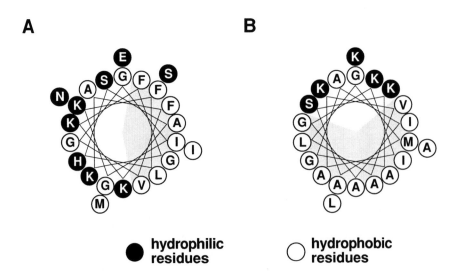

Figure 2. Amphipathic helices formed by antimicrobial peptides. Helical wheel representations of magainin 2 (A) and PGLa (B). The shaded area represents the hydrophobic face.

phase transition temperature of DEPE with two $18:1_{t\Delta 9}$ chains is ~38°C, whereas that of DOPE (di $18:1_{c\Delta 9}$) is −16 °C. The lamellar to hexagonal II phase transition temperature is also decreaseed from about 65 °C to 10 °C by introduction of a *cis* double bond (e.g. (33) and references therein).

Secondary Structures

In the hydrophobic environment of lipid bilayers where no hydrogen-bonding groups exist, antimicrobial peptides form intramolecular or intermolecular hydrogen bonds, folding into amphiphathic secondary structures, i.e. α-helices (Figure 2) or β-sheets. CD and FTIR spectroscopies have been used to obtain rough estimates of peptide conformation. The former technique is easy to use and gives a reliable estimate of helical content (34). For example, magainin 2 contains approx. 70 % helix in acidic phospholipid bilayers (25, 35) and a similar extent of helix formation was reported for PGLa in the presence of PG (23). The latter method also provides quantitative information on β-structures including β-turns if combined with the Fourier self deconvolution curve-fitting procedure (36). The conformation of cecropin P1 in PE/PG bilayers is predominantly α-helical (22). Tachyplesin I (17) and protegrin 1 (2) form antiparallel β-sheet structures in PG-containing bilayers.

High resolution structures can be determined only by NMR. However, there have been very few studies on the conformations of peptides in lipid bilayers because limited motional freedom of the peptides significantly broadens the spectra. The membrane-bound conformations of the venoms melittin (37) and mastoparan X (38) have been determined using the transfer nuclear Overhauser effect. The latter peptide forms a single amphipathic helix, while the former adopts a bent helix conformation with an unordered C-terminal region. It should be noted that the C-terminal conformation and the bent angle in membranes were significantly different from those in so called "membrane-mimetic" media such as detergent micelles or trifluoroethanol.

The backbone conformations as well as the molecular orientations of magainin 2 (14) and PGLa (15) were estimated by solid-state NMR using a series of peptides [15]*N*-labeled at different positions. Both peptides form α-helices, which lie parallel to the membrane surface. Compared with these shorter peptides (< ca. 25 amino acids), which adopt simple secondary structures, the 37-residue cecropin A forms a more complex helix-turn-helix motif in a hexafluoroisopropanol/water mixture (39). Although the mechanisms of action have been discussed based on this conformation, no experimental data are available regarding the membrane-bound structure.

Detection of Peptide-induced Membrane Permeabilization

Lipid bilayers are highly impermeable to polar substances, especially ions, because of the large energy cost of bringing them into the hydrophobic core. Therefore, the release of water-soluble substances entrapped within liposomes has often been used to evaluate the loss of membrane barrier properties. The efflux of small ions such as K^+ and Cl^- can be monitored using an ion-selective electrode (9, 40), an ion-specific fluorescent probe (41) or an isotope (42). The leakage of larger nonelectrolytes such as sugars has been enzymatically or radiometrically detected (43). One of the most common methods utilizes fluorescent ions such as carboxyfluorescein (MW 376), calcein (MW 623), ANTS (MW 427)–DPX (MW 422) pair and FITC-dextrans of varying molecular weights as probes. The leakage of a concentrated, self-quenching dye from the vesicles causes its dilution, resulting in relief from quenching. Therefore, the dye release is quantitatively and continuously determined by

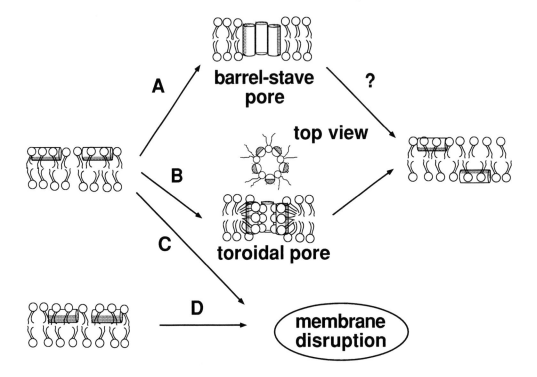

Figure 3. Various modes of membrane permeabilization induced by antimicrobial peptides. If a peptide imposes positive curvature strain on membranes by expanding the polar head group region, it forms either the "barrel-stave" pore (A) or the "toroidal" pore (B), unless strongly inhibited by the bilayer. The peptide can translocate into the inner leaflet upon disintegration of the pore. In the presence of negative curvature-inducing lipids (PS, PA, CL, PE), large amounts of peptides are accumulated on the bilayer surface, eventually leading to irreversible membrane disruption (C). Similar phenomena occur if the peptide imposes negative curvature strain by penetrating deeply into the hydrophobic core of the bilayer (D).

monitoring the enhancement of fluorescence. Another advantage of the quenched fluorescent dye method is that it allows estimation of the lifetime of the dye-permeable membrane defect (44-46).

Potential-sensitive fluorescent dyes are also frequently used to estimate peptide induced membrane permeabilization (11). Addition of valinomycin to K^+-entrapped lipid vesicles suspended in K^+-depleted medium imposes an inside negative transmembrane potential. Peptide-triggered ion flow results in dissipation of the potential, which changes membrane-partitioning of the dye. The absence of any artifacts should be confirmed because binding of the cationic peptide per se could repel the positively charged dye from the bilayer.

Planar lipid bilayers are the most suitable system for detection of ion channels (47). The advantages include easy detection of voltage-dependent channel formation, single channel recording and use of asymmetrical bilayers mimicking biological membranes (48). The disadvantages of these membranes are low reproducibility and the use of a limited number of lipid species because of membrane stability.

Mechanisms of Peptide-induced Membrane Permeabilization

Amphipathic secondary structures can be accommodated in membranes with the hydrophobic amino acid side chains embedded in the hydrocarbon core and the polar side chains interfacing with water. Therefore, the antimicrobial peptides essentially lie parallel to the membrane surface (Figure 3). It is also possible for the peptides to adopt a transmembrane orientation. In this case, to avoid unfavorable exposure of the polar residues to the lipid hydrocarbon chains, the peptides should self-aggregate, forming an aqueous pore in the center. Methods used for the determination of peptide orientation and aggregation have been summarized elsewhere (49).

The hydrophobic face of the surface-lying amphipathic secondary structure interacts with the lipid acyl chains. The peptide penetration depth appears to be dependent on the hydrophobicity of the hydrophobic face. Relatively less hydrophobic magainin 2 showed shallow penetration into the hydrocarbon core. The Trp residue of F5W-magainin 2 is located approximately 1 nm from the bilayer center (50). This location pushes the lipid polar head groups aside, forcing a gap to form in the hydrophobic region, and the membrane locally becomes thinner to fill the gap. An X-ray study revealed that the bilayer thickness decreases almost linearly with increasing peptide concentration (51). The expansion of the polar region implies the local induction of positive curvature strain. Indeed, magainins raise the lamellar-to-hexagonal II phase transition temperature of dipalmitoleoyl phosphatidylethanolamine, indicating that the peptides inhibit formation of the inverted structure of negative curvature by imposing positive curvature strain (52, 53).

The membrane thinning quadratically increases membrane deformation energy, destabilizing the surface-lying state (54). When the energy reaches a critical value, the peptide starts to adopt a transmembrane orientation (Figure 3). The transition from the surface state to the transmembrane state was detected for alamethicin (18) and magainin 1 (19) using oriented circular dichroism, although there is still controversy regarding the orientation of alamethicin (55). Recently, it was suggested that a β-sheet peptide, protegrin 1, follows this kind of orientational transition (56). The critical P/L where the transition occurs depends strongly on lipid composition and the extent of hydration.

Several transmembrane peptides constitute an aqueous pore through which ions can pass. Alamethicin forms the "barrel-stave" pore (Figure 3A), which is constructed solely by a bundle of helices. Although this peptide is not usually classified as a host defense peptide, its pore formation will be described here as an archetype because it is the most extensively studied channel-forming peptide. Refer to (57, 58) for recent reviews. The number of helices of a channel varies from 4 to more than 10, depending on P/L and transmembrane potential. Experimental evidence for this comes from single channel studies using planar lipid bilayers, which show the presence of multi-level bursts of channel opening interspersed with more prolonged closed periods. A neutron diffraction study using oriented multibilayers revealed that the shape of water filling the channel is cylindrical with a diameter of about 1.8 nm at P/L = 1/10 (59). Smaller channels exhibit weak cation selectivity but larger channels have little or no selectivity. It should be noted that under certain conditions, a spectrum of channels of different sizes are always observed but each channel has a defined structure. The voltage-dependence of channel conductance suggests that the helices comprising the bundle align parallel to each other so that the N-terminal end is inserted into the membrane upon application of trans-negative potential by helix dipole–transmembrane electric field interactions.

In the case of magainins, the pore structure is different in that lipids are intercalated between helices (Figure 3B). This may be due to stronger interactions between the positively charged side chains of the peptides and the anionic phospholipid head groups. The "peptide–lipid supramolecular complex pore" model was originally proposed by the author based on the observation that magainin 2 induces rapid lipid flip-flop coupled with pore formation (60). The flip-flop was detected using the dithionite-mediated chemical quenching of NBD-labeled phospholipids. The flip rate is identical to the flop rate and is independent of lipid species. Such phenomena can be explained by lateral diffusion of membrane lipids between the two monolayers connected by the pore. A few months later, Huang and colleagues succeeded in detecting the toroidal structure using neutron scattering (61) and named it the "toroidal (wormhole) pore". Magainins thus exert their cytotoxicity by simultaneously dissipating the transmembrane potential and the lipid asymmetry against membranes of certain lipid compositions. The pore allows the leakage of calcein but is impermeable to FITC-dextran with MW 4400. The pore diameter has been estimated to be 2–3 nm (27). The hypothesis that positive curvature strain imposed by the peptides triggers pore formation was supported by the observation that the incorporation of negative curvature-inducing PE inhibited pore formation of alamethicin (20) and magainin 2 (53). As an extreme case, if the membrane suffers from high intrinsic negative curvature strain, magainin 2 breaks the bilayer organization instead of forming pores (Figure 3C). The peptide effectively forms pores in PG bilayers at low P/L, well below 1/100 with the vesicle morphology intact. In contrast, if PS, PA or CL is used as an acidic phospholipid, membrane permeabilization occurs only at much higher P/L (1/50 to 1/10) with some morphological changes in the liposomes (53). A rapid decrease and a subsequent gradual increase in light scattering were observed. These lipids form inverted phases under conditions of reduced interlipid repulsion, such as at low pH or at high salt concentrations (30-32). The binding of the cationic peptide locally neutralizes the bilayer charge, imposing strong negative curvature strain, which counteracts peptide-induced pore formation. This allows accumulation of a large amount of the peptide in the membrane, finally leading to membrane disruption. The modulation of membrane permeabilization mechanisms by the physicochemical properties of lipid bilayers is extremely important for understanding of the antimicrobial mechanisms of action. Membrane permeabilization by pore formation (Figure 3A and B) can be, at least partially, compensated by ion channels and pumps in bacterial membranes because membrane organization is intact. In contrast, the peptide-induced membrane disruption (Figure 3C and D) is irreversible and unrecoverable, inevitably leading to cell death.

The toroidal pore (Figure 3B) has a positive curvature in the direction of the membrane normal. However, the pore-lining lipids would exhibit negative curvature parallel to the membrane plane. Therefore, peptides with narrower polar faces more effectively form pores. The pore formation rate of PGLa (polar angle ≈ 100°, Figure 2) is much larger than that of magainin 2 (polar angle ≈ 180°, Figure 2) (27). A decrease in magainin polar angle enhances membrane permeabilization activity (52). The balance between positive and negative curvature would depend on the pore size, larger pores having predominantly positive curvature (29). Quantitatively, if the pore diameter exceeds ca. 1.1 nm, positive curvature facilitates pore formation (53).

Amphipathic peptides with large hydrophobicity of the hydrophobic faces are inserted more deeply into the hydrophobic core of lipid bilayers. The Trp residue of a synthetic peptide, 18L, which is localized at the hydrophilic–hydrophobic interface of the helix, shows an emission spectrum very similar to that in hexane, suggesting that the peptide is located in the nonpolar region of the membrane (62). Such a mode of binding expands the hydrophobic core as compared with the head group region. In other words, the peptide imposes negative curvature strain on the membrane, which is supported by a decrease in the T_H of dipalmitoleoylphosphatidylethanolamine (63). This strain finally leads to bilayer disruption, including fusion, instead of forming small pores (Figure 3D). The transient membrane defect allows passage of large solutes (MW 20000). In contrast to the case of magainins, the incorporation of PE facilitates membrane permeabilization.

Shai *et al.* proposed the "carpet-like" mechanism for the mode of action of cecropins and dermaseptins (11, 64 and chapter by Oren and Shai this monograph). In this mechanism, a large amount of peptides covers the membrane surface and disrupts bilayer organization similarly to a detergent. The authors mostly used PS as an acidic phospholipid, which prefers the mechanism shown in Figure 3C. These peptides may follow the scheme shown in Figure 3B in PG membranes.

Translocation of Peptides Across Lipid Bilayers

The pore shown in Figure 3B has a finite lifetime. An increase in the peptide's positive charge reduces pore stability because of enhanced electrostatic repulsion between the side chains (65). We found that magainin 2 stochastically translocates into the inner monolayer upon disintegration of the pores (66). We developed several methods to fluorometrically detect translocation (66, 67). For example, after incubation of a Trp-containing peptide with dansyl-labeled liposomes, the untranslocated peptide remaining on the outer monolayers was removed by either extraction with excess unlabeled liposomes or digestion by addition of trypsin. The amount of the translocated peptide was then quantified on the basis of resonance energy transfer from the Trp residue to the dansyl chromophore.

Translocation is completely coupled to pore formation and lipid flip-flop (60, 66). Therefore, the scheme shown in Figure 3B can also be viewed as the peptide–lipid system constituting a physicochemical signal transduction system; the translocation of an amphipathic peptide quantitatively generates two coupled signals, i.e. ion flow and lipid flow, which short-circuit the two otherwise insulated aqueous phases and two lipid monolayers, respectively.

Pore formation–translocation mechanism is not specific for magainin 2. PGLa (27), mastoparan X (67), melittin (68) and tachyplesin I (69) also follow this scheme. The translocation efficiency is determined by pore formation rate and pore stability. Rapid pore formation and disintegration facilitate peptide internalization without significantly perturbing the barrier properties of the membrane. Buforin 2 from the stomach of the Asian toad *Bufo bufo gargarizans* was suggested to be effectively internalized without significantly permeabilizing membranes (70). The peptide appears to target intracellular nucleic acids.

The pore formation–translocation proceeds in seconds to minutes depending on the P/L used. A reduction in the peptide density in the outer leaflet due to translocation significantly decelerates subsequent pore formation because pore formation is a cooperative process. Therefore, the system reaches a quasi-equilibrium state where pores no longer exist. Since it is in this state that peptide orientation and aggregation have been investigated, it is not surprising that neither transmembrane orientation nor peptide aggregation have been detected.

Peptide–peptide Association

Some peptides hardly or only weakly self-associate in membranes. For example, no aggregation of alamethicin (71) or melittin (72, 73) was detected. Magainin 2 forms a dimer with a small association free energy of 8.8 kJ/mol (27). Other peptides such as paradaxin (74) show significant aggregation. Most antimicrobial peptides belong to the former category. This is partly for the above reason, but more importantly because of their purpose of high selective toxicity. However, it is possible that the formation of a small fraction of aggregates triggers membrane permeabilization.

Association between different peptides has also been reported. Magainin 2 and PGLa of the same origin (*Xenopus* skin) form a 1:1 stoichiometric complex of high potency in membranes with an association free energy of 15 kJ/mol (27), exhibiting marked synergism (10, 27, 75, 76).

Concluding Remarks

Animals and plants utilize physicochemical peptide–lipid interactions to defend themselves against invading microorganisms. This rather simple mechanism provides rapid and broad-spectrum antimicrobial activities. The event leading to cell death is peptide-induced membrane permeabilization, although the detailed molecular mechanisms are strongly dependent on both the peptide and lipid. The driving force appears to be mainly peptide induced stress in bilayer organization, although peptide aggregation may also contribute to membrane perturbation. The author hopes that this short review will be helpful for understanding the mechanisms of action of antimicrobial peptides and the development of novel peptidic antibiotics.

Acknowledgments

This work was supported in part by The Mochida Memorial Foundation for Medical and Pharmaceutical Research, The Kato Memorial Bioscience Foundation and NOVARTIS Foundation (Japan) for the Promotion of Science.

Abbreviations

ANTS	8-aminonaphthalene-1,3,6 trisulfonic acid
CD	circular dichroism
CL	cardiolipin
DEPE	dielaidoyl-L-α-phosphatidylethanolamine
DOPE	dioleoyl-L-α-phosphatidylethanolamine
DPX	*p*-xylene-bis-pyridium bromide
FITC	fluorescein isothiocyanate
FTIR	Fourier transform infrared
LUVs	large unilamellar vesicles
MLVs	multilamellar vesicles
NBD	7-nitrobenz-2-oxa-1,3-diazol-4-yl
PA	phosphatidic acid
PE	phosphatidylethanolamine
PG	phosphatidylglycerol
PS	phosphatidylserine
P/L	peptide-to-lipid ratio
SUVs	small unilamellar vesicles
T_H	bilayer-to-hexagonal II phase transition temperature

References

1. Boman, H.G., Marsh, J. and Goode, J.A., eds. 1994. Antimicrobial Peptides. John Wiley and Sons, Chichester.
2. Waring, A.J., Harwig, S.S.L. and Lehrer, R.I. 1996. Structure and activity of protegrin-1 in model lipid membranes. Protein Peptide Lett. 3: 177–184.
3. Maloy, W.L. and Kari, U.P. 1995. Structure–activity studies on magainins and other host defense peptides. Biopolymers. 37: 105–122.
4. Hancock, R.E.W. and Lehrer, R. 1998. Cationic peptides: a new source of antibiotics. Trends Biotech. 16: 82–88.
5. Matsuzaki, K., Sugishita, K., Harada, M., Fujii, N. and Miyajima, K. 1997. Interactions of an antimicrobial peptide, magainin 2, with outer and inner membranes of Gram negative bacteria. Biochim. Biophys. Acta. 1327: 119–130.
6. Bessalle, R., Kapitkovsky, A., Gorea, A., Shalit, I. and Fridkin, M. 1990. All D-magainin: chirality, antimicrobial activity and proteolytic resistance. FEBS Lett. 274: 151–155.
7. Wade, D., Boman, A., Wåhlin, B., Drain, C.M., Andreu, D., Boman, H.G. and Merrifield, R.B. 1990. All D-amino acid-containing channel forming antibiotic peptides. Proc. Natl. Acad. Sci. USA. 87: 4761–4765.
8. Matsuzaki, K., Harada, M., Handa, T., Funakoshi, S., Fujii, N., Yajima, H. and Miyajima, K. 1989. Magainin 1-induced leakage of entrapped calcein out of negativelycharged lipid vesicles. Biochim. Biophys. Acta. 981: 130–134.
9. Matsuzaki, K., Fukui, M., Fujii, N. and Miyajima, K. 1991. Interactions of an antimicrobial peptide, tachyplesin I, with lipid membranes. Biochim. Biophys. Acta. 1070: 259–264.
10. Vaz Gomes, A., de Waal, A., Berden, J.A. and Westerhoff, H.V. 1993. Electric potential, cooperativity, and synergism of magainin peptides in protein-free liposomes. Biochemistry. 32: 5365–5372.
11. Gazit, E., Boman, A., Boman, H.G. and Shai, Y. 1995. Interaction of the mammalian antibacterial peptide cecropin P1 with phospholipid vesicles. Biochemistry. 34: 11479–11488.
12. White, S.H. and Wimley, W.C. 1994. Peptide in lipid bilayers: structural and thermodynamic basis for partitioning and folding. Curr. Opinion Struct. Biol. 4: 79–86.
13. New, R.R.C., ed. 1990. Liposomes. A Practical Approach. IRL Press, New York.
14. Bechinger, B., Zasloff, M. and Opella, S.J. 1993. Structure and orientation of the antibiotic peptide magainin in membranes by solid-state nuclear magnetic resonance spectroscopy. Protein Sci. 2: 2077–2084.
15. Bechinger, B., Zasloff, M. and Opella, S.J. 1998. Structure and dynamics of the antibiotic peptide PGLa in membranes by solution and solid-state nuclear magnetic resonance spectroscopy. Biophys. J. 74: 981–987.
16. Matsuzaki, K., Shioyama, T., Okamura, E., Umemura, J., Takenaka, T., Takaishi, Y., Fujita, T. and Miyajima, K. 1991. A comparative study on interactions of α-aminoisobutyric acid containing antibiotic peptides, trichopolyn I and hypelcin A with phosphatidylcholine bilayers. Biochim. Biophys. Acta. 1070: 419–428.
17. Matsuzaki, K., Nakayama, M., Fukui, M., Otaka, A., Funakoshi, S., Fujii, N., Bessho, K. and Miyajima, K. 1993. Role of disulfide linkages in tachyplesin–lipid interactions. Biochemistry. 32: 11704–11710.
18. Huang, H.W. and Wu, Y. 1991. Lipid–alamethicin interactions influence alamethicin orientation. Biophys. J. 60: 1079–1087.
19. Ludtke, S.J., He, K., Wu, Y. and Huang, H.W. 1994. Cooperative membrane insertion of magainin correlated with its cytolytic activity. Biochim. Biophys. Acta. 1190: 181–184.

20. Heller, W.T., He, K., Ludtke, S.J., Harroun, T.A. and Huang, H.W. 1997. Effect of changing the size of lipid headgroup on peptide insertion into membranes. Biophys. J. 73: 239–244.

21. Goormaghtigh, E., Cabiaux, V. and Ruysschaert, J.-M. 1993. Polarized attenuated total reflection infrared spectroscopy as a tool to investigate the structure and orientation of amphipathic peptides in a lipid bilayer. In: The Amphipathic Helix. R.M. Epand, ed. CRC Press, Boca Raton. p. 67–86.

22. Gazit, E., Miller, I.R., Biggin, P.C., Sansom, M.S.P. and Shai, Y. 1996. Structure and orientation of the mammalian antimicrobial peptide cecropin P1 within phospholipid membranes. J. Mol. Biol. 258: 860–870.

23. Latal, A., Degovics, G., Epand, R.F., Epand, R.M., Lohner, K. 1997. Structural aspect of the interaction of peptidyl-glycylleucine-carboxyamide, a highly potent antimicrobial peptide from frog skin, with lipids. Eur. J. Biochem. 248: 938–946.

24. Lohner, K., Latal, A., Lehrer, R.I. and Ganz, T. 1997. Differential scanning microcalorimetry indicates that human defensin, HNP-2, interacts specifically with biomembrane mimetic systems. Biochemistry. 36: 1525–1531.

25. Matsuzaki, K., Harada, M., Funakoshi, S., Fujii, N. and Miyajima, K. 1991. Physichochemical determinants for the interactions of magainins 1 and 2 with acidic lipid bilayers. Biochim. Biophys. Acta. 1063: 162–170.

26. Matsuzaki, K., Sugishita, K., Fujii, N. and Miyajima, K. 1995. Molecular basis for membrane selectivity of an antimicrobial peptide, magainin 2. Biochemistry. 34: 3423–3429.

27. Matsuzaki, K., Mitani, Y., Akada, K., Murase, O., Yoneyama, S., Zasloff, M. and Miyajima, K. 1998. Mechanism of synergism between antimicrobial peptides magainin 2 and PGLa. Biochemistry. 37: 15144–15153.

28. Wenk, M.R. and Seelig, J. 1998. Magainin 2 amide interaction with lipid membranes: calorimetric detection of peptide binding and pore formation. Biochemistry. 37: 3909–3916.

29. Epand, R.M. 1998. Lipid polymorphism and lipid–protein interactions. Biochim. Biophys. Acta. 1376: 353–368.

30. Farren, S.B., Hope, M.J. and Cullis, P.R. 1983. Polymorphic phase preference of phosphatidic acid: A ^{31}P and ^{2}H NMR study. Biochem. Biophys. Res. Commun. 111: 675–682.

31. Hope, M.J. and Cullis, P.R. 1980. Effects of divalent cations and pH on phosphatidylserine model membrane: A ^{31}P NMR study. Biochem. Biophys. Res. Commun. 92: 846–852.

32. Seddon, J.M., Kaye, R.D. and Marsh, D. 1983. Induction of the lamellar-inverted hexagonal phase transition in cardiolipin by protons and monovalent cations. Biochim. Biophys. Acta. 734: 347–352.

33. Lohner, K. 1996. Is the high propensity of ethanolamine plasmalogens to form nonlamellar lipid structures manifested in the properties of biomembranes? Chem. Phys. Lipids. 81: 167–184.

34. Yang, J.T., Wu, C.-S.C. and Martinez, H.M. 1986. Calculation of protein conformation from circular dichroism. Methods Enzymol. 130: 208–269.

35. Wieprecht, T., Dathe, M., Beyermann, M., Krause, E., Maloy, W.L., MacDonald, D.L. and Bienert, M. 1997. Peptide hydrophobicity controls the activity and selectivity of magainin 2 amide in interaction with membranes. Biochemistry. 36: 6124–6132.

36. Byler, D.M. and Susi, H. 1986. Examination of the secondary structure of proteins by deconvolved FTIR spectra. Biopolymers. 25: 469–487.

37. Okada, A., Wakamatsu, K., Miyazawa, T. and Higashijima, T. 1994. Vesicle-bound conformation of melittin: Transferred nuclear Overhauser enhancement analysis in the presence of perdeuterated phosphatidylcholine vesicles. Biochemistry. 33: 9438–9446.

38. Wakamatsu, K., Okada, A., Miyazawa, T., Ohya, M. and Higashijima, T. 1992. Membrane-bound conformation of mastoparan X, a G-protein-activating peptide. Biochemistry. 31: 5654–5660.

39. Holak, T.A., Engström, A., Kraulis, P.J., Lindeberg, G., Bennich, H., Jones, T.A., Gronenborn, A.M. and Clore, G.M. 1988. The solution conformation of the antibacterial peptide cecropin A: A nuclear magnetic resonance and dynamical simulated annealing study. Biochemistry. 27: 7620–7629.

40. Matsuzaki, K., Nakai, S., Handa, T., Takaishi, Y., Fujita, T. and Miyajima, K. 1989. Hypelcin A, an α-aminoisobutyric acid containing antibiotic peptide, induced permeability change of phosphatidylcholine bilayers. Biochemistry. 28: 9392–9398.

41. Verkman, A.S. 1990. Development and biological applications of chloride-sensitive indicators. Am. J. Physiol. 259: C375–C388.

42. Pike, M.M., Simon, S.R., Balschi, J.A. and Springer, J.C.S. 1982. High-resolution NMR studies of transmembrane cation transport: Use of an aqueous shift reagent for ^{23}Na. Proc. Natl. Acad. Sci. USA. 79: 810–814.

43. Oku, N., Nojima, S. and Inoue, K. 1980. Selective release of non-electrolytes from liposomes upon perturbation of bilayers by temperature change or polyene antibiotics. Biochim. Biophys. Acta. 595: 277–290.

44. Weinstein, J.N., Ralston, E., Leserman, L.D., Klausner, R.D., Dragsten, P., Henkart, P. and Blumenthal, R. 1984. Self-quenching of carboxyfluorescein fluorescence: uses in studying liposome stability and liposome-cell interaction. In: Liposome Technology. G. Gregoriadis, ed. CRC Press, Boca Raton, FL. p. 183–204.

45. Schwarz, G. and Arbuzova, A. 1995. Pore kinetics reflected in the dequenching of a lipid vesicle entrapped fluorescent dye. Biochim. Biophys. Acta. 1239: 51–57.

46. Ladokhin, A.S., Wimley, W.C. and White, S.H. 1995. Leakage of membrane vesicle contents: determination of mechanism using fluorescence requenching. Biophys. J. 1964–1971.

47. Hanke, W., and Schlue, W.-R. 1993. Planar Lipid Bilayers. Methods and Applications. Academic Press, London.

48. Schröder, G., Brandenburg, K. and Seydel, U. 1992. Polymyxin B induces transient permeability fluctuations in asymmetric planar lipopolysaccharide/phospholipid bilayers. Biochemistry. 31: 631–638.

49. Matsuzaki, K. 1998. Membrane associated peptides. In: Biomembrane Structures. D. Chapman and P. Haris, eds. IOS press, Amsterdam. p. 205–227.

50. Matsuzaki, K., Murase, O., Tokuda, H., Funakoshi, S., Fujii, N. and Miyajima, K. 1994. Orientational and aggregational states of magainin 2 in phospholipid bilayers. Biochemistry. 33: 3342–3349.

51. Ludtke, S., He, K. and Huang, H. 1995. Membrane thinning caused by magainin 2. Biochemistry. 34: 16764–16769.

52. Wieprecht, T., Dathe, M., Epand, R.M., Beyermann, M., Krause, E., Maloy, W.L., MacDonald, D.L. and Bienert, M. 1997. Influence of the angle subtended by the positively charged helix face on the membrane activity of amphipathic, antimicrobial peptides. Biochemistry. 36: 12869–12880.

53. Matsuzaki, K., Sugishita, K., Ishibe, N., Ueha, M., Nakata, S., Miyajima, K. and Epand, R.M. 1998. Relationship of membrane curvature to the formation of pores by magainin. Biochemistry. 37: 11856–11863.

54. Huang, H.W. 1995. Elasticity of lipid bilayer interacting with amphiphilic helical peptides. J. Phys. II (France). 5: 1427–1431.

55. Barranger-Mathys, M. and Cafiso, D.S. 1996. Membrane structure of voltage-gated channel forming peptides by site-directed spin-labeling. Biochemistry. 35: 498–505.

56. Heller, W.T., Waring, A.J., Lehrer, R.I. and Huang, H.W. 1998. Multiple states of β-sheet peptide protegrin in lipid bilayers. Biochemistry. 37: 17331–17338.

57. Bechinger, B. 1997. Structure and functions of channel-forming peptides: magainins, cecropins, melittin and alamethicin. J. Membr. Biol. 156: 197–211.

58. Sansom, M.S.P. 1991. The biophysics of peptide models of ion channels. Prog. Biophys. Mol. Biol. 55: 139–235.

59. He, K., Ludtke, S.J., Worcester, D.L. and Huang, H.W. 1996. Neutron scattering in the plane of membranes: structure of alamethicin pores. Biophys. J. 70: 2659–2666.

60. Matsuzaki, K., Murase, O., Fujii, N. and Miyajima, K. 1996. An antimicrobial peptide, magainin 2, induced rapid flip-flop of phospholipids coupled with pore formation and peptide translocation. Biochemistry. 35: 11361–11368.

61. Ludtke, S.J., He, K., Heller, W.T., Harroun, T.A., Yang, L. and Huang, H.W. 1996. Membrane pores induced by magainin. Biochemistry. 35: 13723–13728.

62. Polozov, I.V., Polozova, A.I., Mishra, V.K., Anantharamaiah, G.M., Segrest, J.P. and Epand, R.M. 1998. Studies of kinetics and equilibrium membrane binding of class A and class L amphipathic peptides. Biochim. Biophys. Acta. 1368: 343–354.

63. Polozov, I.V., Polozova, A.I., Tytler, E.M., Anantharamaiah, G.M., Segrest, J.P., Wooley, G.A. and Epand, R.M. 1997. Role of lipids in the permeabilization of membranes by class L amphiphathic helical peptides. Biochemistry. 36: 9237–9245.

64. Shai, Y. 1995. Molecular recognition between membrane-spanning polypeptides. Trends Biol. Sci. 20: 460–465.

65. Matsuzaki, K., Nakamura, A., Murase, O., Sugishita, K., Fujii, N. and Miyajima, K. 1997. Modulation of magainin 2–lipid bilayer interactions by peptide charge. Biochemistry. 36: 2104–2111.

66. Matsuzaki, K., Murase, O., Fujii, N. and Miyajima, K. 1995. Translocation of a channel forming antimicrobial peptide, magainin 2, across lipid bilayers by forming a pore. Biochemistry. 34: 6521–6526.

67. Matsuzaki, K., Yoneyama, S., Murase, O. and Miyajima, K. 1996. Transbilayer transport of ions and lipids coupled with mastoparan X translocation. Biochemistry. 35: 8450–8456.

68. Matsuzaki, K., Yoneyama, S. and Miyajima, K. 1997. Pore formation and translocation of melittin. Biophys. J. 73: 831–838.

69. Matsuzaki, K., Yoneyama, S., Fujii, N., Miyajima, K., Yamada, K., Kirino, Y. and Anzai, K. 1997. Membrane permeabilization mechanisms of a cyclic antimicrobial peptide, tachyplesin I, and its linear analog. Biochemistry. 36: 9799–9806.

70. Park, C.B., Kim, H.S. and Kim, S.C. 1998. Mechanism of action of the antimicrobial peptide buforin II: Buforin II kills microorganisms by penetrating the cell membrane and inhibiting cellular functions. Biochem. Biophys. Res. Commun. 244: 253–257.

71. Barranger-Mathys, M. and Cafiso, D.S. 1994. Collisions between helical peptides in membranes monitored using electron paramagnetic resonance: Evidence that alamethicin is monomeric in the absence of a membrane potential. Biophys. J. 67: 172–176.

72. Altenbach, C. and Hubbell, W.L. 1988. The aggregation state of spin-labelled melittin in solution and bound to phospholipid membranes: Evidence that membrane-bound melittin is monomeric. Proteins: Struct., Funct., Genet. 3: 230–242.

73. Schwarz, G. and Beschiaschvili, G. 1989. Thermodynamic and kinetic studies on the

association of melittin with a phospholipid bilayer. Biochim. Biophys. Acta. 979: 82–90.

74. Rapaport, D. and Shai, Y. 1992. Aggregation and organization of paradaxin in phospholipid membranes: A fluorescent energy transfer study. J. Biol. Chem. 267: 6502–6509.

75. De Waal, A., Vaz Gomes, A., Mensink, A., Grootegoed, J.A. and Westerhoff, H.V. 1991. Magainins affect respiratory control, membrane potential and motility of hamster spermatozoa. FEBS Lett. 293: 219–223.

76. Westerhoff, H.V., Zasloff, M., Rosner, J.L., Hendler, R.W., de Waal, A., Vaz Gomes, A., Jongsma, A.P.M., Riethorst, A. and Juretic, D. 1995. Functional synergism of the magainins PGLa and magainin-2 in *Escherichia coli.*, tumor cells and liposomes. Eur. J. Biochem. 228: 257–264.

From: *Development of Novel Antimicrobial Agents: Emerging Strategies*
ISBN 1-898486-23-9 © 2001 Horizon Scientific Press, Wymondham, UK.

14

Molecular Mechanism of Cell Selectivity by Linear Amphipathic α-helical and Diastereomeric Antimicrobial Peptides

Ziv Oren and Yechiel Shai

Abstract

Studies described in the last two decades have demonstrated the essential role of antimicrobial peptides in the first line of defense against invading pathogens and their uncontrolled proliferation. Despite numerous studies on the structure and activity of antimicrobial peptides, our knowledge of their mode of action is incomplete and controversial. The most studied group includes the linear, mostly α-helical peptides. Although developed by distant and diverse species such as plants, insects, amphibians and humans, linear antimicrobial peptides share two properties: a net positive charge, and a high propensity to adopt amphipathic α-helical conformation in hydrophobic environments. Numerous studies have shown that peptide-lipid interactions leading to membrane permeation play a major role in their activity. Membrane permeation by amphipathic α-helical peptides has been proposed to occur via one of two general mechanisms: (i) transmembrane pore formation via a "barrel-stave" mechanism; and (ii) membrane destruction/solubilization via a "carpet" mechanism. Critical evaluation of recent studies on linear α-helical antimicrobial peptides is presented in light of these two proposed mechanisms. This chapter, which is focused on representatives of the amphipathic α-helical antimicrobial peptides, supports the "carpet" rather than the "barrel-stave" mechanism. In addition, the different stages of membrane disintegration by antimicrobial peptide will be evaluated based on the recent studies with a novel group of diastereomeric antimicrobial peptides.

Introduction

Antimicrobial peptides are natural antibiotics that constitute a major part of the innate immunity of a wide range of organisms including humans (1). During the last two decades studies have demonstrated the essential role of antimicrobial peptides in the first line of defense against invading pathogens and their uncontrolled proliferation. Despite numerous studies on the structure and activity of antimicrobial peptides, our knowledge of the mode of action by which they select and kill bacterial, fungal, tumor and parasite cells is by far incomplete. The most studied group of antimicrobial peptides includes linear, mostly helical, peptides

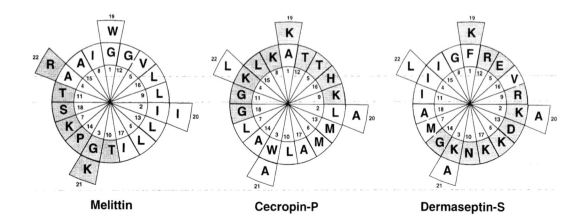

Figure 1. Schiffer-Edmundson wheel projection of the N-terminal 22 amino acids of melittin, cecropin-P and dermaseptin-S. Number 1 represents residue 1 of the peptides. Shaded area indicates hydrophilic amino acids and unshaded area indicates hydrophobic amino acids.

(40 amino acids) that are toxic to bacteria only. Examples are: cecropins, isolated from the cecropia moth (2), magainins (3) and dermaseptins (4), both isolated from the skin of frogs and the human LL-37 (5).

Linear antimicrobial peptides vary considerably in their chain length, hydrophobicity and distribution of charges. Although developed by distant and diverse species such as plants, insects, amphibians and humans, and despite millions of years of evolution, they share common motifs, i.e. they are linear, positively charged and present a large hydrophobic moment and a high propensity to adopt amphipathic α-helical conformation in hydrophobic environments (4, 6-12). In an amphipathic α-helical conformation, polar amino acids are arranged along one side of the helix as a consequence of 1,3 and 1,4 periodicities, and the hydrophobic amino acids along the other side. This structure can be visualized schematically by using the Schiffer and Edmondson wheel projection (13). In this wheel, consecutive amino acids are 100 degrees far from each other such that each cycle contains 3.6 amino acids. Figure 1 shows the wheel projection of three peptides which represent bacteria specific cytolytic peptides, dermaseptin and cecropin, and a non cell selective peptide, melittin.

Mode of Action of Antimicrobial Peptides

Although the exact mechanism by which this family of antimicrobial peptides kills bacteria is not clearly understood, it has been shown that peptide-lipid interaction, rather than a receptor-mediated recognition process, plays a major role in their function. This was demonstrated by the findings that analogues of cecropin and magainin, composed entirely of D-amino acids, possess antibacterial activity indistinguishable from that of the parent molecules (14-16). The interaction of amphipathic α-helical antimicrobial peptides with the bacterial membrane leads to permeation of the bacterial cell wall which leads to the death of the bacteria. The process by which membrane permeation occurs has been described to proceed via either one of the two general mechanisms: (A) transmembrane pore formation via a "barrel-stave" mechanism (17), and (B) membrane destruction/solubilization via a "carpet-like" mechanism (12, 18).

Transmembrane Pore Formation Via a "Barrel-stave" Mechanism

In a classical "barrel-stave" mechanism (17) amphipathic α-helices insert into the membrane and form bundles. In these bundles the hydrophobic surfaces interact with the lipid core of the membrane, and the hydrophilic surfaces point inward, producing a pore (right panel of Figure 2). The interaction of the peptides with the target membrane is driven predominantly by hydrophobic forces because they need to insert into the hydrophobic core of the membrane. As a consequence, they can bind to both zwitterionic and charged phospholipid membranes. Four major steps have been suggested to take place in the "barrel-stave" mechanism: (i) binding of the monomers to the membrane in a α-helical structure; (ii) molecular recognition between membrane-bound monomers which lead to their assembly already at low surface density of bound peptide; (iii) insertion of at least two assembled monomers into the membrane to initiate the formation of a pore, and (iv) progressive recruitment of additional monomers to increase the pore size. A prerequisite condition for this mechanism is that initial assembly of monomers on the surface of the membrane must occur before the peptide is inserted, since it is energetically unfavorable for a single amphipathic α-helix to traverse the membrane as a monomer. In the latter case the low dielectric constant and inability to establish hydrogen bonds, will not allow the fatty acyl region of a lipid bilayer to be in a direct contact with a polar surface of a single amphipathic α-helix. Two examples of cytolytic peptides that insert into membranes via the "barrel-stave" mechanism include alamethicin and pardaxin.

Alamethicin is an amphipathic α-helical peptide containing 20-amino acids and produced by the fungus *Trichoderma viride* (19, 20). The middle of the peptide and the C-terminus possess H-bonding patterns characteristic of 3_{10} helix (21) with a bent around Pro^{14}. Alamethicin has been studied for over 20 years as a model for voltage-gated ion channels. The high concentration dependence of conduction, and the multistep conductances seen in single-channel recordings, were interpreted in terms of a barrel-stave model for the channel pore (for reviews see refs. 22-24). Biophysical studies on the interaction of alamethicin with model membranes also support this mechanism (25).

Pardaxin is an excitatory neurotoxin that has been purified from the Red Sea Moses Sole *Pardachirus marmoratus* (26, 27) and from the Peacock Sole of the Western Pacific *Pardachirus pavoninus* (28), (reviewed in (29)). Pardaxin is composed of 33 amino acids and adopts an amphipathic α-helical structure in hydrophobic environments. The peptide is cytolytic to both bacteria and mammalian cells (27, 30, 31). Similarly to alamethicin, pardaxin has also a helix-hinge-helix structure; the N-helix includes residues 7-11 and the C-helix includes residues 14-26. The helices are separated by a proline residue situated at position 13 similarly to alamethicin (32). Biophysical and functional studies with pardaxin and its analogues revealed that the peptide insert into the membrane and specifically self-associates to form oligomers with different sizes, which support a "barrel-stave" mechanism for its insertion and organization in the membranes (33, 34), (reviewed in ref. 29).

Membrane Destruction/Solubilization Via a "Carpet" Mechanism

The "carpet" mechanism has been proposed initially by Shai and collaborators to describe a situation in which amphipathic α-helical peptides initially bind onto the surface of a mem-

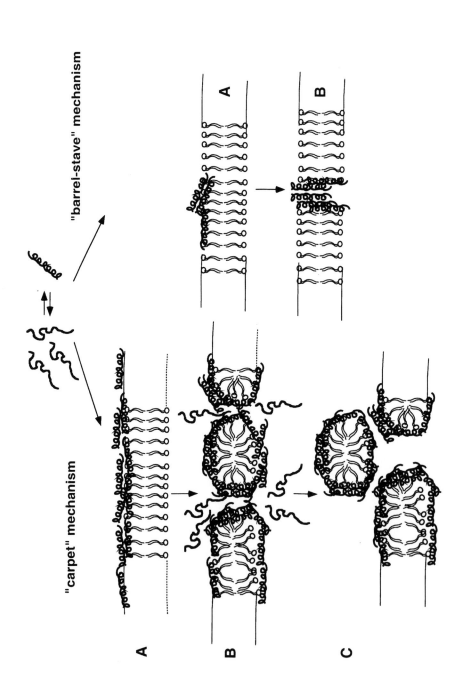

Figure 2. A cartoon illustrating the "barrel-stave" (to the right) and the "carpet" (to the left) models suggested for membrane permeation. In the "carpet" model the peptides are bound to the surface of the membrane with their hydrophobic surfaces facing the membrane and their hydrophilic surfaces facing the solvent (step A). When a threshold concentration of peptide monomers is reached, the membrane breaks into pieces (steps B and C). At this stage a transient pore is formed.

brane in a carpet-like manner and permeate the membrane bilayers without transversing the hydrophobic core of the membrane via a direct contact between the peptides and the aliphatic chains of the lipids (12, 18). The target membrane can be permeated only after a threshold concentration of a peptide has been reached. In contrast to the "barrel-stave" mechanism, the peptide does not insert deeply into the hydrophobic core of the membrane, but rather binds to the interface, near the phospholipid headgroups (left panel in Figure 2). Initial interaction with the negatively-charged target membrane is electrostatically driven, and therefore the active peptides are positively charged. The four steps possibly involved in this model are: (i) preferential binding of positively-charged peptide monomers to the negatively charged phospholipids through electrostatic interaction, (ii) initial formation of a secondary structure and insertion into the interface (iii) laying of amphipathic α-helical monomers in the membrane interface, so that the polar amino acids can interact with the negatively charged phospholipid headgroups or water molecules, and the hydrophobic residues interact with the hydrophobic core of the membrane, and (iv) permeation of the membrane by disrupting the bilayer leading to fragmentation of lipid bilayer into small pieces in a process similar to micellization. An initial step before the collapse of the membrane packing may include transient holes in the membrane. Holes like these may enable the passage of low molecular weight molecules prior to complete membrane lysis. Such holes were described as a toroidal (or wormhole) model (35–37 and chapter by Matsuzaki in this monograph) (Step B in left panel of Figure 2). As seen in Figure 2, these holes may allow the passage of peptide molecules from the outer membrane into the inner membrane of, for example, Gram-negative bacteria, in a process which may be referred to as "self-promoting uptake" (38-41). We will now follow the evidence that support this mechanism.

Steps Involved in Membrane Permeation by Antimicrobial Peptides

Preferential Binding of Positively-charged Peptide Monomers/Oligomers to Negatively Charged Phospholipids

A common feature found in native antimicrobial peptides is their net positive charge contributed by a large number of basic amino acids which are distributed along the hydrophilic face of the amphipathic α-helix (Figure 1). This feature has been proposed to account for their preferential binding to bacteria and not to normal mammalian cells.

The site most likely to be the target of membrane permeating antimicrobial peptides is the inner membrane of bacteria, which typically contains the electron-transport chain and the enzymatic apparatus necessary for oxidative phosphorylation (42). To reach this membrane, the peptides have to traverse the bacterial wall, the outer surface of which contains lipopolysaccarides (LPS) in the case of Gram-negative bacteria, and acidic polysaccharides (teichoic acids) in the case of Gram-positive bacteria, giving the surface of both Gram-positive and Gram-negative bacteria a negative charge (43). Therefore, the net positive charge of the antibacterial peptides facilitates their initial binding to the bacterial surface. In contrast, the outer leaflet of human erythrocytes (representatives of normal mammalian cells) is composed predominantly of zwitterionic phosphatidylcholine (PC) and sphingomyelin phospholipids (44; see also Lohner, this monograph). Studies on the interaction of antimicrobial peptides with model phospholipid membranes revealed low affinity to zwitterionic phospholipids compared to acidic phospholipids. This has been demonstrated with cecropins (10, 12), magainins (45-49) dermaseptins (18, 50) and others (31, 51). The low affinity of antimicrobial peptides to zwitterionic membranes might explain their inability to lyse erythrocytes.

Initial Formation of Secondary Structure and Insertion into the Interface

Little is known about the conformational changes in the insertion process of amphipathic α-helical peptides into membranes. According to one model (52), the secondary structure formation by amphipathic α-helical cytolytic peptides is attained through sequential stages of interfacial binding in an unfolded state, secondary structure formation, and insertion of secondary structure units into the lipid bilayers, leading to disruption of the membrane. The model is supported by a study that examined the conformational changes of the bee venom melittin upon insertion into phospholipids. The results showed that melittin adsorbed on the lipid layer surface contained less α-helix than its counterpart inserted into the lipid layer. As the penetration depth of melittin increased, more ordered structures (α-helix) appeared (53). Before binding to membranes most antimicrobial peptides are present in an aqueous solution in an unfolded state. The initial step of their insertion into the membrane is adsorption of the unfolded form on the surface of the negatively charged phospholipid membrane, as described above. Forces affecting the free energy cost of inserting unfolded peptide into phospholipids membranes include the hydrophobic interactions between the nonpolar amino acids and the phospholipid hydrocarbon core. These forces are countered by the cost of partitioning the polar amino acids and the peptide bond (CONH) (54). When bound to negatively charged membranes, the positive charges of antimicrobial peptides are partially neutralized by the negative charges of the phospholipid headgroups, thus reducing the energy cost of adsorbing the peptide into the membrane. Subsequently, it may allow the intrinsic hydrophobic forces to manifest themselves by forming a stable α-helical structure, driving the peptide further into the interface. However, with the zwitterionic phospholipids the rate limiting step appears to be the initial binding of the antimicrobial peptide to the surface which is very low.

Orientation and Depth of Penetration of Antimicrobial Peptides

The interfaces, defined by the distribution of the water associated with the headgroups, are each about a 15 Å thick and consist of a complex and thermally disordered mixture of water, headgroups, glycerol, carbonyl, and methylenes from the edges of the hydrocarbon core (55). These regions are rich in possibilities for non-covalent interactions with peptides. Several studies which are described in the following paragraphs indicate that antimicrobial peptides are located at the interface, parallel to the surface of the membrane, without deeply inserting into the acyl chain region.

Antimicrobial peptides are located in the interface, preferentially oriented parallel to the surface of phospholipid membranes

Studies using fluorescence (10, 12, 18, 50), and attenuated total reflectance fourier-transform infrared spectroscopy (ATR-FTIR) (56) were used to get insight into the orientation of cecropins and dermaseptins in their membrane-bound state. Dermaseptins and cecropins were labelled selectively at their N-terminal amino acid with the fluorophore NBD (10, 12, 18, 50). NBD fluorescence reflects the environment in which the NBD group is located, displaying a higher quantum yield and a blue shift of the maximal emission wavelength in a more hydrophobic environment. The blue shifts obtained in the presence of vesicles are similar to those observed for an NBD group located at or near the surface of the membrane (emission maximum of ~530 nm) (18, 57, 58). In contrast to these results, peptides known to form channels via a barrel-stave mechanism had their emission maxima at ~520 nm, revealing interaction with the hydrophobic core of the membrane. Examples are pardaxin (33) and the α5 helix from *Bacillus thuringiensis* δ-endotoxin (59). In a further study (56) ATR-FTIR was used to determine the secondary structure and orientation within phospholipid membranes of the mammalian cecropin P1 (CecP). The shape and frequency of the amide I and II absorption peaks of the peptide within acidic PE/PG multibilayers (phosphatidylethanolamine / phosphatidylglycerol) in a 7:3 (w/w) ratio (a phospholipid composition similar to that of many bacterial membranes), indicated that the peptide is predominantly α-helical. Using polarized ATR-FTIR spectroscopy it was found that the peptide is preferentially oriented nearly parallel to the surface of the lipid membranes. In addition, molecular dynamics simulations confirmed that, in response to a cis positive transmembrane voltage difference (56), CecP adopts an orientation parallel to the membrane surface and does not insert into the bilayer. ATR-FTIR and fluorescence spectroscopy studies were used also in the study of the human cecropin like LL-37 indicating that similarly to other antimicrobial peptides, the peptide bind preferentially onto the surface of both negatively-charged and zwitterionic membranes (60).

Solid state NMR measurement of ^{15}N-labeled magainin oriented in multilamellar membranes (61-64), Raman (46), fluorescence (65) and differential scanning calorimetry measurements (66) all indicated that at high and moderate lipid:peptide ratios magainin is oriented parallel to the membrane surface, associated with the lipid head group, and does not significantly disturb the aliphatic chain region. For example, three analogues of magainin 2, each having a Trp residue substituted for Phe at the 5th, 12th, or 16th positions were synthesized and investigated (65). The depths of the Trp residues, which were determined using the n-doxyl phosphatidylcholine quenching technique, were about 10 Å from the bilayer center irrespective of the peptide aggregational state, suggesting that the orientation of the magainin 2 α-helix is parallel to the membrane surface.

Antimicrobial peptides do not affect the acyl chain organization

Linear amphipathic α-helical antimicrobial peptides have little effect on the structure of the acyl chain of lipid bilayers, when inserted in phospholipid membranes. In a Raman study (46), the spectrum of the lipid acyl-chain C-C stretching region was used to indicate the extent of acyl chain disorder induced by bound peptides. Magainin did not disrupt significantly the acyl chains of negatively charged and zwitterionic membranes, even at a high peptide to lipid ratio (w/w) of 1. The effect of magainin 2 on the gel to liquid crystalline phase transition of multilamellar vesicles was measured using differential scanning calorimetry (66). At 3.5 mol % the effect of magainin on the main transition of DPPG (dipalmitoyl phosphatidylglycerol) seemed insignificant, leading to the conclusion that magainin does not penetrate deeply into the acyl chain region. In another study ^{2}H-NMR was used to examine F16W magainin 2 incorporated into palmitoyloleoyl phosphatidylglycerol (POPG) / deuterated palmitoyloleoyl phosphatidylcholine (POPC) (1:3) multilamellar vesicles (MLV) (67). The results showed that the presence of the peptide only slightly decreased the order parameter of the double bond of the PC chain.

Polarized ATR-FTIR spectroscopy was used to determine the effect of cecropin P1 on the multibilayer acyl chain order (56). The incorporation of cecropin P1 did not significantly change the order parameters of the acyl chain of phosphatidylethanolamine (PE) / phosphatidylglycerol PG (7:3 w/w) multibilayers which mimic the phospholipid composition of *E. coli* (68), suggesting that the peptide does not penetrate the hydrocarbon core of the membranes. Similar results were found with LL-37 (60). Contrary to these results, significant effects on the acyl chain were observed with membrane inserted hemolytic peptides such as melittin (69, 70), δ-hemolysin (69), the α4-α5 hairpin helices from *Bacillus thuringiensis* δ-endotoxin (71) and transmembrane segments of Phospholamban (72, 73) and Colicin A (74).

The Membrane Permeation Process

Peptides induced changes in the membrane curvature

The initial step involved in membrane permeation and micellization requires a change in the curvature of the membrane. Membrane lytic peptides can be classified into two groups in terms of their effect on membrane curvature (75): (i) Inducers of a negative curvature strain. A representative example, 18L is a model class L peptide (9). A reciprocal wedge model was suggested by Tytler *et al.* (76) in order to explain the ability of class L peptide to lyse cells. In this model, a class L peptide folds into an amphipathic α-helical structure upon association with phospholipids, such that the polar face of the peptide is associated with the polar head group of the phospholipid. The lipid-associated peptide when viewed in cross section, is an inverted wedge shape in which the hydrophilic face of the helix forms the apex. This structure is expected to force the hydrocarbon chains of phospholipid apart, inducing a negative curvature on the membrane. (ii) Positive curvature inducers. This group of peptides, including Ac-18A-NH_2, a class A peptide (9, 77), cyclic tachyplesin I, and magainin, imposes a positive curvature strain on lipid bilayers. Class A peptides and tachyplesin I have been shown to cause micellization of bilayers (78, 79). In a recent study the effect of magainin 2 on membrane curvature was examined (80). Magainin 2 was found to significantly raise the lamellar to inverse hexagonal (H_{II}) phase transition temperature (T_H) of DPoPE (dipalmitoleoyl-L-α-phosphatidylethanolamine), suggesting that in the absence of electrostatic effects the peptide impose positive curvature strain on the membrane. Phosphatidylserine, phosphatidic acid, and cardiolipin, as opposed to phosphatidylglycerol, are known to form the H_{II} phase under conditions of reduced interlipid electrostatic repulsion (81-83). Permeation of phosphatidylserine, phosphatidic acid, and cardiolipin by magainin 2 occurred at much higher P/L values than phosphatidylglycerol, indicating that lipids that impose higher negative curvature strain on membranes are more resistant to magainin cytolytic activity. Furthermore, addition of a sublytic concentration of cone-shaped LPC (palmitoyl-L-α-phosphatidylcholine), which imposes a positive curvature strain, facilitated leakage of magainin's activity. A high positive curvature in a dimension perpendicular to the bilayer plane is required to stabilize a pore structure toward the formation of a micelle. Two-dimensional ^{1}H NMR experiments in membrane mimetic environment, showed that magainin 2 (84), cecropin A (85) and cecropin P1 (86) are composed of two helical regions with a flexible hinge in between. The ability of amphipathic antimicrobial peptides to induce positive membrane curvature may depend on this-wedge like shape. The role of flexible hinge in antibacterial activity has been previously demonstrated with a series of cecropin-like model peptides (87).

Antimicrobial peptides permeate membranes after a local threshold concentration has been reached

In order to form micelles, a peptide needs to reach high concentration at a certain area of the membrane. In accordance with this assumption, liposome leakage and cytotoxicity experiments suggested the existence of concentration dependent changes in membrane permeating activity of amphipathic antimicrobial peptides. In addition, these studies revealed that initial steps involved in micelles formation include transient holes in the membrane, and changes in the orientation of the peptides. Below a critical concentration, magainin causes only slight leakage, but at higher concentrations magainin causes widespread lysis (42, 45, 46, 66, 88-91). Few examples may demonstrate this effect: (i) The kinetics of magainin 2 induced release of 6-carboxyfluorescein (CF) from phosphatidylserine liposomes, indicating that the fast release of dye is a transient effect resulting from transient destabilization of the bilayer upon initial interaction with the peptide (90). No measurable CF release could be observed until a high level of bound magainin was achieved. (ii) The ability of magainin 2 to decrease the membrane potential in cytochrome oxidase liposomes was investigated by Juretic *et al.* (91). At low concentrations the peptide was almost inactive, but activity was observed when a critical concentration had been reached. (iii) Neutron in-plane scattering detected pores formed by magainin 2 in membranes only when a substantial fraction of the peptide was oriented perpendicular to the membrane (36). (iv) The interaction of magainin 2 and its analog Ala[19]-magainin 2 with lipid was investigated using two lipid photolables. The results indicated a concentration-dependent insertion of the peptide into the lipid bilayer. Higher photolabelling at both shallow and deep photolabels was achieved at high peptide to lipid molar ratio (92).

Membrane permeation combined with membrane binding studies on dermaseptins (18, 50, 93) and cecropins (10, 12) revealed that maximal activity was obtained when the peptides covered the surface of the liposomes. In addition dermaseptins (18, 50), cecropin B, cecropin P (10, 12 and unpublished results) and cecropin A (94) dissipated ion gradients at low concentrations of peptide, while substantially higher concentrations were required to release encapsulated calcein. Further support was obtained by Steiner *et al.*, (7) who found that the stoichiometry of cecropin killing of bacteria suggested that amounts of cecropin sufficient to form a monolayer strongly modified the bacterial membrane. In another study a survival assay was used in order to compare the concentrations of cecropin A required to kill bacteria to the concentration required to release cytoplasmic β-galactosidase (94). The results clearly demonstrated that cecropin A is bactericidal at relatively high concentrations, where it increases membrane permeability to large molecules.

Membrane permeation by antimicrobial peptides revealed by electron microscopy

Examination of *E. coli* treated with cecropins at concentrations below the minimal inhibitory concentrations (MIC) by transmission electron microscopy revealed the formation of patches, whereas total lysis of the bacteria was observed at the MIC concentration (95). These patches were similar to those obtained with non cytolytic derivatives of pardaxin (31) and diastereomers of lytic peptides (96, 97). When magainin was used at a concentration where bacterial viability had decreased about 30 %, similar blebs were observed on the cell surface of the bacteria (49). At higher concentrations magainin 2 caused string-like substances, which are considered to be cellular debris arising from cell lysis. Similar results were obtained when isolated *Heliothis* cecropin and cecropin B were incubated with *E. coli* K12 D31 (98). Large parts of the cell envelope were missing from the bacteria and the outer membrane appeared to be dissociated from the cell. *E. coli* cells contained large lesions with an outer diameter of 9.3 nm and internal pore diameter of 4.2 nm. Scanning and transmission electron microscopy were used to study the morphological changes induced by cecropin B on the bacterial cell of *Klebsiella pneumoniae* (99). The results showed that cecropin B causes bleb-like protrusions on the bacterial cell surface.

A detergent-like effect in membrane permeation by antimicrobial peptides

Comparison between the properties of magainin 2 and the micelles forming detergents Triton-100 and octyl glucoside to induced leakage of carboxyfluorescein (CF) from phosphatidylserine (PS), indicated that their specific activities are similar. The onset of dye leakage took place at a magainin concentration of approximately 3 mol %. Similar results were obtained with Triton-100 and octyl glucoside (90). The addition of magainin at high concentration to phospholipid bilayers resulted in the appearance of large water-filled bilayer disruptions (36). Solid state ^{31}P NMR spectroscopy detected the optical clearing of dense suspensions and the formation of isotropic phases (100).

Diastereomers of Non Cell Selective Cytolytic Peptides: A Novel Group of Non Hemolytic Antimicrobial Peptides

Native Pardaxin Acts Via "Barrel-stave" Mechanism and its Diastereomers Via 'Carpet Like" Mechanism

As previously mentioned, the non cell-selective lytic peptide pardaxin forms channels via a "barrel-stave" mechanism. Previous studies revealed that incorporation of D-amino acids into pardaxin preserved significant affinity to negatively charged phospholipid membranes, but reduced affinity to zwitterionic ones (101). Studies on the structure and organization of the resulting diastereomers in the membrane-bound state revealed that they had reduced α-helical structures, and could not penetrate and assemble in the membrane via a "barrel-stave" mechanism. However, whereas pardaxin was lytic to both bacteria and normal mammalian cells, the diastereomers were devoid of biological activity (102) including antimicrobial activity (Shai, Y. unpublished results). The low net positive charge (+1) of pardaxin, compared to the very high net positive charge, which is characteristic of antimicrobial peptides, might explain the findings that pardaxin diastereomers were devoid of antimicrobial activity. In line with this assumption, we have synthesized pardaxin analogue in which the C-terminal

Table 1. Sequences and designations of lytic peptides and their diastereomers.

Peptide Designation	Sequence [a,b]
TA-paradaxin [c]	GFFALIPKIISSPLFKTLLSAVGSALSSSGGQE-(NH$_2$)$_2$
[D]P^7-pardaxin	GFFALIPKIISSPLFKTLLSAVGSALSSSGGQE-(NH$_2$)$_2$
[D]L^{18}L^{19}-pardaxin	GFFALIPKIISSPLFKTLLSAVGSALSSSGGQE-(NH$_2$)$_2$
[D]P^{7}L^{18}L^{19}-pardaxin	GFFALIPKIISSPLFKTLLSAVGSALSSSGGQE-(NH$_2$)$_2$
melittin [c]	GIGAVLKVLTTGLPALISWIKRKRQQ-NH$_2$
[D]-V^{5,8}I^{17}K^{21}-melittin	GIGAVLKVLTTGLPALISWIKRKRQQ-NH$_2$
[D]L^{3,4,8,10}-K$_3$L$_9$	KLLLLLKLLLLK-NH$_2$
[D]L^{3,4,8,10}-K$_4$L$_8$	KLLLKLLLKLLK-NH$_2$
[D]L^{3,4,8,10}-K$_5$L$_7$	KLLLKLKLKLLK-NH$_2$
[D]L^{3,4,8,10}-K$_7$L$_5$	KKLLKLKLKLKK-NH$_2$

[a] Underlined and bold amino acids were substituted with their D-enantiomers.
[b] E-(NH$_2$)$_2$ stands for glutamic acid in which the two -COOH carboxylate groups were modified to two -CO-NH-CH$_2$CH$_2$-NH$_2$ groups by transamination with diamino ethanol.
[c] Underlined sequences designate the N- and C-helices respectively.

Table 2. Minimal inhibitory concentration (MIC) and hemolytic activity of cytolytic peptides and their diasteriomers.

	Minimal Inhibitory Concentration (μM)				
	E. coli (D21)	*A. calcoaceticus* (Ac11)	*B. megaterium* (Bm11)	*M. luteus* (ATCC 9341)	hemolysis at 50 μM
TA-pardaxin	3	2	0.8	2	100%
[D]P^7-pardaxin	10	5	1.2	5	100%
[D]L^{18}L^{19}-pardaxin	3.5	1.5	0.6	2.5	50%
[D]P^{7}L^{18}L^{19}-pardaxin	6	6	0.9	12.5	3%
melittin	5	20	0.3	ND	100%
[D]-V^{5,8}I^{17}K^{21}-melittin	12	12	0.8	ND	0%
K$_3$L$_9$	9	20	0.7	ND	30%
K$_4$L$_8$	3.5	4	0.4	ND	4%
K$_5$L$_7$	7	20	0.25	ND	0%
K$_7$L$_5$	80	200	1	ND	0%

N.D., not determined

glutamic acid was transaminated and the resulting analogue had a net charge +5 (termed TApar, Table 1) (96). TApar has an α-helical structure and is endowed with high antibacterial activity on Gram-negative and Gram-positive bacteria (Table 2), and with high hemolytic activity on human erythrocytes (Table 2). However, D-amino acids incorporated into TApar seem to reduce its α-helical structure in 40% trifluoroethanol/water as revealed by circular dichroism (CD) spectroscopy (96). This in turn reduced the hemolytic activity of the diastereomeric analogues (Table 2), which indicates the importance of this structure to the cytotoxicity of the peptide to mammalian cells. In contrast, high amphipathic α-helical structure seems not to be crucial for antibacterial activity, since with most of the bacteria tested there was no significant decrease in the antibacterial activity of the peptides (Table 2). The lack of a significant α-helical structure should prevent the diastereomers from inserting and forming a transmembrane pore, and hence a "barrel-stave" mechanism is not favored as their mode of action. The effect of the diastereomers is total lysis of the bacterial cell wall, as revealed by electron microscopy (60, 95 – 97, 103). The properties of the diastereomers suggest that they act via a "carpet-like" mechanism.

Melittin Diastereomers: The Role of Net Positive Charge and Helix Formation in Selective Membrane Binding and Insertion

To test the generality of the diastereomers concept, native melittin was used as a second example (97). Melittin is highly toxic to both bacteria and mammalian cells. The X-ray structure of tetrameric melittin crystallized from aqueous solution (104, 105) and NMR in methanolic solution (106) and dodecylphosphocholine micelles (107-110), indicates that the molecule consists of two α-helical segments (residues 2-10 and 13-26). These segments are connected by a hinge (11-12) to form a bent α-helical rod with the hydrophilic and hydrophobic sides facing opposite directions. D-amino acids were incorporated into the N- and C-terminal helices of melittin (Table 1). Whereas CD spectroscopy revealed very low α-helical structure in TFE/water, NMR study detected an amphipathic α-helix only in the C-terminal region of the diastereomer in TFE/water and methanol solutions and in dodecyl PC/dimyristoyl PG micelles (Sharon, M., Oren, Z., Shai, Y., and Englister, J, unpublished results). As was found with pardaxin, the disruption of the α-helical structure of melittin totally abolished the hemolytic activity of the diastereomers. However, unlike pardaxin, which has low net positive charge, the highly positively charged melittin diastereomer (+6) preserved the antibacterial activity (Table 2).

The interaction of pardaxin, melittin and their diastereomers with model phospholipid membranes was examined in order to elucidate the basis of the selective lytic ability of the diastereomers against bacteria. Pardaxin (29, 30) and melittin (111-115) bind strongly to both zwitterionic and negatively charged membranes. The α-helical regions of both peptides are predominantly hydrophobic and contain only few charged amino acids. Therefore, the binding forces between these helices and phospholipid membranes consist mainly of hydrophobic interactions between the hydrophobic faces of the amphipathic helices and the lipid constituent of the membrane. In contrast to native pardaxin and melittin, their diastereomers (96, 97) bind and permeate strongly only negatively charged phospholipids. Thus, electrostatic interactions appear to play an important role in the initial binding of the diastereomers to negatively-charged membranes.

In a recent study (116) the free energies of transfer of melittin and D-melittin from aqueous phase to the membrane interface of palmitoyloleoyl phosphatidylcholine (POPC) large unilamellar vesicles (LUV) were measured using equilibrium dialysis (54, 117). The partitioning of native melittin into phosphatidylcholine LUV was 20.9 ± 2.9 kJ/mol more favorable than the partitioning of D-melittin. The free energy difference between D-melittin and melittin was consistent with the expected low population of the unfolded membrane-bound state of native melittin, and with the hypothesis that secondary structure formation is mainly driven by a reduction in the free energy of partitioning of peptide bonds caused by hydrogen bonding (54). Rothemund *et al.*, (118) showed that double D-amino acid replacement at the center of an α-helix destabilized the secondary structure by 4.5 kJ/mol. Since more energy has to be invested in the folding of D-melittin it cannot readily form secondary structure in the surface, and lower its free energy of transfer into the membrane interface through the hydrogen bonding of secondary structure formation (116). Hence, the peptide may be released from the zwitterionic membranes before it forms the α-helical structure, which is crucial for insertion into the membrane interface.

Interestingly, despite the disruption of its α-helical structure, the all L-form and the diastereomer of either pardaxin or melittin have similar activities on negatively charged membranes. In the case of melittin, both the wild type and the diastereomer exhibited similar partition coefficients characterized by negative cooperativity, had the same potency to dissipate diffusion potentials, and penetrated to the same depth into the hydrophobic core of PC/PS vesicles, as seen in the tryptophan quenching experiments using brominated phospholipids (97).

Diastereomers of Linear and Short Model Peptides: Modulation of Selective Bacterial Lysis by the Balance Between Positive Charge and Hydrophobicity

To further investigate whether a balance between hydrophobicity and a net positive charge may be a sufficient criterion necessary for selective bacterial lysis, and to gain insight into the mechanism underlying this effect, four diastereomers of linear and short (12 amino acids long) model peptides composed of varying ratios of lysine-to-leucine were synthesized (Table 1) (103). The location of D-amino acids remained constant in all peptides which were constructed for maximum disruption of the α-helical structure.

Similarly to what has been found with the diastereomers of pardaxin and melittin, the short diastereomers exhibited potent antibacterial activity (Table 2) similar to or greater than that of native antibacterial peptides such as dermaseptin S, or the antibiotic drug tetracycline.

Moreover, their hemolytic activity was reduced or abolished (103). Interestingly, [D]-L^{3,4,8,10}-K$_3$L$_9$ had some hemolytic activity despite the lack of α-helical structure. However, increasing the positive charge drastically reduced the hemolytic activity while antibacterial activity was preserved.

The interaction of the model diastereomers with both negatively-charged and zwitterionic phospholipid membranes correlated with their cytolytic activity on erythrocytes and bacteria. Similarly to what has been found with native antimicrobial peptides and diastereomers of melittin and pardaxin, the non hemolytic diastereomers were able to bind strongly only to negatively-charged membranes. Therefore, electrostatic interactions between the positively-charged diastereomers and the negatively-charged phospholipid membranes seem to have an important role in initial interactions and selectivity, but biological activity appears to be driven by the hydrophobic interactions between the nonpolar amino acids and the hydrophobic core of the lipid bilayer. High hydrophobicity may force the immersion of the peptide into the hydrophobic core of the lipid bilayer, regardless of the phospholipid head group, thereby permeating membranes of eukaryotics and prokaryotics, as found in the case of [D]-L^{3,4,8,10}-K$_3$L$_9$. On the other hand high positive charge may prevent efficient immersion step, so that the peptide interacts predominantly with the negatively-charged lipid head group as found in the case of [D]-L^{3,4,8,10}-K$_7$L$_5$ (Table 1).

In contrast to these results, when the α-helical structure of magainin was disrupted by the introduction of three D-amino acids, the resulting diastereomer had no antibacterial activity (119), even though the net positive charge was not altered. Thus, an optimal balance that already exists between the α-helical structure, hydrophobicity and net positive charge of native magainin, allows selective antibacterial activity, and any change in one of these properties could cause a loss in the antibacterial activity of magainin. Contrastingly, hydrophobicity appears to play a major role in compensating for the disruption of α-helical structure by incorporation of D-amino acids in melittin, pardaxin and the diastereomers of linear and short model peptides.

Summary

Functional and structural studies with several antimicrobial α-helical peptides presented in this chapter are in agreement with some or all of the steps involved in the "carpet-like" model (left panel of Figure 2) rather than a "barrel-stave" model for their mode of action. Based on these studies, it is unlikely that the antimicrobial peptides described exert their activity by the formation of transmembrane pores/channels via the "barrel-stave" mechanism. This conclusion is based on the following observations:

(i)   At low peptide concentration the positively charged peptides are electrostatically adsorbed in an unordered form on the surface of the negatively charged phospholipid head groups. They consequently insert into the interface region, primarily in an helical form, parallel to the membrane surface. Occasionally a small number of pores are formed as fluctuation phenomena. Such pores are unstable, and their size and number fluctuate, consistent with the observed ion channel activity, and the leakage experiments. In contrast to channel forming peptides, most antimicrobial peptides do not self-associate in their membrane-bound state unless a threshold concentration has been reached (10, 12, 18, 50, 56). Furthermore, since antimicrobial peptides are highly positively charged amphipathic α-helices, it is energetically unlikely that they will remain monomeric within the hydrophobic core of the lipid membrane.

(ii) Transient holes in the membrane are formed and described by a torodial (or wormhole) model for pore formation, which differs from the barrel-stave model in that the lipid bends back on itself like the inside of a torus (35 – 37 and chapter by Matsuzaki in this monograph). These holes can be described as shown in Figure 2B left panel. The bending requires a lateral expansion in the head group region of the bilayer. Using X-ray diffraction experiments at concentrations below the critical concentration for lysis, the adsorption of magainin 2 in the head group region expands the membrane laterally inducing membrane-thinning, as shown by Ludtke *et al.* (120). Peptide monomers play the role of fillers in the expansion region, thereby stabilizing the pore. The formation of the transient pores was further supported by the large size of the water channel detected by neutron scattering (36). When a transient pore is closed, the participating peptide monomers will again adsorb in the head group region, but they may surface to either side of the membrane. This mechanism was supported by the findings of translocation of magainin across phospholipid bilayer (35). In addition, the peptide dramatically accelerated the flip-flop half-lives of the fluorescent lipids further supporting the pores (37). In addition, step B in the "carpet" model provides a mechanism for peptide translocation across the bilayer accounting for the ability of peptide molecules to reach the target cytoplasmic membrane of Gram-negative bacteria in a process which may be referred as "self-promoting uptake". This mechanism was proposed to be used by defensin macrophage cationic proteins 1, an α-helical cecropin-melittin hybrid peptide and the antimicrobial peptide indolicidin (38, 39, 41).

(iii) At high concentrations amphipathic α-helical peptides cause large-scale membrane permeation due to micellization, which leads to bacterial cell lysis. Antimicrobial peptides including magainin have been shown to impose positive curvature strain on lipid bilayers (80). In agreement with these findings, structural studies show that magainin 2 (84), cecropin A (85) and cecropin P1 (86) adopt a wedge-like shape that may be required for inducing positive curvature in the membrane. The final stages of bacterial lysis, due to changes in membrane curvature and micellization, require high concentration of peptides, as indicated by concentration dependent membrane permeation (42, 45, 46, 66, 88-91), and the stoichiometry of bacterial killing (7, 94). This model is further supported by the extensive bacterial lysis revealed by electron microscopy (49, 98).

Despite the differences in their sequence and structure, diastereomers and native antimicrobial peptides share significant homology in their selectivity, determined mainly by a net positive charge, depth of penetration, orientation, and membrane permeation. Hence, further studies on the properties of the diastereomers can extend our knowledge of the mechanism of native antimicrobial peptides.

The results obtained with pardaxin, melittin and diastereomers of short model peptides indicate that this novel family of antibacterial peptides act on the bacterial membrane via a "carpet" mechanism. Electrostatic interactions between the positively-charged diastereomers and the negatively-charged phospholipid membranes seem to have an important role in initial interactions and selectivity, but biological activity appears to be driven by the hydrophobic interactions between the nonpolar amino acids and the hydrophobic core of the lipid bilayer. Hydrophobicity appears to compensate for the α-helical structure disruption by incorporation of D-amino acids. By modulating the ratio between hydrophobicity and positive charge it is possible to determine the membrane selectivity and antimicrobial activity of the diastereomers. In contrast to these results, an optimal balance between the α-helical structure, hydrophobicity and net positive charge of native antimicrobial peptides, allows selective antimicrobial activity, and any change in one of these properties could cause a loss in their activity. The larger structural and sequence flexibility of the diastereomeric antimicrobial peptides may provide some advantages for the design of a repertoire of potent antimicrobial diastereomeric polypeptides for the treatment of infectious diseases.

Concluding Remarks

Despite the fact that linear antimicrobial peptides were developed by distant and diverse species, during millions of years of evolution, they share common motifs, i.e. they are linear, positively charged and have a high propensity to adopt amphipathic α-helical conformation in hydrophobic environments. The studies presented in this chapter have demonstrated the essential role of these properties in the different stages involved in membrane permeation according to the "carpet" model.

This review focused mainly on cecropins, dermaseptins and magainins as representatives of the amphipathic α-helical antimicrobial peptides. The fact that these peptides vary with regard to their length, amino acid composition, and net positive charge, but act via a common mechanism, may imply that other linear antimicrobial peptides (e.g. LL-37, PGLa, and bombinin-like peptides) exhibiting the same properties, also share the same "carpet" mechanism.

References

1. Boman, H. G. 1995. Peptide antibiotics and their role in innate immunity. Annu. Rev. Immun. 13: 61-92.
2. Steiner, H., Hultmark, D., Engstrom, A., Bennich, H. and Boman, H. G. 1981. Sequence and specificity of two antibacterial proteins involved in insect immunity. Nature. 292: 246-248.
3. Zasloff, M. 1987. Magainins, a class of antimicrobial peptides from Xenopus skin: isolation, characterization of two active forms, and partial cDNA sequence of a precursor. Proc. Natl. Acad. Sci. USA. 84: 5449-5453.
4. Mor, A., Nguyen, V. H., Delfour, A., Migliore-Samour, D. and Nicolas, P. 1991. Isolation, amino acid sequence, and synthesis of dermaseptin, a novel antimicrobial peptide of amphibian skin. Biochemistry. 30: 8824-8830.
5. Agerberth, B., Gunne, H., Odeberg, J., Kogner, P., Boman, H. G. and Gudmundsson, G. H. 1995. FALL-39, a putative human peptide antibiotic, is cysteine-free and expressed in bone marrow and testis. Proc. Natl. Acad. Sci. USA. 92: 195-199.
6. Andreu, D., Merrifield, R. B., Steiner, H. and Boman, H. G. 1985. N-terminal analogues of cecropin A: synthesis, antibacterial activity, and conformational properties. Biochemistry. 24: 1683-1688.
7. Steiner, H., Andreu, D. and Merrifield, R. B. 1988. Binding and action of cecropin and cecropin analogues: antibacterial peptides from insects. Biochim. Biophys. Acta. 939: 260-266.
8. Zasloff, M., Martin, B. and Chen, H. C. 1988. Antimicrobial activity of synthetic magainin peptides and several analogues. Proc. Natl. Acad. Sci. USA. 85: 910-913.
9. Segrest, J. P., De, L. H., Dohlman, J. G., Brouillette, C. G. and Anantharamaiah, G. M. 1990. Amphipathic helix motif: classes and properties. Proteins. 8: 103-117.
10. Gazit, E., Lee, W. J., Brey, P. T. and Shai, Y. 1994. Mode of action of the antibacterial cecropin B2: a spectrofluorometric study. Biochemistry. 33: 10681-10692.

11. Mor, A., Amiche, M. and Nicolas, P. 1994. Structure, synthesis, and activity of dermaseptin b, a novel vertebrate defensive peptide from frog skin: relationship with adenoregulin. Biochemistry. 33: 6642-6650.

12. Gazit, E., Boman, A., Boman, H. G. and Shai, Y. 1995. Interaction of the mammalian antibacterial peptide cecropin P1 with phospholipid vesicles. Biochemistry. 34: 11479-11488.

13. Schiffer, M. and Edmundson, A. B. 1967. Use of helical wheels to represent the structures of protein and to identify segments with helical potential. Biophys. J. 7: 121-135.

14. Bessalle, R., Kapitkovsky, A., Gorea, A., Shalit, I. and Fridkin, M. 1990. All-D-magainin: chirality, antimicrobial activity and proteolytic resistance. FEBS Lett. 274: 151-155.

15. Wade, D., Boman, A., Wahlin, B., Drain, C. M., Andreu, D., Boman, H. G. and Merrifield, R. B. 1990. All-D amino acid-containing channel-forming antibiotic peptides. Proc. Natl. Acad. Sci. USA. 87: 4761-4765.

16. Merrifield, E. L., Mitchell, S. A., Ubach, J., Boman, H. G., Andreu, D. and Merrifield, R. B. 1995. D-enantiomers of 15-residue cecropin A-melittin hybrids. Int. J. Pept. Protein Res. 46: 214-220.

17. Ehrenstein, G. and Lecar, H. 1977. Electrically gated ionic channels in lipid bilayers. Q. Rev. Biophys. 10: 1-34.

18. Pouny, Y., Rapaport, D., Mor, A., Nicolas, P. and Shai, Y. 1992. Interaction of antimicrobial dermaseptin and its fluorescently labeled analogues with phospholipid membranes. Biochemistry. 31: 12416-12423.

19. Meyer, P. and Reusser, F. 1967. A polypeptide antibacterial agent isolated from *Trichoderma viride*. Experientia. 23: 85-86.

20. Pandery, R. C., Cook, J. C. and Rinehart, K. L. 1977. High resolution and field desorption mass spectrometry studies and revised structure of alamethicin I and II. J. Am. Chem. Soc. 99: 8469-8483.

21. Fox, R. O., Jr. and Richards, F. M. 1982. A voltage-gated ion channel model inferred from the crystal structure of alamethicin at 1.5-A resolution. Nature. 300: 325-330.

22. Woolley, G. A. and Wallace, B. A. 1992. Model ion channels: gramicidin and alamethicin. J. Membr. Biol. 129: 109-136.

23. Sansom, M. S. 1993. Alamethicin and related peptaibols model ion channels. Eur. Biophys. J. 22: 105-124.

24. Cafiso, D. S. 1994. Alamethicin: a peptide model for voltage gating and protein-membrane interactions. Annu. Rev. Biophys. Biomol. Struct. 23: 141-165.

25. Rizzo, V., Stankowski, S. and Schwarz, G. 1987. Alamethicin incorporation in lipid bilayers: a thermodynamic study. Biochemistry. 26: 2751-2759.

26. Lazarovici, P., Primor, N. and Loew, L. M. 1986. Purification and pore-forming activity of two hydrophobic polypeptides from the secretion of the Red Sea Moses sole (*Pardachirus marmoratus*). J. Biol. Chem. 261: 16704-16713.

27. Shai, Y., Fox, J., Caratsch, C., Shih, Y. L., Edwards, C. and Lazarovici, P. 1988. Sequencing and synthesis of pardaxin, a polypeptide from the Red Sea Moses sole with ionophore activity. FEBS Lett. 242: 161-166.

28. Thompson, S. A., Tachibana, K., Nakanishi, K. and Kubota, I. 1986. Melittin-like peptides from the shark-repelling defense secretion of the sole *Pardachirus pavoninus*. Science. 233: 341-343.

29. Shai, Y. 1994. Pardaxin: channel formation by a shark repellant peptide from fish. Toxicology. 87: 109-129.

30. Shai, Y., Bach, D. and Yanovsky, A. 1990. Channel formation properties of synthetic pardaxin and analogues. J. Biol. Chem. 265: 20202-20209.

31. Oren, Z. and Shai, Y. 1996. A class of highly potent antibacterial peptides derived from pardaxin, a pore-forming peptide isolated from Moses sole fish *Pardachirus marmoratus*. Eur. J. Biochem. 237: 303-310.

32. Zagorski, M. G., Norman, D. G., Barrow, C. J., Iwashita, T., Tachibana, K. and Patel, D. J. 1991. Solution structure of pardaxin P-2. Biochemistry. 30: 8009-8017.

33. Rapaport, D. and Shai, Y. 1991. Interaction of fluorescently labeled pardaxin and its analogues with lipid bilayers. J. Biol. Chem. 266: 23769-23775.

34. Rapaport, D. and Shai, Y. 1992. Aggregation and organization of pardaxin in phospholipid membranes. A fluorescence energy transfer study. J. Biol. Chem. 267: 6502-6509.

35. Matsuzaki, K., Murase, O., Fujii, N. and Miyajima, K. 1995. Translocation of a channel-forming antimicrobial peptide, magainin 2, across lipid bilayers by forming a pore. Biochemistry. 34: 6521-6526.

36. Ludtke, S. J., He, K., Heller, W. T., Harroun, T. A., Yang, L. and Huang, H. W. 1996. Membrane pores induced by magainin. Biochemistry. 35: 13723-13728.

37. Matsuzaki, K., Murase, O., Fujii, N. and Miyajima, K. 1996. An antimicrobial peptide, magainin 2, induced rapid flip-flop of phospholipids coupled with pore formation and peptide translocation. Biochemistry. 35: 11361-11368.

38. Sawyer, J. G., Martin, N. L. and Hancock, R. E. 1988. Interaction of macrophage cationic proteins with the outer membrane of *Pseudomonas aeruginosa*. Infect. Immun. 56: 693-698.

39. Piers, K. L., Brown, M. H. and Hancock, R. E. 1994. Improvement of outer membrane-permeabilizing and lipopolysaccharide- binding activities of an antimicrobial cationic peptide by C-terminal modification. Antimicrob. Agents Chemother. 38: 2311-2316.

40. Piers, K. L. and Hancock, R. E. 1994. The interaction of a recombinant cecropin/melittin hybrid peptide with the outer membrane of *Pseudomonas aeruginosa*. Mol. Microbiol. 12: 951-958.

41. Falla, T. J., Karunaratne, D. N. and Hancock, R. E. W. 1996. Mode of action of the antimicrobial peptide indolicidin. J. Biol. Chem. 271: 19298-19303.

42. Westerhoff, H. V., Juretic, D., Hendler, R. W. and Zasloff, M. 1989. Magainins and the disruption of membrane-linked free-energy transduction. Proc. Natl. Acad. Sci. USA. 86: 6597-6601.

43. Brock, T. D. (1974). Biology of Microorganisms. 2nd edition, Prentice-Hall Inc., Englewood Cliffs, N.J.

44. Verkleij, A. J., Zwaal, R. F., Roelofsen, B., Comfurius, P., Kastelijn, D. and Deenen, L. V. 1973. The asymmetric distribution of phospholipids in the human red cell membrane. A combined study using phospholipases and freeze-etch electron microscopy. Biochim. Biophys. Acta. 323: 178-193.

45. Matsuzaki, K., Harada, M., Handa, T., Funakoshi, S., Fujii, N., Yajima, H. and Miyajima, K. 1989. Magainin 1-induced leakage of entrapped calcein out of negatively-charged lipid vesicles. Biochim. Biophys. Acta. 981: 130-134.

46. Williams, R. W., Starman, R., Taylor, K. M., Gable, K., Beeler, T., Zasloff, M. and Covell, D. 1990. Raman spectroscopy of synthetic antimicrobial frog peptides magainin 2a and PGLa. Biochemistry. 29: 4490-4496.

47. Gomes, A. V., de Waal, A., Berden, J. A. and Westerhoff, H. V. 1993. Electric potentiation, cooperativity, and synergism of magainin peptides in protein-free liposomes. Biochemistry. 32: 5365-5372.

48. Matsuzaki, K., Murase, O. and Miyajima, K. 1995. Kinetics of pore formation by an antimicrobial peptide, magainin 2, in phospholipid bilayers. Biochemistry. 34: 12553-12559.

49. Matsuzaki, K., Sugishita, K., Harada, M., Fujii, N. and Miyajima, K. 1997. Interactions of an antimicrobial peptide, magainin 2, with outer and inner membranes of Gram-negative bacteria. Biochim. Biophys. Acta. 1327: 119-130.

50. Strahilevitz, J., Mor, A., Nicolas, P. and Shai, Y. 1994. Spectrum of antimicrobial activity and assembly of dermaseptin-b and its precursor form in phospholipid membranes. Biochemistry. 33: 10951-10960.

51. Latal, A., Degovics, G., Epand, R. F., Epand, R. M. and Lohner, K. 1997. Structural aspects of the interaction of peptidyl-glycylleucine- carboxyamide, a highly potent antimicrobial peptide from frog skin, with lipids. Eur. J. Biochem. 248: 938-946.

52. Jacobs, R. E. and White, S. H. 1989. The nature of the hydrophobic binding of small peptides at the bilayer interface:Implications for the insertion of transbilayer helices. Biochemistry. 28: 3421-3437.

53. Sui, S. F., Wu, H., Guo, Y. and Chen, K. S. 1994. Conformational changes of melittin upon insertion into phospholipid monolayer and vesicle. J. Biochem. (Tokyo). 116: 482-487.

54. Wimley, W. C. and White, S. H. 1996. Experimentally determined hydrophobicity scale for proteins at membrane interfaces. Nature Struct. Biol. 3: 842-848.

55. Wiener, M. C. and White, S. H. 1992. Structure of a fluid dioleoyl phosphatidylcholine bilayer determined by joint refinement of x-ray and neutron diffraction data. III. Complete structure. Biophys. J. 61: 434-447.

56. Gazit, E., Miller, I. R., Biggin, P. C., Sansom, M. S. and Shai, Y. 1996. Structure and orientation of the mammalian antibacterial peptide cecropin P1 within phospholipid membranes. J. Mol. Biol. 258: 860-870.

57. Chattopadhyay, A. and London, E. 1987. Parallax method for direct measurement of membrane penetration depth utilizing fluorescence quenching by spin-labeled phospholipids. Biochemistry. 26: 39-45.

58. Rajarathnam, K., Hochman, J., Schindler, M. and Ferguson-Miller, S. 1989. Synthesis, location, and lateral mobility of fluorescently labeled ubiquinone 10 in mitochondrial and artificial membranes. Biochemistry. 28: 3168-76.

59. Gazit, E. and Shai, Y. 1993. Structural and functional characterization of the alpha 5 segment of *Bacillus thuringiensis* delta-endotoxin. Biochemistry. 32: 3429-3436.

60. Oren, Z., Lerman, J. C., Gudmundsson, G. H., Agerberth, B. and Shai, Y. 1999. Structure and organization of the human antimicrobial peptide LL-37 in phospholipid membranes: relevance to the molecular basis for its non-cell selective activity. Biochem. J. 341: 501-513.

61. Bechinger, B., Kim, Y., Chirlian, L. E., Gesell, J., Neumann, J. M., Montal, M., Tomich, J., Zasloff, M. and Opella, S. J. 1991. Orientations of amphipathic helical peptides in membrane bilayers determined by solid-state NMR spectroscopy. J. Biomol. NMR. 1: 167-173.

62. Bechinger, B., Zasloff, M. and Opella, S. J. 1992. Structure and interactions of magainin antibiotic peptides in lipid bilayers: a solid-state nuclear magnetic resonance investigation. Biophys. J. 62: 12-14.

63. Bechinger, B., Zasloff, M. and Opella, S. J. 1993. Structure and orientation of the antibiotic peptide magainin in membranes by solid-state nuclear magnetic resonance spectroscopy. Protein Sci. 2: 2077-2084.

64. Hirsh, D. J., Hammer, J., Maloy, W. L., Blazyk, J. and Schaefer, J. 1996. Secondary structure and location of a magainin analogue in synthetic phospholipid bilayers. Biochemistry. 35: 12733-12741.

65. Matsuzaki, K., Murase, O., Tokuda, H., Funakoshi, S., Fujii, N. and Miyajima, K. 1994.

Orientational and aggregational states of magainin 2 in phospholipid bilayers. Biochemistry. 33: 3342-3349.

66. Matsuzaki, K., Harada, M., Funakoshi, S., Fujii, N. and Miyajima, K. 1991. Physico-chemical determinants for the interactions of magainins 1 and 2 with acidic lipid bilayers. Biochim. Biophys. Acta. 1063: 162-170.

67. Matsuzaki, K. and Seelig, J. 1995 . In Peptide Chemistry 1994. Ohno, M., ed. Protein Research Foundation, Osaka. p. 129-132.

68. Shaw, N. 1974. Lipid composition as a guide to the classification of bacteria. Adv. Appl. Microbiol. 17: 63-108.

69. Brauner, J. W., Mendelsohn, R. and Prendergast, F. G. 1987. Attenuated total reflectance Fourier transform infrared studies of the interaction of melittin, two fragments of melittin, and delta-hemolysin with phosphatidylcholines. Biochemistry. 26: 8151-8158.

70. Frey, S. and Tamm, L. K. 1991. Orientation of melittin in phospholipid bilayers. A polar-ized attenuated total reflection infrared study. Biophys. J. 60: 922-930.

71. Gazit, E., Rocca, P. L., Sansom, M. S. P. and Shai, Y. 1998. The structure and organiza-tion within the membrane of the helices composing the pore-forming domain of *Bacillus thuringiensis* δ- endotoxin are consistent with an "umbrella-like" structure of the pore. Proc. Natl. Acad. Sci. USA. 95: 12289-12294.

72. Arkin, I. T., Rothman, M., Ludlam, C. F., Aimoto, S., Engelman, D. M., Rothschild, K. J. and Smith, S. O. 1995. Structural model of the phospholamban ion channel complex in phospholipid membranes. J. Mol. Biol. 248: 824-834.

73. Tatulian, S. A., Jones, L. R., Reddy, L. G., Stokes, D. L. and Tamm, L. K. 1995. Second-ary structure and orientation of phospholamban reconstituted in supported bilayers from polarized attenuated total reflection FTIR spectroscopy. Biochemistry. 34: 4448-4456.

74. Goormaghtigh, E., De, M. J., Szoka, F., Cabiaux, V., Parente, R. A. and Ruysschaert, J. M. 1991. Secondary structure and orientation of the amphipathic peptide GALA in lipid structures. An infrared-spectroscopic approach. Eur. J. Biochem. 195: 421-429.

75. Epand, R. M., Shai, Y., Segrest, J. P. and Anantharamaiah, G. M. 1995. Mechanisms for the modulation of membrane bilayer properties by amphipathic helical peptides. Biopolymers. 37: 319-338.

76. Tytler, M. E., Segrest, J. P., Epand, R. M., Nie, S. Q., Epand, R. F., Mishra, V. K., Venkatachalapathi, Y. V. and Anantharamaih, G. M. 1993. Reciprocal effects of apolipoprotein and lytic peptide analogs on membranes. J. Biol. Chem. 268: 22112-22118.

77. Polozov, I. V., Polozova, A. I., Tytler, E. M., Anantharamaiah, G. M., Segrest, J. P., Woolley, G. A. and Epand, R. M. 1997. Role of lipids in the permeabilization of membranes by class L amphipathic helical peptides. Biochemistry. 36: 9237-9245.

78. Epand, R. M., Gawish, A., Iqbal, M., Gupta, K. B., Chen, C. H., Segrest, J. P. and Anantharamaiah, G. M. 1987. Studies of synthetic peptide analogs of the amphipathic helix. Effect of charge distribution, hydrophobicity, and secondary structure on lipid association and lecithin:cholesterol acyltransferase activation. J. Biol. Chem. 262: 9389-9396.

79. Matsuzaki, K., Nakayama, M., Fukui, M., Otaka, A., Funakoshi, S., Fujii, N., Bessho, K. and Miyajima, K. 1993. Role of disulfide linkages in tachyplesin-lipid interactions. Bio-chemistry. 32: 11704-11710.

80. Matsuzaki, K., Sugishita, K., Ishibe, N., Ueha, M., Nakata, S., Miyajima, K. and Epand, R. M. 1998. Relationship of membrane curvature to the formation of pores by magainin 2 (In Process Citation). Biochemistry. 37: 11856-11863.

81. Hope, M. J. and Cullis, P. R. 1980. Effects of divalent cations and pH on phosphatidylserine model membranes: a 31P NMR study. Biochem. Biophys. Res. Commun. 92: 846-852.

82. Farren, S. B., Hope, M. J. and Cullis, P. R. 1983. Polymorphic phase preferences of phosphatidic acid: A 31P and 2H NMR study. Biochem. Biophys. Res. Commun. 111: 675-682.

83. Seddon, J. M., Kaye, R. D. and Marsh, D. 1983. Induction of the lamellar-inverted hexagonal phase transition in cardiolipin by protons and monovalent cations. Biochim. Biophys. Acta. 734: 347-352.

84. Gesell, J., Zasloff, M. and Opella, S. J. 1997. Two-dimensional 1H NMR experiments show that the 23-residue magainin antibiotic peptide is an alpha-helix in dodecylphosphocholine micelles, sodium dodecylsulfate micelles, and trifluoroethanol/water solution. J. Biomol. NMR. 9: 127-135.

85. Holak, T. A., Engstrom, A., Kraulis, P. J., Lindeberg, G., Bennich, H., Jones, T. A., Gronenborn, A. M. and Clore, G. M. 1988. The solution conformation of the antibacterial peptide cecropin A: a nuclear magnetic resonance and dynamical simulated annealing study. Biochemistry. 27: 7620-7629.

86. Sipos, D., Andersson, M. and Ehrenberg, A. 1992. The structure of the mammalian antibacterial peptide cecropin P1 in solution, determined by proton-NMR. Eur. J. Biochem. 209: 163-169.

87. Fink, J., Boman, A., Boman, H. G. and Merrifield, R. B. 1989. Design, synthesis and antibacterial activity of cecropin-like model peptides. Int. J. Pept. Protein Res. 33: 412-421.

88. Juretic, D., Chen, H. C., Brown, J. H., Morell, J. L., Hendler, R. W. and Westerhoff, H. V. 1989. Magainin 2 amide and analogues. Antimicrobial activity, membrane depolarization and susceptibility to proteolysis. FEBS Lett. 249: 219-223.

89. Cruciani, R. A., Baker, J. L., Zasloff, M., Chen, H. C. and Colamonici, O. 1991. Antibiotic magainins exert cytolytic activity against transformed cell lines through channel formation. Proc. Natl. Acad. Sci. USA. 88: 3792-3796.

90. Grant, E., Jr., Beeler, T. J., Taylor, K. M., Gable, K. and Roseman, M. A. 1992. Mechanism of magainin 2a induced permeabilization of phospholipid vesicles. Biochemistry. 31: 9912-9918.

91. Juretic, D., Hendler, R. W., Kamp, F., Caughey, W. S., Zasloff, M. and Westerhoff, H. V. 1994. Magainin oligomers reversibly dissipate µH in cytochrome oxidase liposomes. Biochemistry. 33: 4562-4570.

92. Jo, E., Blazyk, J. and Boggs, J. M. 1998. Insertion of magainin into the lipid bilayer detected using lipid photolabels. Biochemistry. 37: 13791-13799.

93. Ghosh, J. K., Shaool, D., Guillaud, P., Ciceron, L., Mazier, D., Kustanovich, I., Shai, Y. and Mor, A. 1997. Selective cytotoxicity of dermaseptin S3 toward intraerythrocytic *Plasmodium falciparum* and the underlying molecular basis. J. Biol. Chem. 272: 31609-31616.

94. Silvestro, L., Gupta, K., Weiser, J. N. and Axelsen, P. H. 1997. The concentration-dependent membrane activity of cecropin A. Biochemistry. 36: 11452-11460.

95. Oren, Z., Hong, J. and Shai, Y. 1999. A comparative study on the structure and function of a cytolytic α- helical peptide and its antimicrobial β- sheet diastereomer. Eur. J. Biochem. 259: 360-369.

96. Shai, Y. and Oren, Z. 1996. Diastereoisomers of cytolysins, a novel class of potent antibacterial peptides. J. Biol. Chem. 271: 7305-7308.

97. Oren, Z. and Shai, Y. 1997. Selective lysis of bacteria but not mammalian cells by diastereomers of melittin: structure-function study. Biochemistry. 36: 1826-1835.

98. Lockey, T. D. and Ourth, D. D. 1996. Formation of pores in *Escherichia coli* cell membranes by a cecropin isolated from hemolymph of *Heliothis virescens* larvae. Eur. J. Biochem. 236: 263-271.

99. Chan, S. C., Yau, W. L., Wang, W., Smith, D. K., Sheu, F. S. and Chen, H. M. 1998. Microscopic observations of the different morphological changes caused by antibacterial peptides on *Klebsiella pneumoniae* and HL-60 leukumeia cells. J. Peptide Sci. 4: 413-425.

100. Bechinger, B. 1997. Structure and functions of channel-forming peptides: magainins, cecropins, melittin and alamethicin. J. Membr. Biol. 156: 197-211.

101. Pouny, Y. and Shai, Y. 1992. Interaction of D-amino acid incorporated analogues of pardaxin with membranes. Biochemistry. 31: 9482-9490.

102. Abu, R. S., Bloch, S. E., Shohami, E., Trembovler, V., Shai, Y., Weidenfeld, J., Yedgar, S., Gutman, Y. and Lazarovici, P. 1998. Pardaxin, a new pharmacological tool to stimulate the arachidonic acid cascade in PC12 cells. J. Pharmacol. Exp. Ther. 287: 889-896.

103. Oren, Z., Hong, J. and Shai, Y. 1997. A repertoire of novel antibacterial diastereomeric peptides with selective cytolytic activity. J. Biol. Chem. 272: 14643-14649.

104. Terwilliger, T. C. and Eisenberg, D. 1982. The structure of melittin. I. Structure determination and partial refinement. J. Biol. Chem. 257: 6010-6015.

105. Terwilliger, T. C. and Eisenberg, D. 1982. The structure of melittin. II. Interpretation of the structure. J. Biol. Chem. 257: 6016-6022.

106. Bazzo, R., Tappin, M. J., Pastore, A., Harvey, T. S., Carver, J. A. and Campbell, I. D. 1988. The structure of melittin. A 1H-NMR study in methanol. Eur. J. Biochem. 173: 139-146.

107. Brown, L. R. and Wuthrich, K. 1981. Melittin bound to dodecylphosphocholine micelles. H-NMR assignments and global conformational features. Biochim. Biophys. Acta. 647: 95-111.

108. Brown, L. R., Braun, W., Kumar, A. and Wüthrich, K. 1982. High resolution nuclear magnetic resonance studies of the conformation and orientation of melittin bound to a lipid-water interface. Biophys. J. 37: 319-328.

109. Inagaki, F., Shimada, I., Kawaguchi, K., Hirano, M., Terasawa, I., Ikura, T. and Go, N. 1989. Structure of melittin bound to perdeuterated dodecylphosphocholine micelles as studied by 2D-NMR and distance geometry calculation. Biochemistry. 28: 5985-5991.

110. Ikura, T., Go, N. and Inagaki, F. 1991. Refined structure of melittin bound to perdeuterated dodecylphosphocholine micelles as studied by 2D-NMR and distance geometry calculation. Proteins. 9: 81-89.

111. Dufourcq, J. and Faucon, J. F. 1977. Intrinsic fluorescence study of lipid-protein interactions in membrane models. Biochim. Biophys. Acta. 467: 1-11.

112. Batenburg, A. M., Hibbeln, J. C. and de, K. B. 1987. Lipid specific penetration of melittin into phospholipid model membranes. Biochim. Biophys. Acta. 903: 155-165.

113. Batenburg, A. M., van, E. J., Leunissen, B. J., Verkleij, A. J. and de, K. B. 1987. Interaction of melittin with negatively charged phospholipids: consequences for lipid organization. Febs Lett. 223: 148-154.

114. Batenburg, A. M., van, E. J. and de, K. B. 1988. Melittin-induced changes of the macroscopic structure of phosphatidylethanolamines. Biochemistry. 27: 2324-2331.

115. Beschiaschvili, G. and Seelig, J. 1990. Melittin binding to mixed phosphatidylglycerol/phosphatidylcholine membranes. Biochemistry. 29: 52-58.

116. Ladokhin, A. S. and White, S. H. 1999. Folding of amphiphatic alpha-helices on membranes:energetics of helix formation by melittin. J. Mol. Biol. 285: 1363-1369.

117. Wimley, W. C. and White, S. H. 1993. Quantitation of electrostatic and hydrophobic membrane interactions by equilibrium dialysis and reverse-phase HPLC. Anal. Biochem. 213: 213-217.
118. Rothemund, S., Krause, E., Beyermann, M., Bienert, M., Sykes, B. D. and Sonnichsen, F. D. 1996. Assignment of the helical structure in neuropeptide Y by HPLC studies of methionine replacement analogues and 1H-NMR spectroscopy. Biopolymers. 39: 207-219.
119. Chen, H. C., Brown, J. H., Morell, J. L. and Huang, C. M. 1988. Synthetic magainin analogues with improved antimicrobial activity. FEBS Lett. 236: 462-466.
120. Ludtke, S., He, K. and Huang, H. 1995. Membrane thinning caused by magainin 2. Biochemistry. 34: 16764-16769.

From: *Development of Novel Antimicrobial Agents: Emerging Strategies*
ISBN 1-898486-23-9 © 2001 Horizon Scientific Press, Wymondham, UK.

15

Molecular Interactions Involved in Bactericidal Activities of Lantibiotics

Ulrike Pag and Hans-Georg Sahl

Abstract

Lantibiotics are produced by a wide range of and mainly act on Gram-positive bacteria. Based on their structures and their mode of action they are currently divided into two distinct groups. The elongated amphiphilic, screw-shaped peptides of the type A-lantibiotics act primarily by pore formation in the bacterial membrane. Additional effects such as inhibition of spore outgrowth and triggering of autolytic enzymes have been observed. The small, globular type-B lantibiotics can be further subdivided: the peptides of the cinnamycin-subtype bind to specific phospholipids and thereby inhibit phospholipases and other enzyme functions; mersacidin and actagardine inhibit the bacterial cell wall biosynthesis by trapping the membrane-bound peptidoglycan precursor lipid II. Recent results indicate that some pore-forming lantibiotics, e.g. nisin and epidermin, also bind to lipid II, using it as a docking molecule for subsequent pore formation. These results demonstrate that at least some lantibiotics, in contrast to many defense peptides, bind with high specificity to particular membrane components. On the other hand, binding to the same membrane target (lipid II) may result in completely different modes of action, i.e. pore formation with nisin and inhibition of cell wall biosynthesis with mersacidin.

Introduction

Lantibiotics contain unusually modified amino acids, particularly dehydroamino acids, and the thioether amino acids lanthionine (Lan) and 3-methyllanthionine (MeLan). The presence of Lan and MeLan residues leads to the formation of characteristic, intra-chain ring structures. Unlike "classical" peptide antibiotics, lantibiotics are ribosomally synthesized as precursur peptides which are subsequently converted into the biologically active peptide by posttranslational modifications. Regarding the enzymes that take part in modifications as well as export and processing, two different classes of lantibiotics can be distinguished (1). Class I lantibiotics are modified by two enzymes, LanB and LanC, catalyzing the dehydration of hydroxyamino acids and the formation of the thioether rings, respectively; proteolytic processing and export from the producing cells are performed by dedicated proteases and ABC-transporters. Class II lantibiotics are modified by only one enzyme, LanM, and possess hybrid ABC-transporters with an additional proteolytic domain at its N-terminus. According to a proposal of Jung (2), based on the structure and mode of action, the lantibiotics are grouped into the elongated, membrane-depolarizing type-A peptides and the globular, enzyme inhibitory type-B lantibiotics. However, in the last decade a significant number of new lantibiotics with intermediate features have been characterized, making a categorization on structural and functional features more difficult. In this chapter different lantibiotics will be discussed with respect to their modes of action.

Type-A Lantibiotics

Pore Formation

Type-A lantibiotics are typically active against a relatively restricted range of Gram-positive bacterial strains. Gram-negative bacteria are only affected when the outer membrane is disrupted e.g. by ion chelators such as EDTA or citrate (3, 4).

The first report on the mechanism of action of the prototype type-A lantibiotic nisin dates back to 1960 when Ramseier (5) observed leakage of UV-absorbing intracellular compounds from treated sensitive cells and suggested a detergent effect (5). Subsequent experiments showed that nisin kills bacterial cells primarily by interference with energy transduction occurring at the cytoplasmic membrane (6, 7). Addition of nisin and other type-A peptides immediately inhibits biosynthesis of macromolecules such as DNA, RNA, protein and polysaccharides (8). Furthermore, bacterial cells are unable to actively take-up amino acids and become leaky for inorganic ions and small metabolites (9). Deenergetized cells are not susceptible to the action of lantibiotics; a membrane potential of at least 50-100 mV is required for the activity of the peptides (3, 10, 11). At pH 5.5, nisin was shown to be active without a potential; under these conditions ΔpH can substitute for $\Delta \psi$ as a driving force for the membrane depolarizing activity of nisin (12). The concept of energization-dependent activity was further supported by experiments with cytoplasmic membrane vesicles. Such membranes are only affected after pre-energization with an appropriate electron donor (3, 7).

Conductance measurements using artificial bilayer membranes (black lipid membranes, BML) were in good agreement with the results obtained with intact cells and physiological membranes (3, 11, 13-15). There was no macroscopic membrane conductivity below a certain threshold potential; the minimum voltage was about 50 mV for epidermin or gallidermin, about 80 mV for nisin, subtilin and Pep5, and around 100 mV for streptococcin A-FF22. In addition, the orientation of the potential has proven to be of importance for some peptides. Pep5 and nisin only function with a *trans*-negative membrane potential, while others such as subtilin, epidermin and streptococcin A-FF22 are active with both voltage orientations. In all cases the current/voltage curves showed a strong hysteresis, indicating that rather the formation than the maintenance of the pore may be the energy-requiring step. Furthermore, analysis of single channels formed in BMLs allowed the determination of the diameter and lifetime of the lantibiotic channels. Streptococcin A-FF22 pores have the shortest lifetimes of only a few milliseconds and a diameter of 0.5-0.6 nm (11). Calculated pore diameters are about 1 nm for nisin and Pep5 and about 2 nm for subtilin; the pores are more stable with a lifetime in the milliseconds range (3, 13, 14). Epidermin and gallidermin pores have lifetimes up to 30 s and their diameter appears to increase with the applied potential (15).

A

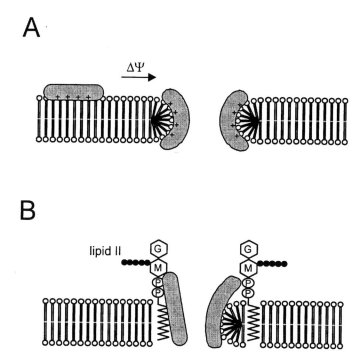

B

Figure 1. Models for the formation of pores by cationic amphiphilic lantibiotics. A: Wedge model as proposed by Moll *et al*. (18) for nisin. B: Model taking into account the demonstrated binding of epidermin and nisin to the cell wall precursor lipid II. Any molecular details on the binding and on the orientation of lipid II in membranes are currently unknown; thus, mechanistic features on the pore formation should not be deduced from the picture.

Based on these results as well as conformational data derived from NMR studies applying nisin in the presence of membrane-mimicking micelles (16, 17), a "wedge model" for pore formation by lantibiotics has been proposed (17, 18) (Figure 1 A). The type-A lantibiotics are rather flexible in aqueous solution and defined structural elements are only identified in small thioether rings (19, 20). Upon contact with the membrane the peptides adopt a helical amphiphilic conformation with the charged residues aligned to one face of the helix and the hydrophobic residues aligned to the other. The cationic peptides interact with the phospholipid head groups by ionic forces causing a locally disturbed bilayer structure, while the hydrophobic side inserts into the membrane (16, 17, 21). Studies with nisin demonstrated that the C-terminal region (22, 23) as well as the overall negative surface charge of the membrane (21, 24-26) are important for binding and pore formation. As lantibiotics are too small peptides to transverse the membrane several times, it is assumed that several molecules have to associate with the membrane in order to form a pore (18). This process leads to dissipation of the membrane potential and promotes a rapid efflux of small metabolites such as amino acids or ATP, which in turn immediately stops all cellular biosynthetic processes (6-9, 15).

Since type-A lantibiotics act on artificial membranes, binding to specific receptors in the cell membrane is not a prerequisite for activity (27). Nevertheless, for nisin a finite number of binding sites and specific antagonization of nisin activity by the inactive N-terminal nisin fragment 1-12 was observed (28), indicating that a particular binding site may be blocked by the fragment. With that respect it is important to recall publications by Linnett & Strominger (29) and Reisinger (30), who reported that nisin inhibits peptidoglycan biosynthesis, and that it binds to the membrane-bound peptidoglycan precursor undecaprenylpyrophosphoryl-

MurNAc(pentapeptide)-GlcNAc, the so-called lipid II. Recent studies applying liposomes supplemented with lipid II demonstrated that nisin and epidermin use lipid II as a docking molecule for specific binding to the bacterial membrane (31) (Figure 1 B). In contrast to a specific receptor molecule, lipid II is not essential for the activity of nisin and epidermin, since both peptides also form pores in the absence of lipid II; however, binding to lipid II seems to significantly decrease the energy necessary for the insertion of the peptides into the membrane. The specific binding of nisin and epidermin to lipid II is presumed to include both the lipid moiety and the disaccharide unit or the peptide side-chain. A pore formation model including a specific docking molecule for binding of the lantibiotics (Figure 1B) may help to explain the striking difference in sensitivity that can be observed even among different strains of a particular bacterial species.

Secondary Effects

Pore formation certainly represents the primary killing activity of nisin and related lantibiotics; under favourable conditions, membranes are depolarized and efflux of small solutes is completed within seconds. Moreover, nisin and Pep5 have been shown to induce autolysis in staphylococci, resulting in massive cell wall degradation, particularly in the area of the septa between dividing daughter cells (32-34). The peptides are able to release two cell wall hydrolysing enzymes, an N-acetylmuramoyl-L-alanine amidase and an N-acetylglucosaminidase, which are strongly cationic proteins and which bind to the cell wall via electrostatic interactions with the negatively charged teichoic and lipoteichoic acids; tight binding to these polymers keeps the autolysins inactive. The cationic peptides displace the enzymes from the cell wall intrinsic inhibitors by a cation exchange-like process, resulting in apparent enzyme activation and rapid cell lysis. Furthermore, nisin and subtilin inhibit the outgrowth of bacterial spores by a mechanism that essentially involves the Dha residue in position 5 (35, 36).

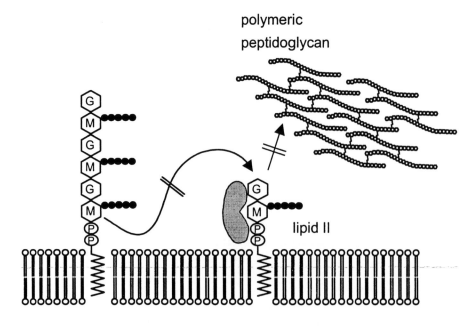

Figure 2. Model for the inhibition of peptidoglycan biosynthesis by the lantibiotics mersacidin and actagardine. The lantibiotics bind to lipid II at the disaccharid-pyrophosphate moiety, thereby blocking the polymerization process (45).

Recently, nisin was shown to autoregulate its own biosynthesis via signal transduction by acting as the extracellular signal for the two-component regulatory system NisKR (37). Through low-level production by uninduced cells nisin accumulates to a threshold concentration and then binds to the membrane bound sensor kinase NisK, which, in response, autophosphorylates a His residue in its intracellular domain. This phosphate residue is then transferred to a conserved Asp residue of the intracellular response regulator NisR, which, after a conformational alteration, activates the transcription of the biosynthetic gene cluster.

Type-B Lantibiotics

With respect to a potential chemotherapeutical application, the type-B lantibiotics mersacidin and actagardine are the most promising antimicrobial agents amongst the lantibiotics. Their activity spectrum covers a variety of Gram-positive bacteria with actagardine being most effective against streptococci an obligate anaerobes (38, 39), while mersacidin is almost equally active against staphylococci, streptococci, bacilli, clostridiae, corynebacteriae, peptostreptococci, and *Propionibacterium acnes* (40-42). Gram-negative bacteria are not susceptible, indicating that the peptides cannot pass the outer membrane of bacteria; neither can they penetrate the eukaryotic cell membrane, since *Listeria monocytogenes*, while being susceptible *in vitro,* is not affected when persisting within eukaryotic cells (43). Although both lantibiotics in vitro exhibit only moderate MIC values, mersacidin has attracted recent attention due to its significant in vivo activity. It effectively cured systemic staphylococcal infections in mice (even those caused by methicillin-resistant *Staphylococcus aureus* (MRSA) strains) (40), as well as subcutaneous staphylococcal abcesses in rats (41). In both cases the lantibiotic equalled or even exceeded the activity of vancomycin.

Mersacidin and actagardine were shown to interfere with the cell wall biosynthesis; they selectively inhibited the incorporation of glucose and D-alanine into cell wall material of *Staphylococcus simulans* 22, whereas DNA, RNA and protein synthesis proceeded unhindered. Recent studies demonstrated that both peptides inhibit peptidoglycan biosynthesis at the level of transglycosylation by forming a complex with the membrane-bound peptidoglycan precursor lipid II (44, 45) (Figure 2). The binding of $[^{14}C]$mersacidin to growing cells, as well as to isolated membranes capable of *in vitro* peptidoglycan synthesis, was strictly dependent on the availability of lipid II, and antibiotic inhibitors of lipid II formation strongly interferred with binding of mersacidin. Furthermore, labeled mersacidin associated tightly with micelles formed from purified and labeled lipid II, and the addition of isolated lipid II to the culture broth efficiently antagonized the bactericidal activity of mersacidin. The molecular target site of mersacidin and actagardine on lipid II differs from that of the glycopeptide antibiotic vancomycin, which binds to lipid II via the C-terminal D-Ala-D-Ala of the pentapeptide side-chain; in contrast, the two lantibiotics rather interact with lipid II by the disaccharide-pyrophosphate moiety (45). Mersacidin and actagardine contain one ring structure that has been almost completely conserved in both molecules, indicating its importance for activity (40, 44, 46).

Lantibiotics of the cinnamycin subtype were found in the course of various screenings and were reisolated several times under different names (for a review see 47). Both duramycin and cinnamycin have demonstrable antimicrobial activity against strains of *Bacillus* species. Treatment of cells with these lantibiotics caused increases in the permeability of the cytoplasmic membrane (48) and impaired both ATP-dependent protein translocation (49) and ATP-dependent calcium uptake (50). In addition, duramycin has also been shown to inhibit chloride transport (51), sodium and potassium ATPases (52) and the proton pump of clathrin-coated vesicles (53). Interestingly, duramycin resistant *B. subtilis* or *B. firmus* have been shown to have membranes which contain an altered lipid composition compared to that of their duramycin-sensitive counterparts; phosphatidylethanolamine is no longer found in the cytoplasmic membrane (50, 54) and, at least in the case of *B. firmus*, is replaced by its plasmalogen form, plasmenylethanolamine (55). In addition to inhibiting the growth of bacilli, duramycin also inhibits a number of metabolic properties of isolated mitochondria and specifically increases the permeability of the inner membrane, the major component of which is phosphatidylethanolamine (56). Similarly, cinnamycin has been shown to induce hemolysis in erythrocytes and phosphatidylethanolamine-containing liposomes (57). Furthermore, the specific binding of the cinnamycin subtype of type-B lantibiotics to phosphatidylethanolamine results in an inhibition of phospholipase A_2, an important enzyme involved in the biosynthesis of prostaglandins and leucotrienes (58, 59, 60, 61). Thus, these type-B lantibiotics may have potential therapeutic applications as immunoregulators.

Acknowledgement

The authors work was supported by the Deutsche Forschungsgemeinschaft (Sa 292/8-1, 8-2), by the Bundesministerium für Forschung und Technologie BMBF (01 KI 9705/8) and by the BONFOR-Programme of the Medical Faculty of the University of Bonn.

References

1. Sahl, H.-G. 1991. Pore formation in bacterial membranes by cationic lantibiotics. In: Nisin and Novel Lantibiotics. G. Jung and H.-G. Sahl, eds. Escom, Leiden.
2. Jung, G. 1991. Lantibiotics - ribosomally synthesized biologically active polypeptides containing sulfide bridges and α,β-didehydroamino acids. Angew. Chem. Int. Ed. Engl. 30: 1051-1068.
3. Kordel, M., Benz, R. and Sahl, H.-G. 1988. Mode of Action of the staphylococcinlike peptide Pep5: voltage-dependent depolarization of bacterial and artificial membranes. J. Bacteriol. 170: 84-88.
4. Stevens, K.A., Sheldon, B.W., Klapes, N.A. and Klaenhammer, T.R. 1991. Nisin treatment for inactivation of *Salmonella* species and other Gram-negative bacteria. Appl. Environ. Microbiol. 57: 3613-3615.
5. Ramseier, H.R. 1960. Die Wirkung von Nisin auf *Clostridium butyricum* Prazm. Arch. Mikrobiol. 37: 57-94.
6. Ruhr, E. and Sahl, H.-G. 1985. Mode of action of the peptide antibiotic nisin and influence on the membrane potential of whole cells and on cytoplasmic and artificial membrane vesicles. Antimicrob. Agents Chemother. 27: 841-845.
7. Sahl, H.-G. 1985. Influence of the staphylococcinlike peptide Pep5 on membrane potential of bacterial cells and cytoplasmic membrane vesicles. J. Bacteriol. 162: 833-836.
8. Sahl, H.-G. and Brandis, H. 1982. Mode of action of the staphylococcin-like peptide Pep5 and culture conditions effecting its activity. Zbl. Bakt. Hyg. A 252: 166-175.
9. Sahl, H.-G. and Brandis, H. 1983. Efflux of low-M_r substances from the cytoplasm of sensitive cells caused by the staphylococcin-like agent Pep5. FEMS Microbiol. Lett. 16: 75-79.
10. Sahl, H.-G. 1991. Pore formation in bacterial membranes by cationic lantibiotics. In: Nisin and Novel Lantibiotics. G. Jung and H.-G. Sahl, eds. Escom, Leiden. p. 347-358.

11. Jack, R.W., Benz, R., Tagg, J.R. and Sahl, H.-G. 1994. The mode of action of SA-FF22, a lantibiotic isolated from *Streptococcus pyogenes* strain FF22. Eur. J. Biochem. 219: 699-705.

12. Gao, T., Abee, T. and Konings, W.N. 1991. Mechanism of the peptide antibiotic nisin in liposomes and cytochrome c oxidase-containing proteoliposomes. Appl. Environ. Microbiol. 57: 2164-2170.

13. Sahl, H.-G., Kordel, M. and Benz, R. 1987. Voltage-dependent depolarization of bacterial membranes and artificial lipid bilayers by the peptide antibiotic nisin. Arch. Microbiol. 149: 120-124.

14. Schüller, F., Benz, R. and Sahl, H.-G. 1989. The peptide antibiotic subtilin acts by formation of voltage-dependent multi-state pores in bacterial and artificial membranes. Eur. J. Biochem. 182: 181-186.

15. Benz, R., Jung, G. and Sahl, H.-G. 1991. Mechanism of channel-formation by lantibiotics in black lipid membranes. In: Nisin and Novel Lantibiotics. G. Jung and H.-G. Sahl, eds. Escom, Leiden. p. 359-372.

16. van den Hooven, H.W., Doeland, C.C.M., van de Kamp, M., Konings, R.N.H., Hilbers, C.W. and van de Ven, F.J.M. 1996. Three-dimensional structure of the lantibiotic nisin in the presence of membrane-mimetic micelles of dodecylphosphocholine and of sodium dodecylsulphate. Eur. J. Biochem. 235: 382-293.

17. van den Hooven, H.W., Spronk, C.A.E.M., van de Kamp, M., Konings, R.N.H., Hilbers, C.W. and van de Ven, F.J.M. 1996. Surface location and orientation of the lantibiotic nisin bound to membrane-mimicking micelles of dodecylphosphocholine and of sodium dodecylsulphate. Eur. J. Biochem. 235: 394-403.

18. Moll, G.N., Roberts, G.C.K., Konings, W.N. and Driessen, A.J.M. 1996. Mechanism of lantibiotic-induced pore formation. Antonie Leeuwenhoek. 69: 185-191.

19. Lian, L.-Y., Chan, W.C., Morley, S.D., Roberts, G.C.K., Bycroft, B.W. and Jackson, D. 1991. NMR studies of the solution structure of nisin A and related peptides. In: Nisin and Novel Lantibiotics. G. Jung and H.-G. Sahl, eds. Escom, Leiden. p. 43-58.

20. van de Ven, F.J.M., van den Hooven, H.W., Konings, R.N.H. and Hilbers, C.W. 1991. The spatial structure of nisin in aqueous solution. In: Nisin and Novel Lantibiotics. G. Jung and H.-G. Sahl, eds. Escom, Leiden. p. 35-42.

21. Demel, R.A., Peelen, T., Siezen, R.J., de Kruijff, B. and Kuipers, O.P. 1996. Nisin Z, mutant nisin Z and lacticin 481 interactions with anionic lipids correlate with antimicrobial activity - a monolayer study. Eur. J. Biochem. 235: 267-274.

22. Breukink, E., van Kraaij, C., Demel, R.A., Siezen, R.J., Kuipers, O.P. and de Kruijff, B. 1997. The C-terminal region of nisin is responsible for the initial interaction of nisin with the target membrane. Biochemistry. 36: 6968-6976.

23. van Kraaij, C., Breukink, E., Noordermeer, M.A., Demel, R.A., Siezen, R.J., Kuipers, O.P. and de Kruijff, B. 1998. Pore formation by nisin involves translocation of its C-terminal part across the membrane. Biochemistry. 37: 16033-16040.

24. Giffard, C.J., Ladha, S., Mackie, A.R., Clark, D.C. and Sanders, D. 1996. Interaction of nisin with planar lipid bilayers monitored by fluorescence recovery after photobleaching. J. Membrane Biol. 151: 293-300.

25. Winkowski, K., Ludescher, R.D. and Montville, T.J. 1996. Physicochemical characterization of the nisin-membrane interaction with liposomes derived from *Listeria monocytogenes*. Appl. Environ. Microbiol. 62: 323-327.

26. El-Jastimi, R. and Lafleur, M. 1997. Structural characterization of free and membrane-bound nisin by infrared spectroscopy. Biochim. Biophys. Acta. 1324: 151-158.

27. Kordel, M., Schüller, F. and Sahl, H.-G. 1989. Interaction of the pore forming-peptide antibiotics Pep5, nisin and subtilin with non-energized liposomes. FEBS Lett. 244: 99-102.

28. Chan, W.C., Leyland, M.L., Clark, J., Dodd, H.M., Lian, L.-Y., Gasson, M.J., Bycroft, B.W. and Roberts, G.C.K. 1996. Structure-activity relationships in the peptide antibiotic nisin: antibacterial activity of fragments of nisin. FEBS Lett. 390: 129-132.

29. Linnett, P.E. and Strominger, J.L. 1973. Additional antibiotic inhibitors of peptidoglycan synthesis. Antimicrob. Agents Chemother. 4: 231-236.

30. Reisinger, P., Seidel, H., Tschesche, H. and Hammes, P. 1980. The effect of nisin on murein synthesis. Arch. Microbiol. 127: 187-193.

31. Brötz, H., Josten, M., Wiedemann, I., Schneider, U., Götz, F., Bierbaum, G. and Sahl, H.-G. 1998. Role of lipid-bound peptidoglycan precursors in the formation of pores by nisin, epidermin and other lantibiotics. Mol. Microbiol. 30: 317-327.

32. Bierbaum, G. and Sahl, H.-G. 1985. Induction of autolysis of staphylococci by the basic peptide antibiotics Pep5 and nisin and their influence on the activity of autolytic enzymes. Arch. Microbiol. 141: 249-254.

33. Bierbaum, G. and Sahl, H.-G. 1987. Autolytic system of *Staphylococus simulans* 22: influence of cationic peptides on activity of *N*-acetylmuramoyl-L-alanine amidase. J. Bacteriol. 169: 5452-5458.

34. Bierbaum, G. and Sahl, H.-G. 1988. Influence of cationic peptides on the activity of the autolytic endo-β-*N*-acetylglucosamidase of *Staphylococcus simulans* 22. FEMS Microbiol. Lett. 58: 223-228.

35. Liu, W. and Hansen, J.N. 1993. The antimicrobial effect of a structural variant of subtilin against outgrowing *Bacillus cereus* T spores and vegetative cells occurs by different mechanisms. Appl. Environ. Microbiol. 59: 648-651.

36. Chan, W.C., Dodd, H.M., Horn, N., Maclean, K., Lian, L.-Y., Bycroft, B.W., Gasson, M.J. and Roberts, G.C.K. 1996. Structure-activity relationships in the peptide antibiotic nisin: role of dehydroalanine 5. Appl. Environ. Microbiol. 62: 2966-2969.

37. Kuipers, O.P., Beerthuyzen, M.M., de Ruyter, P.G.G.A., Luesink, E.J. and de Vos, W.M. 1995. Autoregulation of nisin biosynthesis in *Lactococcus lactis* by signal transduction. J. Biol. Chem. 270: 27299-27304.

38. Arioli, V., Berti, M. and Silvestri, L.G. 1976. Gardimycin, a new antibiotic from *Actinoplanes*. III. Biological properties. J. Antibiot. 29: 511-515.

39. Malabaraba, A., Landi, M., Pallanza, R. and Cavalleri, B. 1985. Physicochemical and biological properties of actagardine and some acid hydrolysis products. J. Antibiot. 38: 1506-1511.

40. Chan, W.C., Leyland, M.L., Clark, J., Dodd, H.M., Lian, L.-Y., Gasson, M.J., Bycroft, B.W. and Roberts, G.C.K. 1996. Structure-activity relationships in the peptide antibiotic nisin: antibacterial activity of fragments of nisin. FEBS Lett. 390: 129-132.

41. Limbert, M., Isert, D., Klesel, N., Markus, A., Seibert, G., Chatterjee, S., Chatterjee, D.K., Jani, R.H. and Ganguli, B.N. 1991. Chemotherapeutic properties of mersacidin *in vitro* and *in vivo*. In: Nisin and Novel Lantibiotics. G. Jung and H.-G. Sahl, eds. Escom, Leiden. p. 448-456.

42. Niu, W.-W. and Neu, H.C. 1991. Activity of mersacidin, a novel peptide, compared with that of vancomycin, teicoplanin and daptomycin. Antimicrob. Agents Chemother. 35: 998-1000.

43. Kretschmar, M., Budeanu, T., Nichterlein, T. and Hof, H. 1993. Wirkung von Mersacidin auf die murine Listeriose. Chemotherapie. J. 2: 176-178.

44. Brötz, H., Bierbaum, G., Reynolds, P.E. and Sahl, H.-G. 1997. The lantibiotic mersacidin inhibits peptidoglycan biosynthesis at the level of transglycosylation. Eur. J. Biochem. 246: 193-199.

45. Brötz, H., Bierbaum, G., Leopold, K., Reynolds, P.E. and Sahl, H.-G. 1998. The lantibiotic mersacidin inhibits peptidoglycan synthesis by targeting lipid II. Antimicrob. Agents Chemother. 42: 54-60.

46. Zimmermann, N., Metzger, J.W. and Jung, G. 1995. The tetracycline lantibiotic actagardine: ^{1}H-NMR and ^{13}C-NMR assignments and revised primary structure. Eur. J. Biochem. 228: 786-797.

47. Sahl, H.-G., Jack, R.W. and Bierbaum, G. 1995. Biosynthesis and biological activities of lantibiotics with unique post-translational modifications. Eur. J. Biochem. 230: 827-853.

48. Racker, E., Riegler, C. and Abdel-Ghany, M. 1983. Stimulation of glycolysis by placental polypeptides and inhibition by duramycin. Cancer. 44: 1364-1367.

49. Chen, L.L. and Tai, P.C. 1987. Effects of antibiotics and other inhibitors on ATP-dependent protein translocation into membrane vesicles. J. Bacteriol. 169: 2372-2379.

50. Navarro, J., Chabot, J., Sherril, K., Aneja, R. and Zahler, S.A. 1985. Interaction of duramycin with artificial and natural membranes. Biochemistry. 24: 4645-4650.

51. Xie, X.S., Stone, D.K. and Racker, E. 1983. Determinants of clathrin-coated vesicle acidification. J. Biol. Chem. 258: 14834-14838.

52. Stone, D.K., Xie, X.S. and Racker, E. 1984. Inhibition of clathrin-coated vesicle acidification by duramycin. J. Biol. Chem. 259: 2701-2703.

53. Nakamura, S. and Racker, E. 1984. Inhibitory effect of duramycin on partial reactions catalyzed by (Na^+, K^+) adenosine-triphosphatase from dog kidney. Biochemistry. 23: 385-389.

54. Dunkley, E.A.Jr., Clejan, S., Guffanti, A.A. and Krulwich, T.A. 1988. Large decreases in membrane phosphatidylethanolamine and diphosphatidylglycerol upon mutation to duramycin resistance do not change the protonophore resistance of *Bacillus subtilis*. Biochim. Biophys. Acta. 943: 13-18.

55. Clejan, S., Guffanti, A.A., Cohen, M.A. and Krulwich, T.A. 1989. Mutations of *Bacillus firmus* OF4 to duramycin resistance results in substantial replacement of membrane lipid phosphatidylethanolamine by its plasmalogen form. J. Bacteriol. 171: 1744-1746.

56. Sokolove, P.M., Westphal, P.A., Kester, M.B., Wierwile, R. and Van Meter, K.S. 1989. Duramycin effects on the structure and function of heart mitochondria. Biochim. Biophys. Acta. 983: 15-22.

57. Choung, S.-Y., Kobayashi, T. and Inoue, J. 1988. Haemolytic activity of a cyclic peptide Ro 09-0198. Biochim. Biophys. Acta. 940: 171-179.

58. Märki, F. and Franson, R. 1986. Endogenous suppression of neutral-active and calcium-dependent phospholipase A_2 in human polymorphonuclear leucocytes. Biochim. Biophys. Acta. 879: 149-156.

59. Märki, F., Hänni, E., Fredenhagen, A. and van Oostrum, J. 1991. Mode of action of the lanthionine-containing peptide antibiotics duramycin, duramycin B and C and cinnamycin as indirect inhibitors of phospholipase A_2. Biochem. Pharmacol. 42: 2027-2035.

60. Fredenhagen, A., Fendrich, G., Märki, F., Märki, W., Gruner, J., Raschdorf, F. and Peter, H.H. 1990. Duramycins B and C, two new lanthionine-containing antibiotics as inhibitors of phospholipase A_2. J. Antibiot. 43: 1403-1412.

61. Fredenhagen, A., Märki, F., Fendrich, G., Märki, W., Gruner, J., van Oostrum, J., Raschdorf, F. and Peter, H.H. 1991. Duramycin B and C, two new lanthionine-containing antibiotics as inhibitors of phospholipase A_2 and structural revision of duramycin and cinnamycin. In: Nisin and Novel Lantibiotics. G. Jung and H.-G. Sahl, eds. Escom, Leiden. p. 131-140.

From: *Development of Novel Antimicrobial Agents: Emerging Strategies*
ISBN 1-898486-23-9 © 2001 Horizon Scientific Press, Wymondham, UK.

16

Cyclic Cystine-knot β-Stranded Antimicrobial Peptides: Occurrence, Design and Synthesis

James P. Tam, Yi-An Lu, Jin-Long Yang and Qitao Yu

Abstract

Amphipathicity of antimicrobial peptides is an important attribute to their membranolytic actions. However, the relationship of amphipathicity to membranolytic selectivity that dissociates cytotoxicity from antimicrobial activity remains poorly understood. Analog study using rigid preorganized amphipathic structures may provide insight for selective interactions with microbial rather than eukaryotic membrane. Cyclic cystine-knot peptides with two or three β strands, referred as $cc_3\beta2$ and $cc_3\beta3$ peptides respectively, represent novel and highly constrained scaffoldings of antimicrobial peptides containing 18 to 33 amino acid residues. This report describes their natural occurrence in higher organisms as well as our efforts in designing and developing new synthetic methods for $cc_3\beta2$, $cc_3\beta3$ peptides and their analogs. The rigidity imparted by the close-ended amide backbone and the tricystine constraints of $cc_3\beta2$ and $cc_3\beta3$ peptides also facilitates developing therapeutic useful peptide antibiotics of β-stranded defensins, tachyplesins and protegrins that are membrane-selective, salt-insensitive and low cytoxicity.

Introduction

Antimicrobial peptides are important components of the host defense against microbial infections (1-6, see also Ganz and Lehrer this monography). These peptides are usually cationic, open-chained and possess broad-spectrum activities against bacteria and fungi through killing mechanisms such as altered permeability and pore formation that are directed against microbial membranes. Furthermore, they are known to have the propensity to fold as amphipathic structures with clusters of hydrophobic and charge regions. This structural property appears to be closely related to their membranolytic activity, including their activity spectrum and hemolytic selectivity.

Modulation of amphipathicity and hydrophobicity has been used for designing analogs and for correlating mechanisms of action (7-19). However, it is often difficult to dissociate the sequence changes from concomitant conformational changes that affect membranolytic activity. Furthermore, the activity of some antimicrobial peptides such as defensins is abro-

Table 1. Classification, structures and occurrence of different cyclic cystine-rich β-strand peptides.

Classification	End-to-end lactam	β-strand	No of Cystine	Amino acid	Occurrence
cc₃β3	Yes	3	3	27-31	plants
cc₂β3	Yes	3	2	33	Synthetic defensin
ccβ3	Yes	3	1	33	Synthetic defensin
cβ3	Yes	3	0	33	Synthetic defensin
cc₃β2	Yes	2	3	18	RTD-1 and synthetic PG and TP
cc₂β2	Yes	2	2	18	Synthetic PG and TP
ccβ2	Yes	2	1	18	Synthetic PG and TP
cβ2	Yes	2	0	18	Synthetic PG and TP

gated under physiological conditions containing 100-150 mM NaCl (hereafter referred to as high-salt conditions) that limit their therapeutic applications (5-6). To overcome these limitations, our laboratory has studied the design of rigid, compact, and pre-organized amphipathic structures to minimize conformational changes and salt sensitivity in analog studies. In particular, we are interested in the design of cyclic, cystine-stabilized β-stranded peptides based on defensins, tachyplesins, and protegrins so as to improve their activity and selectivity under high-salt conditions.

Although cyclic peptides are produced naturally in microbes, they are usually restricted to relatively small peptides of about 10 amino acids. In 1999, two novel types of large cyclic, tricystine β-stranded antimicrobial peptides were identified from higher organisms (20, 21). A common feature of these antimicrobial peptides is their conformational rigidity imparted by a cystine-knot and end-to-end cyclic peptide backbone framework. For convenience, these tricystine peptides are classified structurally as the cc₃β2 and cc₃β3 families containing two or three strands with a cyclic cystine-knot motif, respectively (Table 1). We have also prepared cyclic analogs of these constrained peptides containing zero to three disulfide constraints and have referred to them (e.g. cc₃β2 series) as cβ2 (no cystine), ccβ2 (one cystine), cc₂β2 (two cystine) and cc₃β3 (three cystine), respectively. Herein, we report the occurrence, syntheses, conformation and activity of these cyclic and cystine-rich peptides.

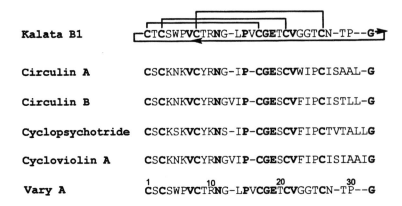

Figure 1. Amino acid sequences of cc₃β3 plant peptides of kalata, circuline A, B, cyclopsychotride, cycloviolin A and vary A. The conserved residues are indicated in bold. Sequence numbers are arbitrarily based on the linear precursors for chemical syntheses.

The Cyclic Cystine-knot Three β-Stranded (cc₃β3) Antimicrobial Peptides

Occurrence

The cc₃β3 peptides are large end-to-end cyclic peptides of 29 to 31 amino acids from plants of the *Rubiaceae* and *Violaceae* families (22-26). They are different from the known plant defensin-like antimicrobial peptides which are open-chained, larger and significantly more basic in character (1-6). Thus far, 17 members of cc₃β3 peptides with significant sequence identity have been identified and most of which in the past several years. All contain six cysteines which can be aligned perfectly (Figure 1). These include circulin A and B (CirA, CirB) from the African tropical tree *C. parvifolia* (1), cyclopsychotride (Cpt) from the South American tropical plant *Psychotria longipes* (2), kalata from the African plant *Oldenlandia afinis* (3), violapeptide-1 and vary peptides A-H from *V. arvensis* Murray, and cycloviolins A-D from *Lenonia cymosa C.* As a group, they are reported to be the largest cyclic peptides found in nature.

The ccβ3 peptides contain a cystine-knot motif commonly observed in protease inhibitors and toxins (27). This knotted motif contains a disulfide bonding pattern of Cys I-IV, II-V and III-VI, with the Cys III-VI disulfide threading through the other two. Although the primary sequence of kalata was known since 1970s, the constraints contributed by the cystine-knot disulfides and cyclic amide backbone that confer resistance to enzymatic proteolysis complicates the elucidation of their disulfide connectivity by traditional methods. The initial evidence tentatively suggesting the cystine-knot arrangement of these peptides was found in kalata using NMR methods and distance calculation in 1995 (28). Subsequently, the cystine-knot motif of circulins and other cc₃β3 members are determined by partial acid or enzymatic hydrolysis and mass spectral determination (29). Recently, we have confirmed the cystine-knot motifs of kalata, CirA, CirB and Cpt through total chemical synthesis and comparison with native samples obtained from plant extracts (21, 30, 31).

Structures

The structures of kalata and CirA determined by 2D NMR (28, 32) display compact and fairly rigid structures with a similar global fold of distorted triple stranded β-sheets (Figure 2). The rigidity is contributed by the peptide backbone that is extensively folded back on itself and by the cross bracing of the three-disulfide bonds. The cystine-knot occupies most of the interior spaces and forms a sulfur-rich core. Based on the sequence homology and the conserved cystine-knot motif, other cc₃β3 peptides are likely similar well-ordered three β-sheet structures.

Antimicrobial Activity

Computer modeling of kalata and CirA structures show that they display clusters of cationic charges and hydrophobic amino acids on their surfaces, characteristics of antimicrobial peptides. The hypothesis that these cc₃β3 peptides may contain antimicrobial activity has been confirmed in four cc₃β3 peptides (kalata, Cpt, Cir A and B) prepared by chemical synthesis and tested against nine strains of microbes. Kalata and CirA are specific for the Gram-positive *S. aureus* with a minimum inhibition concentration (MIC) of about 0.2 μM. They are relatively ineffective against Gram-negative bacteria such as *E. coli* and *P. aeruginosa*. However, CirB and Cpt are active against both Gram-positive and negative bacteria. In particular,

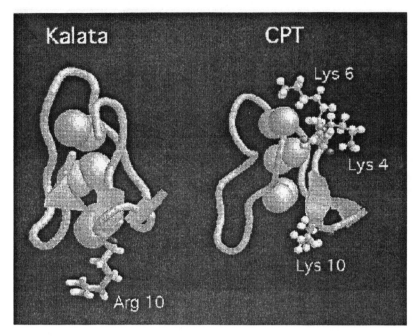

Figure 2. Computer modeling of kalata and cyclopsychotride. The backbone structures with cationic amino acids (Arg and Lys) in ball-and-stick figures. Arg[10] located at the bottom is used as reference point.

CirB showed potent activity against *E. coli* with a MIC of 0.41 µM. All four cyclic peptides are moderately active against two strains of fungi, *C. kefyr and C. tropicalis* but are inactive against *C. albicans*.

The plant $cc_3\beta3$ peptides are not highly cationic and contain −1 to +2 net positive charges (26). Thus, it is not surprising to find that their activity spectra are relatively narrow and selective when compared to the highly basic β-stranded antimicrobial peptides such as defensins. Furthermore, their antimicrobial activity appears to depend on their cationic charge for initial recognition and binding (21). Modifying the sole cationic Arg residue in kalata, which has no net positive charge, with a keto aldehyde significantly reduces its activity against *S. aureus* while blocking the Arg in CirA, which has +2 net positive charge, produces no significant effect. The two-disulfide $cc_2\beta2$ variants exhibit antimicrobial profiles and potency similar to their native peptides. However, in high-salt assays (100 mM of NaCl), few of these macrocyclic peptides, natives or analogs, retain antimicrobial activity. These results show that the macrocyclic peptides possess specific and potent antimicrobial activity that is salt-dependent and their initial interactions with the microbial surfaces may be electrostatic, an effect commonly found in defensin antimicrobial peptides. Furthermore, their $cc_3\beta3$ structure end-to-end cyclic structure with a cystine-knot motif represents a novel natural-occurring scaffolding of antimicrobial peptides and is different from the known, open-chained plant defensins that are larger in size.

CirA and CirB, identified through the anti-HIV screening program, possess anti-viral activity (33). Both kalata and Cpt were also discovered through drug screening programs. Kalata shows uterotonic activity (34) and hemolytic activity (35) whereas Cpt inhibits neurotensin binding to its receptor (36). Although we have shown that these plant $cc_3\beta3$ peptides possess antimicrobial activity and may participate in defence mechanism, their true physiological function remains to be determined.

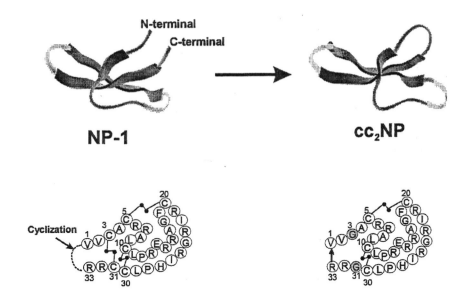

Figure 3. Amino acid sequence of cc₂NP-1 and circularized cNP-1. Open circles indicate unchanged amino acid residues with respect to NP-1. Disulfide linkages are shown by connecting lines. Arrow indicates the NH₂ to COOH direction of peptide bonds that link the NH₂ and COOH termini of NP-1.

Design of Synthetic Salt-insensitive α-Defensins with cβ3, ccβ3 and cc₂β3 Motifs

Defensins are a family of open-chained, β-stranded antimicrobial peptides that range in size from 3-5 kDa and contain three or four intramolecular disulfide bonds (1-6, 37-38). Based on their cystine pairings and homology to known structures, defensins fall into two structural classes: the α- and β-defensins of vertebrates, and the plant/insect-defensins. α-Defensins were first identified in granules of leukocytes (5,6), and later found in small intestinal Paneth cells (38). β-Defensins are expressed by many epithelial cells and glands and in some leukocytes (38-40). Their widespread distribution equips α- and β-defensins to participate in systemic and mucosal host defense.

Collectively, defensins exhibit antimicrobial activity towards many bacteria, yeast, fungi and enveloped viruses. The antibacterial activity of α- and β-defensins is greatly affected by the ambient concentration of NaCl, and links between local salt-dependent inactivation of defensins and the chronic pulmonary infections in patients with cystic fibrosis have been proposed (41). However, to be clinically relevant, antimicrobial activity must be retained under physiological conditions of 120-150 mM NaCl which is simulated by the high-salt assay.

The constrained cc₃β3 plant peptides provide a template for designing such α-defensin analogs to improve their salt insensitivity. The folded structures of α- and β-defensins are very similar, despite differences in disulfide pairing. α-Defensins contain three disulfide bonds and are circularly permutated by their end-to-end (CysI:CysVI) disulfide bond, which permits end-to-end cyclization by an amide bond to form cc₂NP (Figure 3).

Seven cyclic defensin analogs of a typical rabbit α-defensin, NP-1 representing cβ3, ccβ3 and cc₂β2 motifs as well as a dimer of ccβ3 have been prepared (42). These analogs retain defensin-like architecture after the characteristic end-to-end (Cys I:Cys VI) α-defensin disulfide bond is replaced by a backbone amide bond. The cyclic structure of ccβ3 and cβ3 designs with one or no cross-bracing disulfide bond, respectively, suffice for antimicrobial activity as NP-1 but are less cytotoxic. For example, all analogs retain activity against *Escherichia coli* in NaCl concentrations that render NP-1 inactive. A contributing factor may be the end-to-end cyclization that clusters positive charges resembling those in protegrins (43) and tachyplesins (44). We have also designed a retro-isomer of NP-1 and which is as active as the parent compound, suggesting that overall topology and amphipathicity govern its antimicrobial activity. Synthetically, the cyclic α-defensin analogs with cβ3 to $cc_2β3$ motifs minimize problems encountered during the oxidative folding of three-disulfide defensins. Furthermore, the ccβ3 design have shown and should permit the synthesis of dimers and higher oligomers of α-defensins which are believed to be the membrane-active forms.

The Cyclic Cystine-knot Two β-Stranded ($cc_3β2$) Antimicrobial Peptides

RTD –Theta Defensin

This prototypic cyclic tricystine peptide, RTD-1 (rhesus theta defensin, Figure 4), was identified in 1999 from monkey leukocytes (20). It represents the first known cyclic peptide isolated from mammal. RTD-1 contains 18 amino acids and is classified structurally in this report as a $cc_3β2$ peptide because it also contains a "cystine knot"-like motif but on a two-β–strand framework with three parallel cross-bracing disulfides. However, the disulfide connectivity of the cystine motif in RTD is Cys I-VI, II-V and III-IV that differs from the Cys I-IV, II-V and III-VI pattern found in the $cc_3β3$ cystine-knot peptides. In contrast to the solvent-shielded sulfur core of $cc_3β3$ peptide, the sulfur core of $cc_3β2$ peptides is located on one face of the strand and exposed to solvent. Such a structural motif provides an invariable hydrophobic sulfur-rich cluster to form an amphipathic molecule. The cyclic RTD-1 peptide

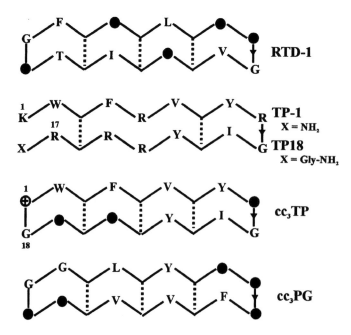

Figure 4. Amino acid sequences and topology of charged amino acid in RTD-1, TP-1, TP18, cc_3TP, cc_3PG; cationic amino acid ● represents Arg, ⊕ represents Lys; - - - - represents a disulfide bond; arrow indicates the N to C direction of peptide bonds.

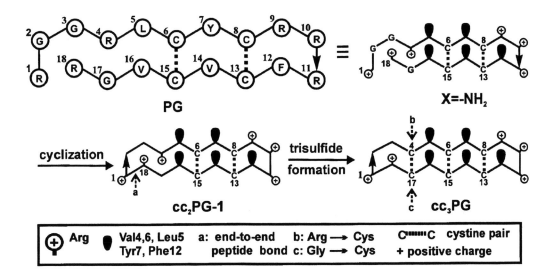

Figure 5. Amino acid sequences and design of cyclic tricystine cc₃PG.

is generated by a novel posttranslational head-to-tail ligation of two truncated -defensin-like gene products (20). Unlike other members of the defensin family, RTD-1 is salt-insensitive regarding its activity towards Gram-positive bacteria and fungi.

Amphipathic Design of Cyclic Cystine-knot $cc_3β2$ forms of Tachyplesin and Protegrin

Our need for rigid peptides to correlate conformational effects and salt-sensitivity in antimicrobial peptides has also led to the development of amphipathic scaffoldings of $cc_3β2$ peptides based on protegrins (45) and tachyplesins (46, 47) that bear structural similarity to RTD-1. The size and structural constraint elements of the synthetic $cc_3β2$ peptides, cc_3PG and cc_3TP (Figure 4), are similar to the recently discovered RTD-1. Protegrin-1 (PG-1), a prototypic member of the broad-spectrum cationic antimicrobial peptide family of protegrins, consists of 18 amino acids and forms an ordered antiparallel β-strand structure stabilized by two cystine bonds, but has disordered structures at its N- and C-termini. Such a structure is amenable to the $cc_3β2$ design because the close proximity of their N- and C-termini. Thus, cc_3PG is formed (45), first by cyclization of PG-1 by a peptide bond linking the N- and C-termini to afford cc_2PG, and then adding a third disulfide $Cys^{4,7}$ by replacing the corresponding cross Arg^4 and Gly^{17} residues (Figure 5).

Similarly, cc_3TP (46) is designed based on the tachyplesin-1 (TP-1) isolated from the Japanese horseshoe crab (*Tachypleus tridentatus*). cc_3TP contains the whole length of the 17-amino-acid TP+8 sequence and an additional Gly at its COOH-terminus (TP+8). End-to-end cyclization of the lengthened TP+8 to an 18-residue affords cc_2TP with an even-numbered, symmetrical antiparallel stranded cyclic peptide template with two cross-braced disulfide bonds. cc_3TP is obtained by a third cross-strand disulfide bond by replacing the $Arg^{5,14}$ pair positioned in the middle of the antiparallel β strands with a $Cys^{5,14}$ pair. Although cc_3TP gains two additional constraints through an end-to-end cyclic peptide bond, it contains only four cationic amino acids when compared to TP-1 which contains six cationic amino acids. The loss of these two cationic charges leads to altered activity spectrum of cc_3TP.

Table 2. A comparison of antimcrobial activity of RTD-1, cc3TP and cc3PG.

Organism	Media Salt	TP-1	TP-18	PG-1	MIC(μM) RTD-1	cc₃TP	cc₃PG
Gram-negative							
E. coli	L	0.3	0.4	0.9	2.0	3.0	0.6
	H	0.4	0.1	0.8	28.4	6.9	1.4
P. aeruginosa	L	0.9	0.5	1.2	2.0	5.0	1.2
	H	0.5	0.2	2.0	5.2	5.2	3.6
P. vulgaris	L	0.7	0.4	2.4	10.2	6.0	1.1
	H	1.0	0.7	2.8	50.4	14.4	0.9
K. oxytoca	L	0.2	0.3	0.7	2.0	2.1	28.6
	H	0.5	0.2	0.9	12.8	7.8	53.2
Gram-positive							
S. aureus	L	0.4	0.4	0.7	1.1	1.0	0.7
	H	0.5	0.5	0.6	0.9	0.8	0.7
M. luteus	L	1.0	0.1	0.3	0.5	0.5	0.3
	H	1.1	0.1	0.8	0.4	0.4	0.6
E. faecalis	L	0.3	0.3	0.3	1.2	1.0	5.2
	H	0.4	0.4	0.7	1.2	0.9	5.8
Fungi							
C. albicans	L	0.7	0.2	1.3	4.0	5.1	8.0
	H	0.9	0.4	1.0	5.2	17.2	26.4
C. kefyr	L	0.9	0.3	1.8	1.8	3.9	2.1
	H	1.3	0.4	1.4	1.2	4.1	1.7
C. tropicalis	L	0.5	0.4	1.0	2.8	5.2	10.2
	H	1.0	0.8	1.5	1.0	1.1	19.8

*Experiment were performed in radial diffusion assay with underlay gel containing 1% agarose, 10 mM phosphate buffer with (high-salt) or without (low-salt) 100 mM NaCl. Activity against multiple strains are expressed as the minimum inhibitory concentration (MIC, μM).

A common feature shared by these naturally occurring and synthetic $cc_3\beta2$ peptides is their pseudosymmetry which is contributed by a closed-chained, two-β-stranded structure with three evenly spaced disulfide bonds. From a minimalistic perspective, they approximate a "β tile"- or "β-tape"-like structure consisting of four consecutively fused cyclic hexapeptides. The up-and-down side-chain arrangements on these rigid two-β-strand frameworks produce a top face (arbitrarily assigned) containing clusters of hydrophobic and positive charged amino acids and a sulfur-rich hydrophobic bottom face formed by three cystine pairs (Figure 6).

Topologically, the distributions of hydrophobic and charged clusters of these $cc_3\beta2$ peptides are different. RTD-1 has an alternating hydrophobic and charged motif whereas the synthetic cc_3TP and cc_3PG contain a central hydrophobic cluster and two charged clusters at the two four-residue reverse turns (Figure 5). Despite their differences in sequences, circular dichroism (CD) measurements have confirmed that these highly rigidified $cc_3\beta2$ peptides display similar ordered β-sheet structures (20, 45-47).

Comparison of Antimicrobial Activity of RTD, cc₃PG and cc₃TP

TP-1 and TP18 exhibit MICs < 0.9 μM in low-salt assays and 0.4 to 1.3 μM in high-salt assays against ten organisms (45-47). PG-1 also exhibits similar activity profile but displays higher potency against Gram-positive bacteria with MICs <0.7 μM and is less active against fungi than TP-1 and TP18 (Table 2). As a group, the three $cc_3\beta2$ peptides, RTD-1, cc_3PG and

cc_3TP, are less potent than PG-1 and TP-1, particularly against some Gram-negative bacteria and fungi. Among these $cc_3\beta2$ peptides, cc_3PG is most potent and display activity profile very similar to PG-1. In comparing these highly constrained $cc_3\beta2$ antimicrobial peptides, the ratio and topology of cationic and hydrophobic amino acids appear to be important for their activity. cc_3TP contains an extra cystine pair that replaces a charged Arg pair in their sequences based on TP-1. As a result, they are less cationic (by one positive charge) and also less active against Gram-negative bacteria than RTD-1 or cc_3PG. However, cc_3TP is as potent as TP-1 or PG-1 against the three test Gram-positive bacteria in low- and high-salt conditions.

These results are consistent with the general observation that antimicrobial activity, particularly against Gram-negative bacteria, increases with increasing numbers of cationic charges as those found in the defensin family (5-6, 37-38). The decrease in cationic charges coupled with the increase in hydrophobic amino acids also render cc_3TP relatively insoluble under assay conditions containing phosphate buffers. In particular, the topology of cc_3TP containing a continuous stretch of seven hydrophobic amino acids Trp^2-Tyr^8 occupying one entire β strand favors aggregation in aqueous solutions and is reminiscent of the highly insoluble β-amyloid peptides (48). Thus, our current assay conditions using the overlay gel diffusion method may not accurately portray their antimicrobial activity profile. However, it is also

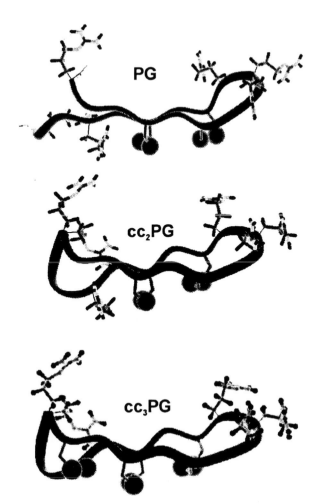

Figure 6. Computer modeling of protegrin (PG), cc_2PG and cc_3PG. Molecular models are based on reported structure of PG. Backbone structure are shown as ribbons, cationic amino acid (Arg) in sticks and the disulfide pairs in ball- and –sticks.

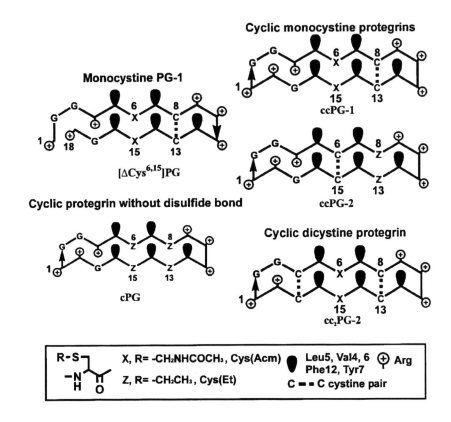

Figure 7. Design of cPG. Amino acid sequences of a monocyctine protegrin [ΔCys ^{6,15}]PG, cPG, ccPG, cc$_2$PG with varying number of cross-strand disulfide constraints.

likely that increasing the constraints of cc$_3$TP limits its conformational spaces, which results in narrowing its activity spectrum.

Correlation Conformational Rigidity of cc$_3$PG and Cyclic Protegrins to Antimicrobial Activity

To determine an optimal balance between conformational constraint and antimicrobial potency, we have studied the correlation of conformational rigidity with antimicrobial activity and cytotoxicity using cyclic protegrin analogs (45). These cyclic analogs are prepared by linking their N- and C-termini as end-to-end peptide bonds together with varying numbers (zero to three) of cross-strand disulfide constraints (cPG, ccPGs, cc$_2$PGs and cc$_3$PGs, Figure 5 and 7). Antimicrobial assays against ten organisms in high- and low-salt conditions show that all cyclic protegrins retain broad-spectrum antimicrobial activity against Gram-positive and Gram-negative bacteria, fungi and HIV-1. Increasing conformational constraints also correlate proportionally in enhanced membranolytic selectivity of 2 to 30 fold and significant decreases of 2 to 10 fold decrease in cytoxicity against human cells when compared to PG-1. However, increasing constraints also lead to decrease potency against several organisms such as *K. oxytoca, E. faecalis* and *C. albicans*. The triple-constrained cc$_2$PG-1 or cc$_2$PG-2 displays activity spectra and potency similar to or better than PG-1 in eight of the tested organisms. cPG, a single-constrained cyclic protegrin with no disulfide bond, and the double-constrained ccPG-1 and ccPG-2, two cyclic mimics of PG-1 with one disulfide bond, exhibit activity spectra, potency and cytoxicity similar to PG-1. In contrast, the most constrained cc_3PG retains similar activity as PG-1 in only six of the test ten organisms. Circular dichoism measurements indicate that cyclic protegrins constrained with one to three disulfide bridges display β-strand structures in water or water-trifluoroethanol (TFE) mixtures, which are not seen in the cyclic protegrin analogs without any disulfide constraint (45). These results show that cyclic structures are useful in the design of antimicrobial peptides and cc_2PG suffices for optimal activity. However, an increase in their rigidity of the cyclic β2 design based on the PG-1 sequence decreases potency in some organisms but enhances membranolytic selectivity for interactions with microbial rather than eukaryotic membranes.

Topological Charge Pattern and Activity Profile of cc_3TP

The variable top-face topology of cc_3TP and RTD-1 is different even though both contain the $cc_3\beta2$ design. We have compared these two topological categories (46), using two groups of ccTP **2**-like analogs **2-3** with the RTD-like analogs **4-5** (Figure 8). cc_3TP **1** is less cationic than RTD-1 due to the removal of the $Arg^{5,14}$ pairs of cationic charges by the $Cys^{5,14}$ disulfide constraint that results in a continuous cluster of eleven hydrophobic amino acids. Six of these are located on the variable face, with Trp^2, Phe^4, Val^6 and Tyr^8 on one β-strand and Ile^{11} and Tyr^{13} on the opposite strand together with five half-cystines on the invariable face. The four

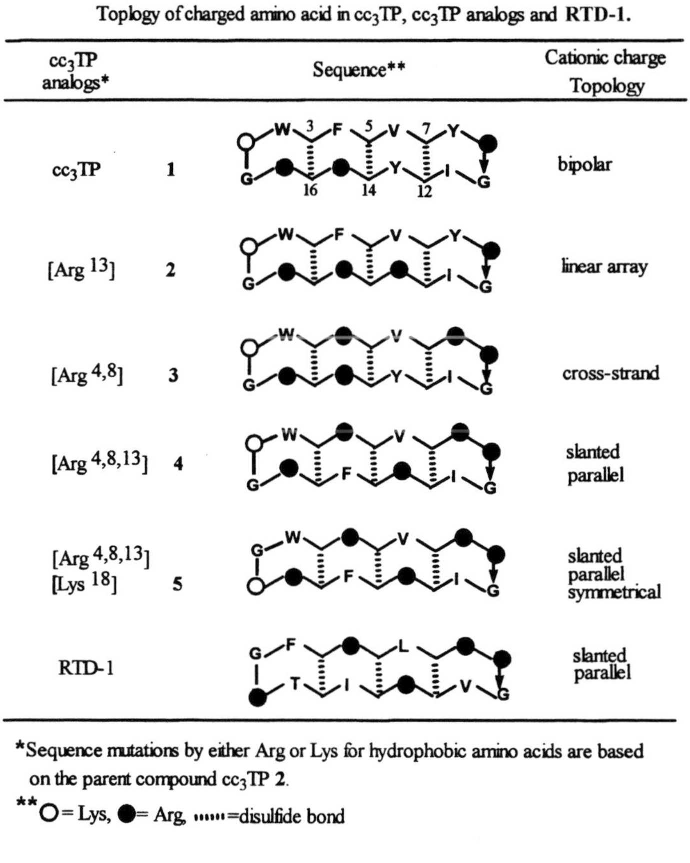

Toplogy of charged amino acid in cc_3TP, cc_3TP analogs and RTD-1.

cc_3TP analogs*		Sequence**	Cationic charge Topology
cc_3TP	1		bipolar
[Arg ^{13}]	2		linear array
[Arg ^{4,8}]	3		cross-strand
[Arg ^{4,8,13}]	4		slanted parallel
[Arg ^{4,8,13}] [Lys ^{18}]	5		slanted parallel symmetrical
RTD-1			slanted parallel

*Sequence mutations by either Arg or Lys for hydrophobic amino acids are based on the parent compound cc_3TP **2**.

**O = Lys, ● = Arg, ⸝⸝⸝⸝⸝⸝ = disulfide bond

Figure 8. Topology of charged amino acids in cc_3TP, cc_3TP analogs and RTD-1.

cationic charges are clustered near both ends, with a linear array of three cationic charges at one corner and the lone remaining Arg at the opposite corner (Figure 8). This bipolar topology of a dense hydrophobic cluster may favor aggregation in aqueous solutions. Thus, subsequent analogs are designed to disperse cationic charges and to segregate hydrophobic clusters. The linear array design [Arg^{13}]cc$_3$TP **2** and the cross-strand design [Arg^{4,8}]ccTP **3** contain aromatic amino acid replacements, respectively, by one or two Arg residues. Analog **2** with Tyr13→Arg replacement preserves the single continuous hydrophobic strand of seven amino acids, Trp^2-Tyr^8, and the single linear array of three cationic charges found in cc$_3$TP **1**. Analog **3,** with a dual replacement of Tyr8,13→Arg breaks up this continuous stretch and shifts the cross-strand Arg^{4,8} pair to near the middle of the variable face. The slanted parallel design of analog **4** and **5** are patterned topologically with cationic charges and hydrophobic clusters in an alternating slanted parallel pattern, with analog **5** forming a nearly perfectly symmetrical alternating cluster pattern. This alternating parallel pattern is also found in the naturally occurring cyclic peptide RTD-1.

Comparing with TP-1, all five cc33 tachyplesins and RTD-1 retain comparable activity of 0.3 to 1.9 µM, particularly under high-salt conditions, against the three Gram-positive bacteria, *S. aureus*, *M. luteus* and *E. faecalis*, but are less active against two of the four Gram-positive bacteria, *K. oxytoca* and, in particular, *P. vulgaris*, as well as one of the test fungi, *C. albicans*. The RTD-like group consisting of RTD-1 and cc3TP analogs 4 and 5 display relatively similar activity spectrum activity against *E. coli*, *P. aeruginosa*, *C. kefyr* and *C. tropicalis*. However, their activities in low- and high-salt conditions are markedly different. RTD-1 is generally more active in low- than in high-salt conditions, whereas the activity of cc3TP analogs 4 and 5 is reversed. The most dramatic difference of this activity reversal is observed against *E. coli* in which the MIC of RTD-1 is 2.0 µM in the low-salt assay and 28.4 µM, a 14 fold decrease, in the high-salt assay. In contrast, the MICs of cc3TP analogs 4 and 5 are 5.5 and 6.1 µM, under low-salt conditions but improve about five fold to 1.2 and 1.2 µM, respectively, under high-salt conditions. This unusual behavior in the high-salt condition against *E. coli* is also observed in the cc3TP-like group consisting of cc3TP analogs of 1-3. In general, the antimicrobial profiles of cc3TP 1 and RTD-1 are similar despite their differences in topological arrangement of cationic charges. These results show that the basic design of ccβ2 scaffold appears to be rather specific for Gram-positive bacteria and some fungi. However, selective modification of variable top face may lead to broader antimicrobial activity spectrum.

Correlation of Conformational Rigidity and Activity of Cyclic Tachyplesins

To confirm the results of analog study based on cyclic protegrin peptides, a similar study using a series of 11 peptides including cyclic tachyplesins with graded decrease from four to no constraints is designed to evaluate conformational rigidity in relation to antimicrobial activity (47). This series retains the charge and hydrophobic amino acids of tachyplesin, but contains zero to four covalent constraints (Figure 9). In general, an increase or decrease in conformational constraints in this series of peptides containing six cationic charges produces surprisingly little overall effect in their activity spectra under low-salt conditions compared with TP-1, the parent compound. TP-1 and TP18 show MICs of 0.2 to 1.3 µM and are generally more active against the test organisms in low-salt than in high-salt assays. Except for linear peptides assayed under high-salt conditions, peptides with increased or decreased conformational constraints retain broad activity spectra with small variations in potency of 2 to 10 fold compared to tachyplesin. Comparing with TP and TP18, the triple-constrained cyclic peptides cc$_2$TP-2 displays similar potency and activity spectrum in low- and high-salt conditions. In contrast, the most constrained cc_3TP and another triple-constrained cc_2TP-1, both of which are less cationic by two positive charges are less active than TP-1 against Gram-negative bacteria and fungi in low- and high-salt conditions. However, cc_3TP is as active as TP-1 against three test Gram-positive bacteria *S. aureus, M. luteus* and *E. faecalis* with MICs at 1.0, 0.5 and 1.0 μM, respectively in low-salt and 0.4 to 0.9 μM in high-salt assays. The cc_2TP-1 displays an activity profile similar to cc_3TP with MICs ranging from 0.6 to 4.6 μM against Gram-positive bacteria. Compared to cc_3TP, the loss of the cystine constraint in cc_2TP-1 led to >5 fold loss of potency against *K. oxytoca, E. faecalis* and *C. albicans*.

Similarly, no large changes in the potency or activity spectrum of the double- or single-constrained peptides ccTP-1, ccTP-2 and [ΔCys^{3,16}]TP18 and the linear peptides, can be discerned in low-salt conditions. In high-salt conditions, there is an increasing loss of potency with decreasing conformational constraints, as observed in peptides [ΔCys^{3,16}]TP18, LTP and rLTP. These results suggest that the primary sequence of TP with six bulky amino acids and six cysteines provides the threshold of hydrophobicity for interaction with microbial membranes without any constraint under low-ionic conditions. However, increasing constraint provides enhanced activity under high-salt conditions.

The importance of conformation constraint can be illustrated by the linear and cyclic monocystine TP analogs. TP and TP18 as well as the homologous family of protegrin peptides maintain their conformations by two disulfide bridges. The inner disulfide bond (Cys^{7,12}) has been known to be important in maintaining the β-strand structure. Indeed, the monocystine open-chained [ΔCys^{3,16}]TP18 and the ccTP-2 without the outer Cys^{3,16}-disulfide bond retain comparable activity as TP in low-salt assays. In contrast, removal of the outer disulfide bond in the ccTP-2 produced 2 to 23 lower potencies in high-salt conditions compared to TP against all four test Gram-negative organisms, *P. vulgaris, K. oxytoca, P. aeruginosa* and *E. coli.*

Test of Amphipathicity Based on [Gly$_6$]cc$_3$TP

To test whether the combination of rigidity, disulfide motifs and pseudosymmetry in the $cc_3\beta2$ template can maintain amphipathicity through the variable top face and the invariable hydrophobic sulfur-rich bottom face, global and local alterations are used as design elements in the analog study of cyclic TPs (47). The 17-residue TP contains six bulky hydrophobic acids, four of which are aromatic. Contributions to conformational stability and amphipathicity by single substitutions of these hydrophobic amino acids have been shown to be important to the antimicrobial activity of TP-1 (49). Gly, the smallest of the 20 genetically coded amino acids, generally disrupts ordered structures in peptides (50). By replacing six bulky hydrophobic amino acids of TP (Trp^2, Phe^4, Val^6, Tyr^{8,13} and Ile^{11}) with six Gly, significant changes in physiochemical properties are anticipated, including the disruption of amphipathic structures and the total loss of the hydrophobic amino acid cluster.

The previously described series of cyclic and linear TP peptides, which retains the charge and hydrophobic amino acids of tachyplesin, but contains zero to four covalent constraints, is used for comparing a parallel series of [Gly$_6$]-analogs with all six bulky hydrophobic amino acids in their sequences replaced by Gly (47). Such a [Gly$_6$]-TP series would test the importance of rigidity and a cystine core in maintaining the amphipathic design. Furthermore, the sequences of [Gly$_6$]-analogs are simplified to three types of amino acids and are particularly rich in Gly residues. The most constrained [Gly$_6$]cc$_3$TP contains eight Gly, six Cys and four cationic amino acids and no other types of amino acids. Circular dichroism measurements show that [Gly$_6$]cc$_3$TP and [Gly$_6$]cc$_2$TP display well ordered β-sheet structures while the less constrained [Gly$_6$]-analogs are disordered.

Compared to previous series, the global change of replacing six bulky hydrophobic residues in the [Gly₆]-series sharply decreases potency and produces large variations of activity spectra and potency that roughly correlates with the decreases in conformational constraints. Except against *E. coli,* the Gly-rich analogs with two or less covalent constraints are largely inactive under high-salt conditions. Remarkably, the most constrained [Gly₆]cc ₃TP retains a broad activity spectrum against all ten test microbes in both low- and high-salt assays. [Gly₆]cc ₃TP is as potent as the parent cc₃TP under low-salt conditions. In ten test organisms, the activity of [Gly₆]cc ₃TP against seven organisms in low-salt conditions was similar or higher than those cc₃TP. In high-salt conditions, peptide [Gly₆]cc ₃TP retains similar or exhibits higher activity in six test organisms as compared to cc₃TP. However, other [Gly₆]-analogs are far less active than [Gly₆]cc ₃TP. With the exceptions of [Gly₆]cc₃TP and [Gly₆]cc₂TP-1, their activities are abrogated in high-salt assays.

These results show that conformational rigidity increases activity under high-salt conditions while the combination of conformational rigidity and cystine pairs in the [Gly₆]-analogs can maintain amphipathicity and broad-spectrum antimicrobial activity under high-salt conditions. Increased constraints in the [Gly₆]-analogs also lead to specific recognition of microbial envelopes. The two most constrained analogs [Gly₆]cc₃TP and [Gly₆]cc₂TP-1 are

1 5 14 18			
Lys-*Trp*-Cys-*Phe*-Arg-*Val*-Cys-*Tyr*-Arg-Gly-*Ile*-Cys-*Tyr*-Arg-Arg-Cys-Arg-Gly-NH₂ (TP18)			

Number of constraints	Schematic[a] structure	Compound name/sequence	[Gly₆]-analogs[b]
IV		cc₃TP	[Gly₆]cc₃TP
III		cc₂TP-1	[Gly₆]cc₂TP-1
		cc₂TP-2	[Gly₆]cc₂TP-2
II		TP18	[Gly₆]TP18
	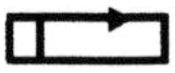	ccTP-1	[Gly₆]ccTP-1
	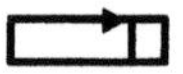	ccTP-2	[Gly₆]ccTP-2
I		[ΔCys^{3,16}]TP18	
		[Gly₄]cTP	[Gly₁₀]cTP
0		LTP/KWC*FRVC*YR-GIC*YRRC*RG-X	[Gly₈]LTP
		rLTP/GYRGIGYRRGR-GKWGFRV-X	[Gly₁₂]rLTP
		[Gly₄]rLTP/GYRGIGYR-RGRGKWGFRV-X	

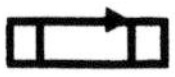

Figure 9. Two analog series of TP18. Sequence of TP 18 shown on top of figure. Closed structures indicate end-to-end peptide bond between Lys[1] and Gly[18]. Arrow indicates the N to C direction of peptide bonds. Cysteine replacements are shown in compound name. Linear peptide analogs contain the following sequence: LTP, KWC(Acm)FRVC(Acm)YRGI C(Acm)YR RC(Acm)RG-X, rLTP and [Gly₄]rLTP GYRGIGYRRGRGKWGFRV-X, where X represents thioester –CO-SCH₂CH₂CONH₂ in rLTP, -NHOH in [Gly₄]rLTP. [Gly₆]-analogs contain six Gly replacements Trp[2], Phe[4], Val[6], Tyr[8,13] and Ile[11] shown as italic on top of figure. Amino acid sequences of [Gly₆]LTP, KGC(Acm)GRG C(Acm)GRGG C(Acm)GRRC(Acm)RG-X and [Gly₆]rLTP GGRGGGGGRRGRGKGGGRG-X. X represents -SCH₂CH₂CONH₂.

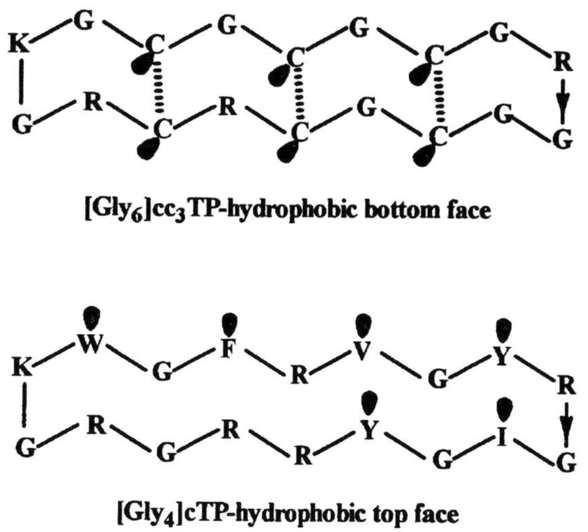

[Gly₆]cc₃TP-hydrophobic bottom face

[Gly₄]cTP-hydrophobic top face

Figure 10. Sequence and topology of [Gly₆]cc₃TP and [Gly₄]cTP with opposite hydrophobic face; black dot represents a hydrophobic amino acid.

active against *M. luteus* at 0.7 μM or less, even under high-salt condition. Radiolabeling these peptides would be useful diagnostically to distinguish infection from inflammation. Collectively, these results show that [Gly₆]cc₃TP could serve as a template for further analog study to improve potency and specificity through single or multiple replacements of hydrophobic or unnatural amino acids.

Unexpectedly, the Gly-rich tachyplesin analogs with or without constraints also display specific activity under low-salt conditions against *E. coli*. Seven [Gly₆]-TP analogs show moderate but consistent antimicrobial activity against *E. coli* with MICs ranging from 2 to 4 μM under low-salt condition, irrespective of the numbers of their constraints. The more constrained cc₃β2 and cc₂β2 [Gly₆]-analogs can retain their activity under high-salt condition with MIC of 1.7 M. The origin of this specificity is not clear but could be related to the spatial clustering of cationic charges of these cationic Gly-rich sequences (Figure 9). Their cyclic constrained nature forms a cluster of cationic charges in the **RCRGK** sequence. The R**RGRGK**GGR sequence is found in [Gly₆]cTP and the retro-linear peptide [Gly₆]rLTP but not in the normal linear peptide, [Gly₆]LTP, which is considerably less active (MIC 17.2 μM). It is interesting to note that Gly-rich sequences with weakly cationic clusters are found in microcins (51) which are oxazole-thiazole peptide antibiotics with strings of Gly repeats.

An interesting comparison can be made for two amphipathic peptides with different degrees of constraints (47). Peptides [Gly₄]cTP and [Gly₆]cc₃TP (Figure 10) share some common global design changes. They contain Gly replacements for either the cystine pairs or six hydrophobic amino acids. [Gly₄]cTP with four Gly replacing two cystine pairs retains a cyclic structure without any disulfide constraint. Such a design places both cationic charges and hydrophobic amino acids on the top face but is devoid of the bottom-face sulfur-rich cystine pairs. The [Gly₆]cc₃TP has the reverse topological amphipathic design which replaces the top-face hydrophobic clusters with Gly but retains the bottom-face sulfur rich cluster. Thus, it is interesting to find that [Gly₄]cTP and [Gly₆]cc₃TP display comparable activity profiles and potencies in low-salt conditions. However, [Gly₆]cc₃TP which is more amphipathic based on reverse-phase HPLC analysis and more constrained than [Gly₄]cTP, retains its activity in high-salt conditions. Moreover [Gly₆]cc₃TP (EC₅₀ >450 μM) is far less hemolytic than [Gly₄]cTP (EC₅₀ 142 μM), and it has a higher therapeutic index that is favorable for clinical considerations. These results suggest that a highly constrained peptide such as [Gly₆]cc₃TP could maintain broad-spectrum activity in both low-and high-salt conditions even without any bulky amino acids in its sequence.

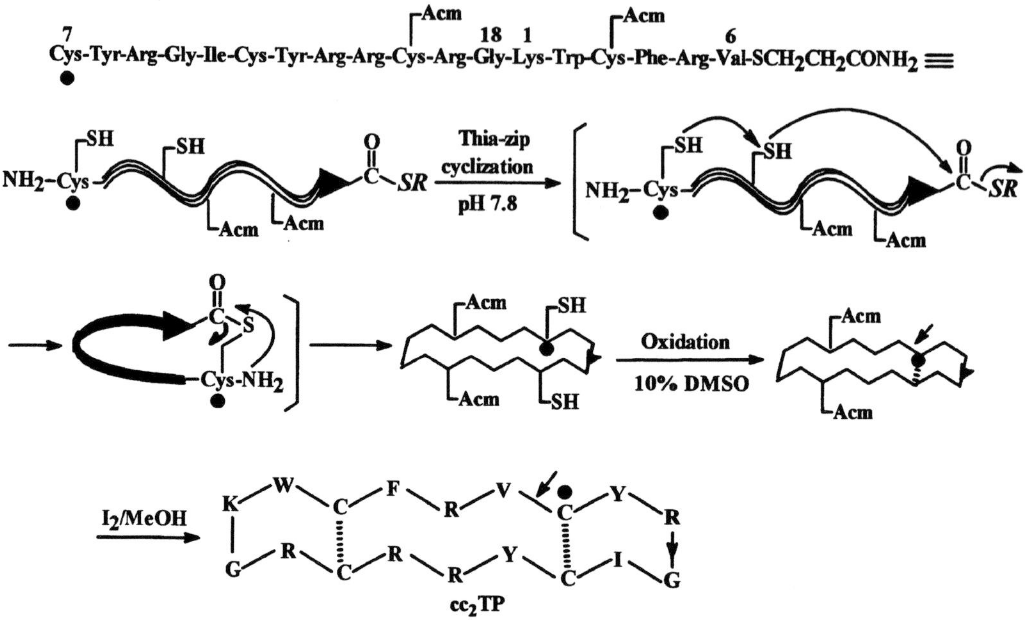

Figure 11. A representative synthetic scheme for end-to-end cyclization and disulfide formation of cc₂TP .Thioester peptide was obtained by stepwise synthesis and the cyclic peptide from thia zip-assisted cyclization. Disulfide bond is first formed by DMSO oxidation. The Cys(Acm) protected disulfide is formed by I₂/CH₃OH. SR represents -SCH₂CH₂CONH₂

New Methodology in the Synthesis and Characterization of Cyclic Cystine-knot Peptides

Syntheses of cyclic cystine-knot peptides and their analogs present challenges using the conventional methods of peptide synthesis because multi-tiered protecting groups are required to achieve selective formation of various constraints of a cyclic amide with a mono- di- or tri-cystine structure. We have therefore developed new orthogonal cyclization methods to facilitate their synthesis. Two orthogonal ligation methods used in this report are Cys-thioester and Ag⁺ ion-assisted cyclizations (21, 30-31, 45-47, 51-54). Both cyclization methods employ linear peptide thioesters as precursors obtained by a stepwise solid phase method and then the precursors are cyclized to form end-to-end peptide bonds in aqueous solutions to afford the desired products. The advantage of these methods is their simplicity, particularly since cyclic peptides can be obtained without undergoing a deprotection step after the cyclization reaction (52). End-group determination coupled with enzymatic digestion are used to confirm the end-to-end cyclic structures (42, 45-47, 51-54).

Cys-thioester Cyclization

The requirement for a Cys-thioester cyclization is an amino terminal (NT)-Cys and a carboxyl terminal (CT)-thioester in the linear precursor sequence (Figure 11). The NT-Cys specifically recognizes the CT-thioester to undergo an intramolecular ligation to form a Xaa-Cys bond. Since the intramolecular ligation is usually much faster than the intermolecular ligation, no oligomerization or polymerization is observed. The Cys-thioester cyclization is performed in a phosphate buffer at pH 7.8 without the use of any coupling reagent (21, 30-31, 51). The process is further assisted by a thia-zip reaction involving a series of thiol-assisted intramolecular rearrangements (54). Ultimately, an amino terminal thiolactone is formed, leading to a spontaneous ring contraction through an S, N-acyl isomerization to form the end-to-end peptide bond (54). The cyclization rates of the cyclic cystine peptides including those cβ2, ccβ2, cc$_2$β2 cc$_3$β2 and the corresponding cc$_3$β3 peptides are found to be facile and completed in <16 h as monitored by reverse phase-high performance liquid chromatography (RP-HPLC).

Ag$^+$-Assisted Cyclization

This cyclization method is used for preparing cβ2 and cβ3 peptides without any Cys in their sequences (Figure 12). Similar to the previously described Cys-thioester cyclization method, an unprotected thioester precursor is required. Unlike the Cys-thioester method, it does not require an NT-Cys present in the precursor peptide. Cyclization of an unprotected peptide thioester obtained by solid-phase synthesis is performed in an aqueous solution assisted by Ag$^+$ ion which coordinates the reactive N- and C-terminal thioester ends of an unprotected peptide thioester to form a cyclic peptide (53). Based on steric consideration, an unhindered amino acid such as Gly is often selected as a carboxyl-terminal amino acid of the precursor thioester peptides. Cyclization assisted by Ag$^+$ ion is performed at pH 5.6 to afford cyclic peptide often in >70% yield.

Disulfide-bond Formation

After the cyclization by the orthogonal methods, two methods of disulfide bond formation have been used. The first method is achieved essentially in a one-pot reaction by adding 10% DMSO (dimethylsulfoxide) to the peptide solution where all disulfides are randomly formed (55). This one-pot reaction is usually complete in 12 h as monitored by HPLC and mass spectrometry (MS). The DMSO-mediated oxidation method is convenient and has been successfully applied to cc$_2$β2, cc$_2$β3 and cc$_3$β2 peptides of the RTD, PG, TP and cyclic α-defensin

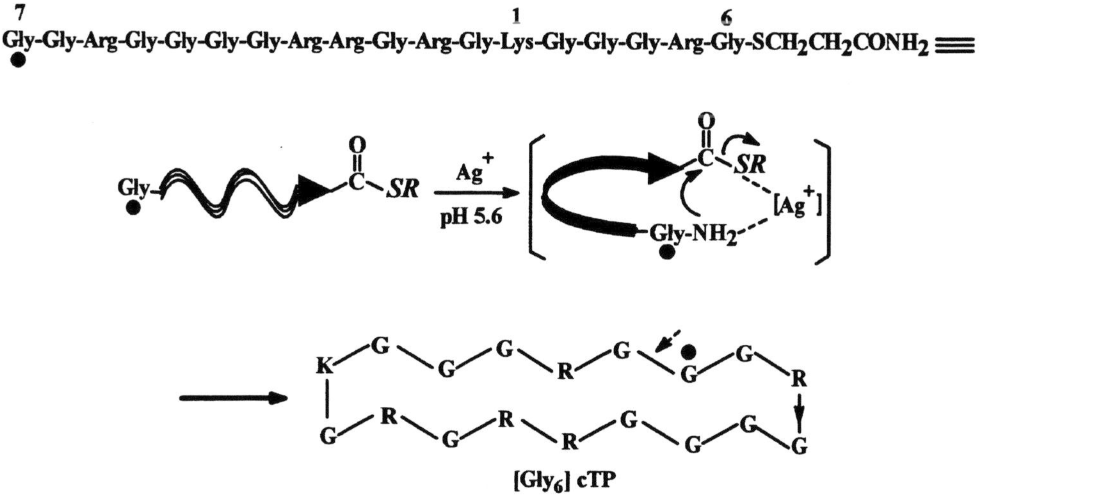

Figure 12. Orthogonal cyclization to form end-to-end cyclic peptide [Gly$_6$]cTP by Ag+ ion-assisted. Thioester peptide is obtained by stepwise synthesis. SR represents -SCH$_2$CH$_2$CONH$_2$.

Table 3. Antimicrobial activity of protegrin and its cyclic analogs

Organism	Media Salt	MIC(μM)					
		[ΔCys⁶,¹³]PG	cc₂PG-1	cc₂PG-2	ccPG-1	ccPG2	cPG
Gram-negative							
E. coli	L	1.1	0.7	0.9	0.5	0.4	0.8
	H	0.8	0.7	1.4	0.8	1.8	0.8
P. aeruginosa	L	2.3	1.0	4.1	2.3	6.2	1.1
	H	2.5	2.3	3.2	1.1	4.8	1.6
P. vulgaris	L	1.2	2.9	1.5	1.8	2.0	2.3
	H	1.7	2.8	5.0	1.4	0.8	1.7
K. oxytoca	L	1.1	16.8	4.4	0.8	17.8	4.0
	H	1.3	14.8	4.0	1.2	13.2	1.6
Gram-positive							
S. aureus	L	0.7	0.5	0.5	0.4	0.5	0.4
	H	1.1	0.4	0.5	0.5	0.8	0.5
M. luteus	L	0.2	0.4	0.3	0.3	0.3	0.6
	H	0.8	0.6	0.5	0.3	0.5	0.7
E. faecalis	L	0.5	1.6	0.5	0.4	0.4	0.9
	H	0.7	1.4	0.5	0.8	0.6	0.8
Fungi							
C. albicans	L	1.3	2.3	2.0	0.5	1.9	1.3
	H	2.0	3.4	8.2	0.7	6.6	7.2
C. kefyr	L	1.9	1.4	1.7	1.2	1.9	1.3
	H	2.1	1.2	1.5	0.8	2.1	0.8
C..tropicalis	L	2.0	1.5	1.7	0.6	1.2	4.0
	H	1.6	1.9	1.2	1.3	1.4	1.5

*Experiment were performed in radial diffusion assay (see Table 2)

series. However, it has not been successful with the cc₃β3 series of plant peptides such as CirA, CirB and Cpt. For these peptides, a two-step chemoselective oxidative disulfide formation has been employed in which a specific disulfide pair is protected by S-acetamidomethyl (Acm) group (56). The first step uses the DMSO oxidation to form the two disulfide pairs. Removal of the Cys(Acm) is performed by adjusting the aqueous solution to pH 4 with acetic acid/0.1 N HCl and treatment with I₂/methanol to form the desired third cystine pair (55).

Characterization of Disulfide Bond Connectivity of Cyclic Cystine-knot Peptides

Two fragmentation methods coupled with mass-spectral analysis have been used to characterize disulfide bond connectivity of ccβ2 and ccβ3 peptides. The first fragmentation method is based on partial acid hydrolysis with a relatively weak acidic mixture such as 0.25 M oxalic acid at 100 °C for 3 to 5 hr (29-31). Disulfide bonds are stable under these conditions of partial acid hydrolysis. This is an advantage because most proteases have maximum activity in neutral and alkaline conditions, which are also the conditions under which disulfide bond may undergo disulfide rearrangements. Factors in hydrolytic conditions such as the time, temperature or pH of the acidic mixture can be varied to modulate the degree of fragmentation. The fragmentation is particularly facile with certain "hot spots" in the peptide sequences. These include acid-susceptible dipeptide sequences of Xaa-Ser/Thr, Asp/Glu-Xaa, and Asn/Gln-Xaa, particularly when Xaa is Gly, Ser, Thr, Asp, Glu, Asn, Gln and Pro.

Hydroxylic side chain of Ser and Thr can isomerize in acidic conditions through N,O-acyl transfer to esters which then can be readily hydrolyzed that favors hydrolysis at the amino side of the Xaa-Ser/Thr dipeptide. Anchimeric assistance of carboxylic side chain of Asp and Glu through either a 5 or 6 member transition state accelerates hydrolysis at the carboxyl side of Asp/Glu-Xaa dipeptide. Similarly, amide side chains of Asn and Gln are also known to undergo aspartimide or glutarimide function at the carboxyl side in the degradation. In general, the five member formation of Asp-Asn formation is about 100 fold faster than the corresponding reactions by Glu/Gln. Other than side chain anchimeric assistance in the acid catalyzed proteolysis steric factor and bond strains also contribute to the susceptibility of peptide bonds. Gly is favored in the hydrolytic pathway because it offers little steric hindrance. On the contrary, amino acids with stericly hindered β-branched side chains such as Ile are resistant to acid hydrolysis. For example, Ile-Val has led to incomplete hydrolysis.

The more popular fragmentation approach of $cc_2\beta3$ and $cc_3\beta3$ peptides is by enzymatic hydrolysis (20, 26, 45-47). Depending on the disulfide connectivity, singly or combination of endopeptidases such as trypsin or endo-Lys for basic amino acids, chymotrypsin for aromatic amino acids, and staph. aureus V8 for acidic amino acids has been used together with matrix assisted laser desorption ionization mass spectrometry (MALDI/MS) analysis for "On-target" disulfide analysis (57).

Circular Dichroism Measurements

CD measurements provide a convenient method to correlate structure and functions (45-47). However, there is insufficient information to deconvolute CD spectra of these highly constrained two- or three β-strand cyclic cystine-stabilized structures with unambiguity due to the various contributions by β-turns, aromatic residues, disulfides and β-sheet structures. Nevertheless, it is possible to show that the constrained cyclic peptides and their corresponding [Gly_6]-analogs display some degree of ordered structures in aqueous and high-salt solutions. For example, in our study of different $cc_3\beta2$ peptides, three types of CD spectra are

Table 4. Hemolytic activity and membranolytic selective index of RTD-1, ccβ2,cc₂β2 and cc₃β3.

Peptide	EC_{50} or EC_{25}* (µM)	Membranolytic selective index					
		E. coli	P. aeruginosa	S. aureus	M. luteus	C. albicans	C. tropicalis
TP-1	91	228	182	182	83	101	91
PG-1	81	98	40	130	102	79	56
RTD-1	2350*	83	452	2611	5875	452	2260
cc₃TP	3800*	1267	760	3800	7600	745	731
cc₃PG	799	555	221	1109	1288	30	40
cc₂TP	159	318	795	318	145	177	159
cc₂PG-1	137	201	60	390	220	47	73
cc₂PG-2	123	101	43	285	279	15	100
ccTP	141	1410	705	235	282	201	235
ccPG-1	130	171	122	270	406	176	104
ccPG-2	139	70	26	162	228	20	98

Hemolytic activity of peptides was determined using human erythrocytes with phosphate-buffered saline. Peptide concentrations causing 50% (EC_{50}) or 25% (EC_{25}*) hemolysis were derived from the dose-response curves. The membranolytic selective index is the ratio of EC_{50} or EC_{25} (µM) to MIC (µM) of antimicrobial activity of peptides under physiological conditions.

obtained in methanol, water and phosphate-buffered high salt solution with 100 mM NaCl at pH 7. The type I spectra include the cystine-stabilized peptides of the open-chained PG-1, TP-1, TP18, monocystine TP and PG. They show two positive bands near 200 and 230 nm and a weak negative band near 208 nm. The positive band (π-π^* transition) around 200 nm and a weaker negative band (n-π^* transition) around 207 nm have been associated with ordered β-sheet peptides connected by a reverse turn (58) and are consistent with the reported CD spectra of PG and TP (59, 60). The strong positive band near 230 nm is usually not observed in β-strand proteins but has been observed in cystine-knot -strand peptides (61). The type II spectra displayed by the less constrained peptides that include cyclic TP and PG without disulfide constraints, linear peptides and the [Gly$_6$]cTP analogs containing only a negative band near 200 nm characteristic of predominantly unordered structures in aqueous solution. Similar CD spectra of linear tachyplesin analogs have been previously reported by others (49, 60). The type III spectra displayed by ccβ2 structures such as RTD, cc$_3$PG, cc$_3$TP as well as their cc$_3\beta$2 and cc$_2\beta$2 analogs could be considered as non-classical CD spectra of β-strand structures. These cyclic peptides retain the negative band near 208 nm and the positive band near 230 nm, but show variations in ellipticity below 200 nm. Furthermore, there are substantial differences in their magnitudes as measured by ellipticity and band shifts in their minima and maxima in methanol, water or phosphate-buffered high-salt condition. For example, the CD spectra of cc$_3$TP and its [Gly$_6$]-analog, which are determined in methanol to avoid peptide aggregation, showed a negative band at 203-208 nm, a strong positive band at 228-230 nm, but a weak positive band < 200 nm. In water or high-salt condition, cc$_3$TP shows a slightly different spectrum with a blue shifted minimum to 202 nm and the absence of the positive band < 200 nm whereas the [Gly$_6$]-analog shows a strong positive band < 200 nm. Tachyplesin is known to behave as a soluble monomeric peptide without any tendency to aggregate in aqueous solutions (59) and our results also confirm that there are no significant changes in the CD spectra of TP or TP18 in methanol or water. The differences in CD spectra observed in water and in high-salt conditions or methanol suggest the possibility of aggregation of the highly hydrophobic and amphipathic cc$_3$TP in low ionic aqueous environments.

Antimicrobial and Hemolytic Activities

A two-stage radial diffusion assay in agarose gels (62) is employed in our study to test the cc$_3\beta$3 and cc$_3\beta$2 peptides against ten organisms under both low- and high-salt conditions, without and with 100 mM NaCl, respectively. Four Gram-negative bacteria include *Escherichia coli* (ATCC #25922), *Pseudomonas aeruginosa* (#27853), *Klebsiella oxytoca* (#49131), and *Proteus vulgaris* (#49132). The three Gram-positive bacteria are *Staphylococcus aureus* (#29213), *Micrococcus luteus* (#49732) and *Enterococcus faecalis* (#29212). The three fungi are *Candida albicans* (#37092), *Candida kefyr* (#37095), and *Candida tropicalis* (#37097). Each peptide is tested over a range of 5,000 fold concentrations and activities are expressed as the minimum inhibitory concentrations (MICs) of 0.1 to >500 µM. Selected MICs of cc$_2\beta$2 and cc$_3\beta$3 peptides are shown in Table 2 and 3.

The hemolytic activity of cyclic peptides on human erythrocytes has also been determined (63) (Table 4). The EC$_{50}$ of TP produced 50% hemolysis of human erythrocytes is 91 µM. Most cyclic peptides are less hemolytic than TP-1 or PG-1, especially those highly constrained with their EC$_{50}$'s >450 µM. Table 4 also compares the membranolytic selectivity indexes (EC$_{50}$/MIC) of some cc$_2\beta$2 and cc$_3\beta$2 peptides to show their improvements in cytoxicity over TP-1 or PG-1 under high-salt conditions. Some cc$_3\beta$3 peptides such as cc$_3$TP analogs (46-47) are highly nonhemolytic to human erythrocytes and the conventional EC$_{50}$ could not be determined because it would require unrealistic high concentrations of peptides.

However, their effective concentrations (EC_{25}) causing 25% red blood cell hemolysis are found to range from 590 µM in cc_3TP analog **2** to 3,900 µM in cc_3TP analog **5** as compared to 29 µM for TP-1. Based on their EC_{25}, we have shown that it is possible to dissociate antimicrobial activity from hemolytic activity using the cc_2β3 design to create analogs with high selectivity index (EC_{25}/MIC) as high as 13,000. For example cc_3TP analogs **2** to **5** show selective indexes from 311 to 13000 ranging as compared to 26 to 97 in TP-1, against the three test Gram-positive organisms.

Concluding Remarks

The cc_3β2 and cc_3β3 peptides or their cc_2β2 and cc_2β3 analogs rigidified by three or two cross-braced disulfide bonds, respectively, and an end-to-end cyclic peptide bond provide novel amphipathic scaffoldings to develop salt-insensitive and selective antimicrobial peptides. Of the two, the smaller scaffoldings of the 18-residue cc_3β2 and cc_2β2 peptides show promise for analog study for developing less toxic therapeutic antibiotics. Both contain two-β-stranded tile-like structures and amphipathicity due to positive charges tethered above the plane and a hydrophobic sulfur core of cystine pairs below the plane. The robustness of the cc_3β2 and cc_2β2 design has been tested on protegrin and tachyplesin peptides with global and/or local changes. These include different degrees of constraints of cβ2, ccβ2 and cc_2β2 peptides and selected [Gly_6]-analogs in which all bulky hydrophobic amino acids are replaced by Gly. Of particular interest is the globally altered [Gly_6]-analog of cc_3β2 peptide such as [Gly_6]cc_3TP that maintain an amphipathic β tile-like framework. CD measurements in aqueous and high-salt solutions also support that such a highly modified [Gly_6]cc_32β peptide displays ordered structures with characteristics of β-sheet peptides. Thus, [Gly_6]ccTP could serve as a rigid amphipathic template for further analog study to improve potency and specificity through single or multiple replacements of hydrophobic or unnatural amino acids.

Although the generally recognized mechanisms of killing microbes by antimicrobial peptides include their inherent property of membranolytic activity, several antimicrobial peptides have been shown to act on specific targets. Some α-defensins are inhibitors of adrenocorticotrophin receptors and block the production of immuno suppressive adrenal steroid hormones during acute infection (64). TP analogs and defensins are also known to be chemoattractants (65). Tachyplesin analogs are inhibitors of CXCR4 receptors and are effective as anti-HIV infective agents (66). Several β-defensins have now been linked to signal cellular immunity as chemotactic factors to both immature dendritic cells and memory T cells through the CCR6 chemokine receptor (67). Insect defensins bind to the Toll family of receptors to mediate antimicrobial response defence (68-70). The Gly-rich antimicrobial peptide microcin B17 has been shown to be a DNA gyrase inhibitor (71) and the bactenecin family of peptides with Arg-Pro rich sequences is believed to have DNA-binding properties (72). Recently, it has also been shown that the antimicrobial peptide nisin Z is specific for the membrane-anchored cell-wall precursor Lipid II (73, as well as Pag and Sahl this monography), which is also the target of the conventional antibiotic vancomycin. However, nisin Z binds to Lipid II with high affinity, rendering the plasma membrane more permeable. These specific modes of actions by antimicrobial peptides in addition to their nonspecific membranolytic activity provide a new incentive for developing highly rigid antimicrobial peptides selective for targeting specific host or microbial components.

Acknowledgement

This work was supported in part by US Public Health Service NIH Grant CA36544, AI46164, and GM57145.

References

1. Boman, H.G. 1995. Peptide antibiotics and their role in innate immunity. Ann. Rev. Immunol. 13: 61-92.
2. Nicolas, P. and Mor, A. 1995. Peptides as weapons against microorganisms in the chemical defense system of vertebrates. Annu. Rev. Micro. 49: 277-304.
3. Zasloff, M. 1992. Antibiotic peptides as mediators of innate immunity. Curr. Opin. Immunol. 4: 3-7.
4. Hancock, R.E. 1997. Peptide antibiotics. Lancet. 349: 418-422.
5. Lehrer, R.I., Lichtenstein, A.K. and Ganz, T. 1993. Defensins: antimicrobial and cytotoxic peptides of mammalian cells. Ann. Rev. Immunol. 11: 105-128.
6. Lehrer, R.I. and Ganz, T. 1999. Antimicrobial peptides in mammalian and insect host defence. Curr. Opin. Immunol. 11: 23-27.
7. Maloy, W.L. and Kari, U.P. 1995. Structure-activity studies on magainins and other host defense peptides. Biopolymers. 37: 105-122.
8. Wade, D., Boman, A., Wahlin, B., Drain, C.M., Andreu, D., Boman, H.G. and Merrifield, R.B. 1990. All-D amino acid-containing channel-forming antibiotic peptides. Proc. Natl Acad. Sci. USA. 87: 4761-4765.
9. Merrifield, E.L., Mitchell, S.A., Ubach, J., Boman, H.G., Andreu, D. and Merrifield, R.B. 1995. D-enantiomers of 15-residue cecropin A-melittin hybrids. Int. Peptide Protein Res. 46: 214-220.
10. Kondejewski, L.H., Farmer, S.W., Wishart, D.S., Kay, C.M., Hancock, R.E. and Hodges, R.S. 1996. Modulation of structure and antibacterial and hemolytic activity by ring size in cyclic gramicidin S analogs. J. Biol. Chem. 271: 25261-25268.
11. Subbalakshmi, C., Krishnakumari, V., Nagaraj, R. and Sitaram, N. 1996. Requirements for antibacterial and hemolytic activities in the bovine neutrophil derived 13-residue peptide indolicidin. FEBS Letters. 395: 48-52.
12. Hong, S.Y., Oh, J.E. and Lee, K.H. 1999. Effect of D-amino acid substitution on the stability, the secondary structure, and the activity of membrane-active peptide. Biochem. Pharmacol. 58: 1775-1780.
13. Andreu, D., Ubach, J., Boman, A., Wahlin, B., Wade, D., Merrifield, R.B. and Boman, H.G. 1992. Shortened cecropin A-melittin hybrids. Significant size reduction retains potent antibiotic activity. FEBS Letters. 296: 190-194.
14. Hamuro, Y., Schneider, J.P. and DeGrado, W.F. 1999. Design of antibacterial –peptides. J. Am. Chem. Soc. 121: 12200-12201.
15. Yokum, T.S., Elzer, P.H. and McLaughlin, M.L. 1996. Antimicrobial alpha, alpha-dialkylated amino acid rich peptides with *in vivo* activity against an intracellular pathogen. J. Med. Chem. 39: 3603-3605.
16. Kondejewski, L.H., Jelokhani-Niaraki, M., Farmer, S.W., Lix, B., Kay, C.M., Sykes, B.D., Hancock, R.E. and Hodges, R.S. 1999. Dissociation of antimicrobial and hemolytic activities in cyclic peptide diastereomers by systematic alterations in amphipathicity. J. Biol. Chem. 274: 13181-13192.

17. Dathe, M., Wieprecht, T., Nikolenko, H., Handel, L., Maloy, W.L., MacDonald, D.L., Beyermann, M. and Bienert, M. 1997. Hydrophobicity, hydrophobic moment and angle subtended by charged residues modulate antibacterial and haemolytic activity of amphipathic helical peptides. FEBS Letters. 403: 208-212.
18. Blondelle, S.E. and Houghten, R.A. 1992. Design of model amphipathic peptides having potent antimicrobial activities. Biochemistry. 31: 12688-12694.
19. Dathe, M., Schumann, M., Wieprecht, T., Winkler, A., Beyermann, M., Krause, E., Matsuzaki, K., Murase, O. and Bienert M. 1996. Peptide helicity and membrane surface charge modulate the balance of electrostatic and hydrophobic interactions with lipid bilayers and biological membranes. Biochemistry. 35: 12612-12622.
20. Tang, Y.Q., Yuan, J., Osapay, G., Osapay, K., Tran, D., Miller, C.J., Ouellette, A.J. and Selsted, M.E. 1999. A cyclic antimicrobial peptide produced in primate leukocytes by the ligation of two truncated alpha-defensins. Science. 286: 498-502.
21. Tam, J.P., Lu, Y.-A., Yang, J.L. and Chiu, K.W. 1999. An unusual structural motif of antimicrobial peptides containing end-to-end macrocycle and cystine-knot disulfides. Proc. Natl. Acad. Sci. USA. 96: 8913-8918.
22. Cardellina, J.H 2d., Munro, M.H., Fuller, R.W., Manfredi, K.P., McKee, T.C., Tischler, M., Bokesch, H.R., Gustafson, K.R., Beutler, J.A. and Boyd, M.R. 1993. A chemical screening strategy for the dereplication and prioritization of HIV-inhibitory aqueous natural products extracts. J. Nat. Prod. 56: 1123-1129.
23. Gran, L. 1973. Oxytocic principles of Oldenlandia affinis. Lloydia. 36: 174-178.
24. Goransson, U., Luijendijk, T., Johansson, S., Bohlin, L. and Cleaeson, P. 1999. Seven novel macrocyclic polypeptides from *Viola arvensis*. J. Nat. Prod. 62: 283-286.
25. Claeson, P., Goransson, U., Johansson, S., Luijendijk, T. and Bohlin, L. 1998. Fractionation protocol for the isolation of polypeptides from plant biomass. J. Nat. Prod. 61: 77-81.
26. Hallock, Y.F., Sowder, II R.C., Pannell, L.K., Hughes, C.B., Johnson, D.G., Gulakowski, R.G., Cardellina, J.H. and Boyd, M.R. 2000. Cycloviolins A-D, anti-HIV macrocyclic peptides from Leonia cymosa. J. Org. Chem. 65: 124-128.
27. Pallaghy, P.K., Nielsen, K.J., Craik, D.J. and Norton, R.S. 1994. A common structural motif incorporating a cystine knot and a triple-stranded beta-sheet in toxic and inhibitory polypeptides. Protein Science. 3: 1833-1839.
28. Saether, O., Craik, D.J., Campbell, I.D., Sletten, K., Juul, J. and Norman, D.G. 1995. Elucidation of the primary and three-dimensional structure of the uterotonic polypeptide kalata B1. Biochemistry. 34: 4147-4158.
29. Derua, R., Gustafson, K.R. and Pannell, L.K. 1996. Analysis of the disulfide linkage pattern in circulin A and B, HIV-inhibitory macrocyclic peptides. Biochem. Biophys. Res. Commun. 228: 632-638.
30. Tam, J.P. and Lu, Y.-A. 1997. Synthesis of large cyclic cystine-knot peptide by orthogonal coupling strategy using unprotected peptide precursor. Tetrahedron Lett. 38, 5599-5602.
31. Tam, J.P. and Lu, Y.-A. 1998. A biomimetic strategy in the synthesis and fragmentation of cyclic protein. Protein Science. 7: 1583-1592.
32. Daly, N.L., Koltay, A., Gustafson, K.R., Boyd, M.R., Casas-Finet, J.R. and Craik, D.J. 1999. Solution structure by NMR of circulin A: a macrocyclic knotted peptide having anti-HIV activity. J. Mol. Biol. 285: 333-345.
33. Gustafson, K.R., Sowder, II R.C., Henderson, L.E., Parsons, I.C., Kashman, Y., Cardellina, II J.H., McMahon, J.B., Buckheit, Jr R.W., Pannell, L.K. and Boyd, M.R. 1994. Circulines A and B: novel HIV-inhibitory macrocyclic peptides from the tropical tree *Chassalin parvifolia*. J. Am. Chem. Soc. 116: 9337-9338.

34. Gran, L. 1973. On the effect of a polypeptide isolated from "Kalata-Kalata" (*Oldenlandia affinis* DC) on the oestrogen dominated uterus. Acta Pharmacol. Toxicol. 33: 400-408.

35. Daly, N.L., Love, S., Alewood, P.F. and Craik, DJ. 1999. Chemical synthesis and folding pathways of large cyclic polypeptides: studies of the cystine knot polypeptide kalata B1. Biochemistry. 38: 10606-10614.

36. Witherup, K.M., Bogusky, M.J., Anderson, P.S., Ramjit, H., Ransom, R.W., Wood, T. and Sardana, M. 1994. Cyclopsychotride A, a biologically active, 31-residue cyclic peptide isolated from *Psychotria longipes*. J. Nat. Prod. 57: 1619-1625.

37. Meister, M., Lemaitre, B. and Hoffmann, J.A. 1997. Antimicrobial peptide defense in Drosophila. Bioessays. 19: 1019-1026.

38. Ouellette, A.J. and Selsted, M.E. 1996. Paneth cell defensins: endogenous peptide components of intestinal host defense. FASEB J. 10: 1280-1289.

39. Porter, E.M., van Dam, E., Valore, E.V. and Ganz, T. 1997. Broad-spectrum antimicrobial activity of human intestinal defensin 5. Infect. Immun. 65: 2396-2401.

40. Valore, E.V., Park, C.H., Quayle, A.J., Wiles, K.R., McCray, P.B Jr. and Ganz, T. 1998. Human beta-defensin-1: an antimicrobial peptide of urogenital tissues. J. Clin. Invest. 101: 1633-1642.

41. Goldman, M.J., Anderson, G.M., Stolzenberg, E.D., Kari, U.P., Zasloff, M. and Wilson, J.M. 1997. Human beta-defensin-1 is a salt-sensitive antibiotic in lung that is inactivated in cystic fibrosis. Cell. 88: 553-560.

42. Yu, Q., Lehrer, R.I. and Tam, J.P. 2000. Engineered salt-insensitive alpha-defensins with end-to-end circularized structures. J. Biol. Chem. 275: 3943-3949.

43. Kokryakov, V.N., Harwig, S.S., Panyutich, E.A., Shevchenko, A.A., Aleshina, G.M., Shamova, O.V., Korneva, H.A. and Lehrer, R.I. 1993. Protegrins: leukocyte antimicrobial peptides that combine features of corticostatic defensins and tachyplesins. FEBS Lett. 327: 231-236.

44. Nakamura, T., Furunaka, H., Miyata, T., Tokunaga, F., Muta, T., Iwanaga, S., Niwa, M., Takao, T. and Shimonishi, Y. 1998. Tachyplesin, a class of antimicrobial peptide from the hemocytes of the horseshoe crab (*Tachypleus tridentatus*). Isolation and chemical structure. J. Biol. Chem. 263: 16709-16713.

45. Tam, J.P., Wu, C. and Yang, J.L. 2000. Membranolytic selectivity of cystine-stabilized cyclic protegrins. Eur. J. Biochem. 267. 3289-3300.

46. Tam, J.P., Lu, Y.-A. and Yang, J.L. 2000. Marked increase in membranolytic selectivity of novel cyclic tachyplesins constrained with an antiparallel two-beta strand cystine knot framework. Biochem. Biophys. Res. Commun. 267: 783-790.

47. Tam, J.P., Lu, Y.-A. and Yang, J.L. 2000. Design of salt-insensitive glycine-rich antimicrobial peptide with cyclic tricystine structures. Biochemistry. 39: 7159-7169.

48. Hilbich, C., Kisters-Woike, B., Reed, J., Masters, C.L. and Beyreuther, K. 1991. Aggregation and secondary structure of synthetic amyloid beta A4 peptides of Alzheimer's disease. J. Mol. Biol. 218: 149-163.

49. Rao, A.G. 1999. Conformation and antimicrobial activity of linear derivatives of tachyplesin lacking disulfide bonds. Arch. Biochem. Biophy. 361: 127-134.

50. Richardson, J.S. and Richardson, D.C. 1988. Amino acid preferences for specific locations at the ends of alpha helices Science. 240: 1648-52.

51. Zhang, L. and Tam, J.P. 1997. Synthesis and application of unprotected cyclic peptides as building blocks for peptide dendrimers. J. Am. Chem. Soc. 119: 2363-2370.

52. Tam, J.P., Yu, Q. and Miao, Z. 1999. Orthogonal ligation strategies for peptide and protein. Biopolymers. 51: 311-332.

53. Zhang, L. and Tam, J.P. 1999. Lactone and lactam library synthesis by silver ion-assisted

orthogonal cyclization of unprotected peptides. J. Am. Chem. Soc. 121: 3311-3320.

54. Tam, J.P., Lu, Y.-A. and Yu, Q. 1999. Thia zip reaction for synthesis of large cyclic peptides: mechanisms and applications. J. Am. Chem. Soc. 121: 4316-4324.

55. Tam, J.P., Wu, C.R., Liu, W. and Zhang, J.W. 1991. Disulfide bond formation in peptides by dimethyl sulfoxide. Scope and application. J. Am. Chem. Soc. 113: 6657-6662.

56. Yang, Y., Sweeney, W.V., Schneider, K., Chait, B.T. and Tam, J.P. 1994. Two-step selective formation of three disulfide bridges in the synthesis of the C-terminal epidermal growth factor-like domain in human blood coagulation factor IX. Protein Science. 3: 1267-1275.

57. Kussmann, M., Lassing, U., Sturmer, C.A., Przybylski, M. and Roepstorff, P. 1997. Matrix-assisted laser desorption/ionization mass spectrometric peptide mapping of the neural cell adhesion protein neurolin purified by sodium dodecyl sulfate polyacrylamide gel electrophoresis or acidic precipitation. J. Mass Spectrom. 32: 483-493.

58. Woody, R.W. 1995. Circular dichroism. Methods in Enzymology. 246: 34-71.

59. Oishi, O., Yamashita, S., Nishimoto, E., Lee, S., Sugihara, G. and Ohno, M. 1997. Conformations and orientations of aromatic amino acid residues of tachyplesin I in phospholipid membranes. Biochemistry. 36: 4352-4359.

60. Tamamura, H., Ikoma, R., Niwa, M., Funakoshi, S., Murakami, T. and Fujii, N. 1993. Antimicrobial activity and conformation of tachyplesin I and its analogs. Chem. Pharm. Bull. 41: 978-980.

61. Daly, N.L., Love. S., Alewood, P.F. and Craik, D.J. 1999. Chemical synthesis and folding pathways of large cyclic polypeptides: studies of the cystine knot polypeptide kalata B1. Biochemistry. 38: 10606-10614.

62. Lehrer, R.I., Rosenman, M., Harwig, S.S., Jackson, R. and Eisenhauer, P. 1991. Ultrasensitive assays for endogenous antimicrobial polypeptides. J. Immunol. Methods. 137: 167-173.

63. Fehlbaum, P., Bulet, P., Michaut, L., Lagueux, M., Broekaert, W.F., Hetru, C. and Hoffmann, J.A. 1994. Insect immunity. Septic injury of Drosophila induces the synthesis of a potent antifungal peptide with sequence homology to plant antifungal peptides. J. Biol. Chem. 269: 33159-33163.

64. Solomon, S., Hu, J., Zhu, Q., Belcourt, D., Bennett, H.P., Bateman, A. and Antakly, T. 1991. Corticostatic peptides. J. Steroid Biochem. Mol. Biol. 40: 391-398.

65. Chertov, O., Michiel, D.F., Xu ,L., Wang, J.M., Tani, K., Murphy, W.J., Longo, D.L. and Taub, D.D. 1996. Oppenheim J.J. Identification of defensin-1, defensin-2, and CAP37/azurocidin as T-cell chemoattractant proteins released from interleukin-8-stimulated neutrophils. J. Biol. Chem. 271: 2935-2940.

66. Xu, Y., Tamamura, H., Arakaki, R., Nakashima, H., Zhang, X., Fujii, N., Uchiyama, T. and Hattori, T. 1999. Marked increase in anti-HIV activity, as well as inhibitory activity against HIV entry mediated by CXCR4, linked to enhancement of the binding ability of tachyplesin analogs to CXCR4. AIDS Research & Human Retroviruses. 15: 419-427.

67. Yang, D., Chertov, O., Bykovskaia, S.N., Chen, Q., Buffo, M.J., Shogan, J., Anderson, M. Schroder, J.M., Wang, J.M., Howard, O.M. and Oppenheim, J.J. 1999. Beta-defensins: linking innate and adaptive immunity through dendritic and T cell CCR6. Science. 286: 525-528.

68. Hoffmann, J.A., Kafatos, F.C., Janeway, C.A. and Ezekowitz, R.A. 1999. Phylogenetic perspectives in innate immunity. Science. 284: 1313-1318.

69. Medzhitov, R. and Janeway, C.A Jr. 1998. Self-defense: the fruit fly style. Proc. Natl. Acad. Sci. USA. 95: 429-430.

70. Williams, M.J., Rodriguez, A., Kimbrell, D.A. and Eldon, E.D. 1997. Thr 18-wheeler mutation revesls complex antibacterial gene regulation in *Drosophila* host defense. EMBO J. 16: 6120-6130.

71. Yorgey, P., Lee, J., Kordel, J., Vivas, E., Warner, P., Jebaratnam, D. and Kolter, R. 1994. Posttranslational modifications in microcin B17 define an additional class of DNA gyrase inhibitor. Proc. Natl. Acad. Sci. USA. 91: 4519-4523.

72. Skerlavaj, B., Romeo, D. and Gennaro, R. 1990. Rapid membrane permeabilization and inhibition of vital functions of gram-negative bacteria by bactenecins. Infect. Immun. 58: 3724-3730.

73. Breukink, E., Wiedemann, I,. van Kraaij, C., Kuipers, O.P., Sahl, H. and de Kruijff, B. 1999. Use of the cell wall precursor lipid II by a pore-forming peptide antibiotic. Science. 286: 2361-2364.

From: *Development of Novel Antimicrobial Agents: Emerging Strategies*
ISBN 1-898486-23-9 © 2001 Horizon Scientific Press, Wymondham, UK.

17

Synthetic Combinatorial Libraries: An Emerging Approach Toward the Identification of Novel Antibiotics

Sylvie E. Blondelle

Abstract

Development of novel therapeutics for the treatment of bacterial infection has become a clinical imperative. The greatest threat to current antibiotic coverage is the rapid evolution and spread of drug-resistance, which has now been reported against every currently available antibiotics. A solution to this dilemma is to develop a broad range of lead compounds available for clinical trials. Synthetic combinatorial libraries made up of hundreds to millions of small organic molecules, peptidomimetic compounds, and peptides have been successfully developed and used to discover new antimicrobial leads. The strength of combinatorial libraries relies on: 1) the rapid identification of highly active compounds from large pools or arrays of individual compounds, 2) the ongoing development of solid-phase approaches for the generation of small molecule libraries in an efficient and reproducible manner, and 3) the generation of many non-support bound libraries, which allow the ready performance of cell-based assays. Although not thoroughly exploited in the antibiotic research area, combinatorial chemistry is anticipated to greatly decrease the time it takes to develop novel antibiotic leads.

Introduction

Infectious diseases are nowadays the leading cause of death world-wide (1), and the third leading cause of death in the United States (2). Pathogens antimicrobial resistance is a major concern in infectious diseases such as acute respiratory infections, tuberculosis, and diarrheal diseases. In particular, new opportunistic pathogens are emerging and disseminating in a growing immune system-debilitated host population, and common or resurgent pathogens are becoming resistant to standard antibiotics in community-acquired infection (3). For example, the increased isolation of methicillin-resistant *Staphylococcus aureus* (MRSA) strains represents significant problems in clinical practice (4, 5), with the recent reports of three separate cases of *S. aureus* infection by methicillin-resistant strains having intermediate resistance to vancomycin (6-8). Similarly to the considerable increase in incidence of drug-resistance in Gram-positive infections in the hospital setting, notably among *S. aureus* and enterococci species, drug-resistance in Gram-negative organisms such as *Pseudomonas*, *Serratia*, and *Acinetobacter* species remains an ongoing problem (9). The emergence of multiple drug resistant *Mycobacterium tuberculosis* strains in the AIDS, drug addict, and prison populations represents another example of potential threat caused by such phenomena in terms of the spread of infection (10). Continued development of novel therapeutics for the treatment of bacterial infection has therefore become an important clinical imperative.

The difficulty in designing new antimicrobial agents relies in the inability to predict which infectious diseases will be untreatable 8 to 10 years from now (typical time for a new therapeutic substance to be introduced into the market). Indeed, drug-resistant strains are aggressively spreading and the changes in pathogens are difficult to predict. This dilemma may be dealt in part through the development of a wide range of new lead compounds available for clinical trials. An expending approach in both academic and industrial laboratories is the development and use of synthetic combinatorial libraries (SCLs) for the identification of novel biologically active agents. The SCL approaches enable the synthesis and ready use of libraries composed of collections of thousands to millions of compounds for drug discovery. These libraries represent chemical collections of heterocyclic and small organic compounds, peptidomimetics, and peptides and have been successfully used in a variety of biological assays (reviewed in refs. 11-13). Although not yet fully exploited in the antibiotic research area (only five separate groups have to date reported their studies), combinatorial chemistry is anticipated to greatly decrease the time it takes to develop novel antibiotics.

Table 1. SCLs screened for antimicrobial and/or antifungal activity.

Library Chemical Class	Target Microorganism	Total Compounds	Activity of Lead Compound[b]	Reference
Tetrapeptide	*Staphylococcus aureus*	10,185,728	MIC = 3-4 µg/ml	(38)
Hexapeptide	*Staphylococcus aureus*	34,012,224	MIC = 3-7µg/ml	(30)
Hexapeptide	*Staphylococcus aureus*	52,128,400	IC_{50} = 5µg/ml	(50,58)
Hexapeptide	*Escherichia coli*	52,128,400	MIC = 16-32µg/ml	(59)
Hexapeptide	*Candida albicans*	52,128,400	IC_{50} = 28µg/ml	(60)
Hexapeptide	Fungi broad spectrum[a]	52,128,400	MIC = 5-36µg/ml	(20)
Decapeptide	*Candida albicans*	40,353,607	MIC = 0.78µg/ml	(21)
18-mer peptide	*Staphylococcus aureus*	130,321	MIC = 2-3µg/ml	(44)
Permethylated hexapeptide	*Staphylococcus aureus*	34,012,224	MIC = 11-15µg/ml	(22)
Alkylated tetrapeptide	*Staphylococcus aureus*	7,311,616	N/A	(23)
Pentamine	*Staphylococcus aureus*	7,311,616	N/A	(23)
N-alkyl aminocyclic urea/thiourea	*Candida albicans*	118,400	MIC = 8-16µg/ml	(24)
Bicyclic guanidine	*Candida albicans*	102,459	MIC = 3-4g/ml	(25)
Polyazapyridinophane	*Escherichia coli imp⁻ Streptococcus pyogenes*	1,500	MIC = 5-10µM	(27)
Pyridinopolyamine	*Escherichia coli imp⁻ Streptococcus pyogenes*	1,638	MIC = 1-3µM	(26)
Oxyamine	*Escherichia coli imp⁻ Streptococcus pyogenes*	405	MIC = 1-5µM	(28)
Triazine	*Staphylococcus aureus Bacillus subtilis*	46,656	MIC = 4-16µg/ml	(29)

[a]*Fusarium oxysporum, Rhizoctonia solani, Ceratocystis fagacearum, Pythium ultimum.* [b]MIC = minimum inhibitory concentration and IC_{50} = concentration necessary to inhibit 50% growth as determined by a microdilution assay.

L-amino acid peptides

Ac-RRWWCR-NH$_2$
Ac-RRWWRF-NH$_2$
Ac-RRWCKR-NH$_2$
Ac-FRWLLF-NH$_2$
Ac-FRWWHR-NH$_2$

KKVVFKVKFK-NH$_2$
KKVVVKVKFK-NH$_2$
KKVVFKFKFK-NH$_2$

YKLLKLLPKLKGLLFKL-NH$_2$
YKLLKLLLPKLKPLLFKL-NH$_2$

N-alkylated peptides

pm[LFIFFF]
pm[FFIFFF]
pm[FFFFFFF]

N-benzyl aminocyclic ureas

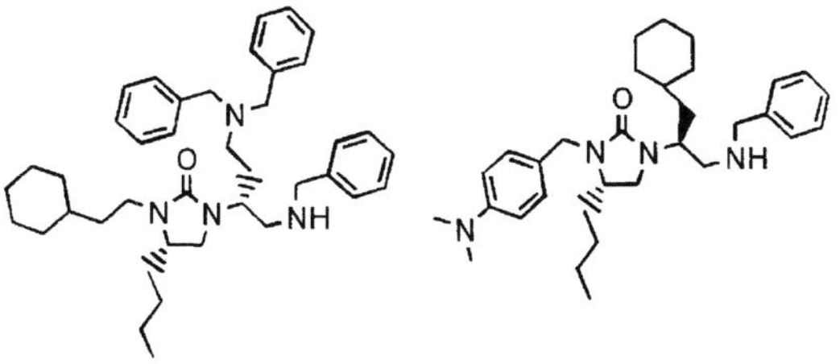

Bicyclic guanidines

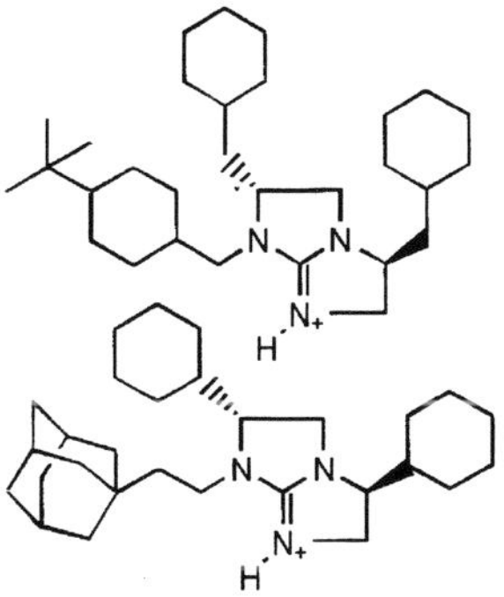

Triazines

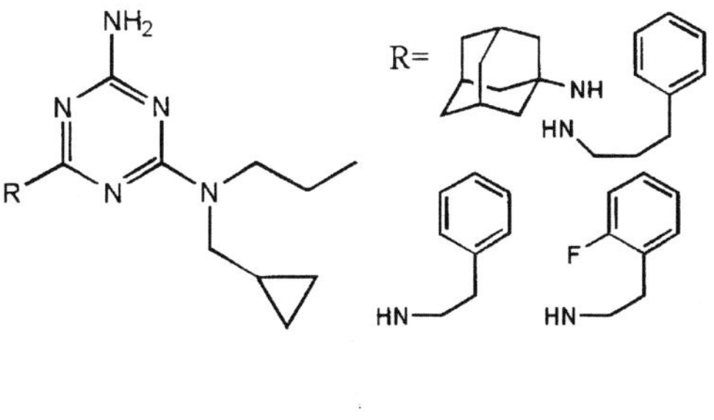

D-amino acid peptides

frlkfh-NH$_2$
frlhf-NH$_2$

L-, D-, unnatural amino acid containing peptides

(αFmoc-εLys)Wfl-NH$_2$
(αFmoc-εLys)WKW-NH$_2$
(αFmoc-εLys)WYr-NH$_2$
(αFmoc-εLys)cir-NH$_2$

Oxyamines

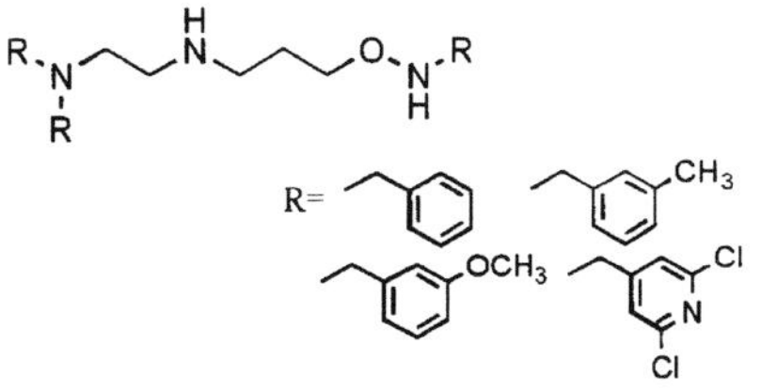

Pyridinopolyamines

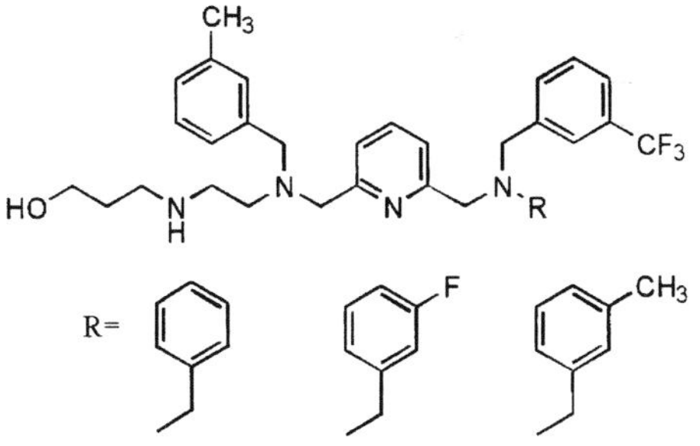

Polyazapyridinophanes

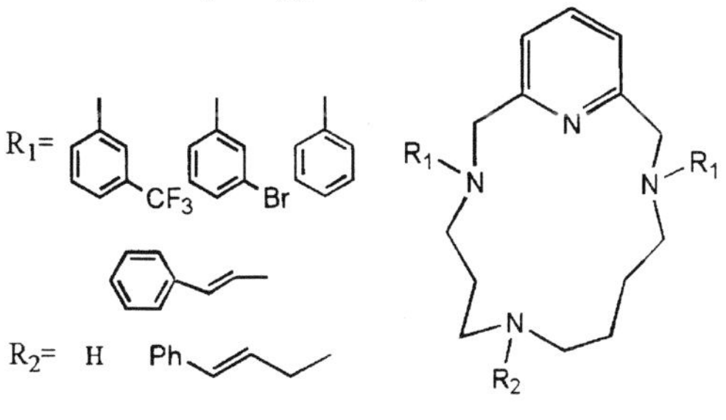

Figure 1.Structure of lead antimicrobial and/or antifungal compounds identified from combinatorial libraries.

Combinatorial Libraries: Vast Source of Antibiotic Candidates

Novel drug candidates are typically discovered through large scale screening of individual synthetic compounds and natural products, as well as through synthesis programs aimed at chemically modifying existing drugs. These candidates are therefore found in large banks of existing known compounds, or are isolated from natural sources such as soil samples, marine waters, insects, and tropical plants (reviewed in refs. 14-16). A better understanding of the mechanisms of resistance also enables the definition of novel inhibition targets and pathways to be explored (17), and in turn, permits the chemical design of modified versions of existing antimicrobial compounds. The major classes of existing antibiotics include tetracyclines, glycopeptides, β-lactams, quinolones, macrolides, oxazolidinones, aminoglycosides, cyclic peptides, polypeptides, and isolated natural products (reviewed in ref. 18). Although, these various compounds can be referred as a library or large collection of drug candidates, the identification of such compounds results from 50 years of effort by synthetic chemists at various academic and pharmaceutical organizations.

Synthetic combinatorial chemistry approaches have created a vast new source of molecular diversity for the potential identification of lead compounds. This revolutionary field enables hundreds to thousands of times more compounds to be synthesized and screened relative to traditional approaches. Antimicrobial and/or antifungal compounds identified through combinatorial chemistry approaches include peptides of various length (19-21), N-alkylated peptides (22,23), pentamines (23), N-alkyl aminocyclic ureas and thioureas (24), bicyclic guanidines (25), pyridinopolyamines (26), polyazapyridinophanes (27), pyridinopolyamines (26), oxamines (28), and triazines (29) (Table 1 and Figure 1).

Combinatorial Library Classification

Library Definition

As mentioned earlier, large collection of synthetically prepared compounds or compounds isolated from natural sources have been referred as library of compounds for high throughput screening purposes. However, these compound collections differ from the synthetic combinatorial library repertoire in that a given SCL is composed of individual compounds of the same chemical class that are simultaneously generated in a systematic manner (30). Thus, a SCL is built around a given pharmacophore by generating all possible combinations of selected sets of functional groups at a defined number of positions. Each SCL then differs from one to the other by its chemical class (i.e., from peptides to organic molecules), size (i.e., total number of individual compounds present in the library), number and location of the diversity positions, and format driving the deconvolution process toward the identification of the active components (i.e., single compound arrays, tagged arrays, resin-bound one bead/one compound, mixtures of compounds free in solution, etc.).

Library Generation

Two approaches have been taken for the generation of combinatorial libraries: 1) molecular biology in which libraries of peptides are expressed on fusion phage or other vector systems (31); 2) solution and solid-phase chemical synthesis (12). Since the identification of novel antimicrobial lead compounds has only been reported from studies using synthetically prepared combinatorial libraries, the molecular biology approach will not be discussed in this review.

Most of the SCLs reported to date have been generated by solid-phase synthetic methods, while the preparation of SCLs by solution phase has been relatively unexplored. The primary advantages of solid-phase methods over solution phase synthesis rely on the capability of driving reactions on polymer supports to completion (often >99.8%), the ability to readily remove excess reagents or starting materials, and ease of automation. In particular, in the past five years, a number of standard organic reactions have been adapted to solid phase chemistry [reviewed in (32)], and these synthetic approaches and concepts have now been broadly applied to the synthesis of heterocyclic and other classic organic compounds (12,33-35). These advances in solid-phase chemistry have greatly contributed to the fast growing development of non-peptide SCLs. On the other hand, solution phase synthesis allows purification steps to be performed following any chemical reaction, which is required for reactions that cannot be driven to completion using chemical pathways known to date, as well as the synthesis of larger quantities of library material.

Chemical approaches to the generation of combinatorial libraries allow for the insertion of D-amino acids (36,37), non-proteinogenic amino acids (38), and carboxylic acids (39) as building blocks, for the post-synthetic chemical modification of peptide libraries as a means of generating peptidomimetic libraries (22), and for the stepwise synthesis of small organic molecule libraries (12). Typically, SCLs containing several positions of diversity are synthesized by the consecutive incorporation of multifunctional building blocks with orthogonal protecting groups. In the case of the first building block, the solid phase support serves as a protecting group for one functionality following incorporation. Upon deprotection of the orthogonal protecting groups, subsequent building blocks are similarly incorporated until all positions of diversity are added. Two synthetic approaches involving either the mixing of multiple resins (30,40) or the use of mixtures of incoming reagents (41-43) are now widely used to generate mixture-based SCLs where multiple functionalities are simultaneously incorporated at diverse positions within a combinatorial library. The principal differences between the various synthetic combinatorial technologies are the solid support, means of incorporating the building blocks to generate mixtures, and screening conditions (support-bound or in solution).

Table 2. Single compound arrays vs mixture screening.

Library format[a]	# of samples	# of compounds per sample	# of plates[b]
Individual array			
OOO	27,000	1	282
Iterative[c]			
OXX	30	900	1
iterations			
$@_1OX$	30	30	1
$@_1@_2O$	30	1	1
Positional Scanning			
OXX / XOX / XXO	90	900	1
deconvolution			
$@_1@_2@_3$	1 to 250	1	1 to 3

[a]A library of 27,000 compounds made up using 30 different building blocks at each of the three diversity positions is used in this example. O represents one of the 30 functional groups, X represents an equimolar mixture of the 30 functional groups at the given diversity position, and @ represents a defined functional group that has been selected to carry out the deconvolution process (see Figures 3 and 4). [b]A single data point per sample and the use of all 96 wells in each assay plate have been considered to calculate the number of plates. [c]A single mixture is selected at each step of the deconvolution process in this example.

Table 3. Natural *vs* synthetic mixture screening.

	Natural product extracts	Mixture-based SCLs
Direct screening of compound mixtures	Yes	Yes
Discovery of highly active compounds	Yes	Yes
Equal concentrations of compounds	No	Yes
Known chemical structure	No	Yes
Known synthetic pathway	No	Yes

Individual Compound Arrays vs Mixture-based SCLs

The robotic synthesis of very large individual compound arrays and the synthesis of mixtures of compounds have both pragmatic long-term utility. Robotically prepared individual compound arrays and classical high throughput screening systems provide thorough information since all compounds in a given class are individually examined. In contrast, while not providing complete information on every single compounds present in the library, mixture-based libraries have the advantage of greatly decreasing the economics and time constraints relevant to compound array systems. For example, a substantial cost reduction in using libraries formatted as mixtures is found in decreased amounts of reagents and materials required for screening (up to hundred fold - Table 2). Furthermore, when using mixture-based libraries, screening of millions of compounds can be accomplished even in assays that are not formatted for conventional high throughput, such as tissue or *in vivo* assays, as well as by any academic laboratory having low screening capacity.

The use of mixture-based SCLs is comparable to the screening of natural product extracts and bacterial broths in that these diversity sources are composed of a complex mixture of compounds. However, mixture-based SCLs have a number of clear advantages (Table 3). For example, the structural nature of the compounds making up the SCLs is known, the concentration of every individual compounds within the libraries is approximately equal, and no synthetic hurdles exist once an individual active compound has been identified.

Generic vs Biased SCLs

The majority of combinatorial libraries developed to date are built around a given pharmacophore that is unrelated to any known antibiotics. These generic libraries are actually often generated for screening in various biological assays instead of being targeted to a single biological assay. The premise behind the selection of the library scaffolds is that the broader are the chemical diversity and number of compounds for each class of structures generated, the greater is the probability of identifying ligands having chemical characteristics that are different from native ligands. Such unrelated ligands are therefore more prone to circumvent existing drug-resistance than analogs of existing antibiotics.

On the other hand, once a lead compound has been identified from screening a generic SCL or traditional structure-activity relationship (SAR) studies, a biased combinatorial library can be generated around this lead compound. This allows the synthesis of thousands of analogs of a given compound in a time frame that could not be previously manageable using traditional chemical methodologies. For example, a SCL was built around a 18-mer peptide that was found using traditional SAR studies to exhibit antimicrobial and antifungal activity (44). Analogs having 10-fold higher anti-staphylococcus activity, combined with a 4- to 9-fold lower hemolytic activity than the original sequence were identified from a biased library containing over 130,000 peptide analogs in which 4 out of the 18 positions were randomized.

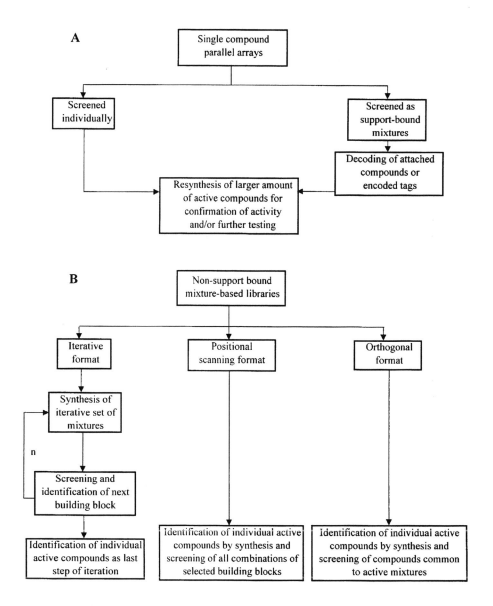

Figure 2. Flow chart of deconvolution processes. **A** Libraries formatted as single compound parallel arrays; **B** Mixture-based libraries.

Deconvolution Processes Toward the Identification Active Library Components

Single Compound Parallel Arrays

The deconvolution processes to active library components are linked to the library format (Figure 2A). For obvious raisons, no further deconvolution steps are necessary when screening single compound array libraries (i.e., each compound, either in solution or support bound, is screened separately from the others). Resynthesis of the active compounds may however be required for further testing in those cases when the libraries are robotically generated in small quantities.

On the other hand, one-bead-one-compound libraries screened as mixture of beads per test sample, although viewed as parallel arrays with each individual bead screened as a single entity, requires further deconvolution steps. In the case of peptide libraries, deconvolution of the active peptides from such libraries is carried out by colorimetric methods to visualize the bead to which a soluble antibody, receptor, enzyme, etc. binds, followed by microsequencing (40). A modification of this approach uses encoded "tags" to allow the rapid identification of peptides which sequences cannot be directly determined from individual beads. These tags include polynucleotides, which are then decoded by PCR (45), polypeptides, decoded by sequencing (46,47), or halogenated aromatic molecules, decoded by electron capture capillary gas chromatography (48).

The presence of immobilizing supports however increases the difficulty of using support-bound libraries in cell-based assay systems, such as traditional broth assays involving whole bacteria or fungi, that require soluble interactants. A novel assay format has been developed for the screening of support-bound libraries (49) and has recently been imple-

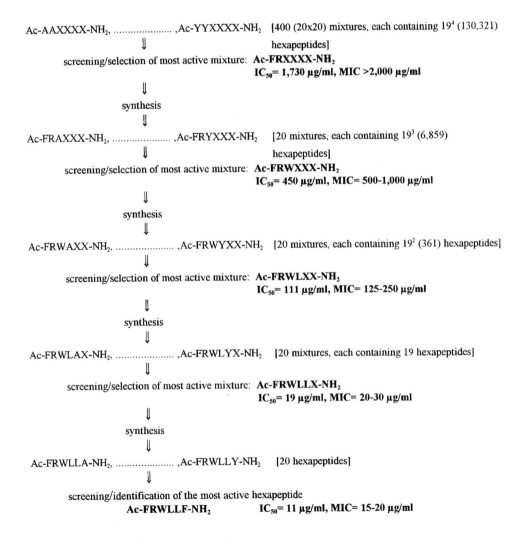

Figure 3. Identification of antistaphylococcal compounds from an iterative mixture-based SCL (50).

mented for detection of antimicrobial activity (29). This assay referred as lawn format assay consists in spreading beads on agar plates, photocleaving small quantities of each compound from its respective bead, detecting the diffusion zone, and sequencing the relevant beads as described above. This method was initially validated with the detection of penicillin V on *Bacillus subtilis* agar plates from a mixture of unmodified beads, and then applied to the screening of over 46,000 triazines.

Non-support Bound Mixture-based Libraries

Two primary approaches for the identification of individual active compounds from mixture-based libraries have been developed, namely the iterative process (30) and the positional scanning approach (43) (Figure 2B). The iterative process of selection and synthesis results in the identification, at a given position, of residue(s) or functional group(s) of the active sequences present within a mixture per assay and synthesis cycle. The iterative process is illustrated in Figure 3 for the identification of antistaphylococcal hexapeptides from a SCL containing over 50 million sequences (50). This iterative library has six positions of diversity and is composed of 400 separate mixtures. Each compound within a given mixture has a common defined residue at two positions, which differentiate each mixture from the others (designated OOXXXX, where O represents one of 20 L-amino acid, and X represents a close to equimolar mixture of 19 L-amino acids). Following the screening of the 400 mixtures, the iterative process is carried out in which the subsequent four X positions of the most active OOXXXX mixture(s) are successively defined (i.e., OOOXXX, OOOOXX, OOOOOX, and OOOOOO). The selection criteria at each deconvolution step are based on the biological activity of interest and on the chemical and structural nature of the functional groups defining the most active mixtures. The most active sequence in this library is thus directly identified by the systematic arrangement of the library and the continually defining nature of the iterative process. Variations of the original iterative process have been proposed and include the subtractive pooling strategy in which mixtures are characterized by a group of missing functionalities (51), and bogus coin strategy in which mixtures are generated using various ratios of building blocks (52). However, none of these methods have been applied for antimicrobial research.

The positional scanning approach uses a positional scanning SCL (PS-SCL), in which the functional group(s) at each position of the active molecule(s) can be determined directly from the initial screening data. The positional scanning process is illustrated in Figure 4 for the identification of N-benzyl aminocyclic ureas having anti-candidal activity from a PS-SCL containing over 100,000 compounds (24). As shown in this example, when using a library having three diversity positions, a complete PS-SCL consists of three sublibraries (designated OXX, XOX, and XXO), each of which has a defined functionality at one position and a mixture of functional groups at each of the other two positions. The location of the defined functionality varies from one sublibrary to the others. In the present example, 40, 37, and 80 separate functional groups were included at positions 1, 2, and 3, respectively. Each sublibrary within a PS-SCL, while addressing a separate position of the sequence, represents the same collection of individual compounds (e.g., 40x37x80=118,400 individual). When used in concert, the data derived from each sublibrary (i.e., each of the separate sets of mixtures) yield information about the key functionalities for every position. Following the screening of the PS-SCL, the most active mixtures are selected from each of the sublibraries. Individual compounds that represent all possible combinations of the functionalities defining those selected mixtures are then synthesized to confirm the screening data and determine the relative activities of the generated individual compounds. The PS-SCL approach relies on the

Position 1 [40 separate mixtures, each containing 37x80 (i.e., 2,960) aminocyclic ureas]

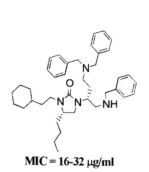

MIC = 250-500 μg/ml

Position 2 [37 separate mixtures, each containing 40x80 (i.e., 3,200) aminocyclic ureas]

MIC = 125-250 μg/ml

Position 3 [80 separate mixtures, each containing 40x37 (i.e., 1,480) aminocyclic ureas]

MIC = 250-500 μg/ml

1. Screening of the three sublibraries
2. Identification of the most active mixture(s) from each sublibrary
3. Synthesis of individual compound(s) representing all combinations of functionalities defining the most active mixtures
4. Screening of the individual compounds
5. Indentification of the most active compound

MIC = 16-32 μg/ml

Figure 4. Identification of anticandidal compounds from a positional scanning mixture-based SCL (24). R_X represents an equimolar mixture of all of the functional groups at the corresponding diversity position

connectivity between the functionalities defining the most active mixtures at each position being addressed. If the functional groups defining the most active mixtures from each of the sublibraries are indeed found to be connected to each other (i.e., are part of the same active compound(s)), then combinations of these groups will lead to active individual compounds. The advantage of the PS-SCL format is that iterations are not required, and only a single synthesis is needed to obtain individual compounds.

Variations to the positional scanning deconvolution process that have not yet been used to identify novel antibiotics include: 1) the indexed pooling strategy in which mixtures are screened in a matrix format (53); 2) the "library of libraries" approach directed toward the identification of structural motifs necessary for a given biological activity (54); and 3) a deletion deconvolution in which each mixture is characterized by a deleted functionality (55). Finally, a separate method is the orthogonal deconvolution which compares two libraries generated in such a format that each mixture of a library has a single common compound with any mixture of the second library (56). The common compounds to the active mixtures are then separately synthesized to determine and confirm their activities. However, no use of orthogonal libraries has been reported for antimicrobial research.

Screening and Deconvolution Criteria

As mentioned earlier for the screening of support-bound libraries on agar plates (29), development or optimization of an assay system is necessary to the screening of large number of samples. The main goals in optimizing or developing an assay are to obtain the highest and most reproducible signal-to-background ratio possible, and to be able to screen in parallel all or significant number of the library samples. Most of the antimicrobial studies using SCLs have been tested in broth assay systems with existing standardized parallel screening protocols (e.g., 96-well plate format for microdilution assay with known intra- and inter-variability). No optimization was therefore required prior to screen the libraries in such assay systems. When using mixture-based SCLs, the initial concentration used to screen a library depends on the library diversity as well as the presence or not of organic co-solvent. In particular, nonpeptide SCLs by their very nature are often not highly water soluble and solubility is enhanced by the presence of 1 to 10% DMF or DMSO or other organic co-solvent.

Table 4. Importance of dose-response curve determination over single concentration screening.

Defined amino acid[a]	% inhibition at 2.5mg/ml[b]	IC_{50} (μg/ml)[b]
α-Fmoc-ϵ-L-lysine	100	44
D-tryptophan	94	212
L-tryptophan	96	322
L-norleucine	100	357
7-amino heptanoic acid	98	360
D-phenylalanine	99	430
p-nitro-L-phenylalanine	100	447
L-phenylalanine	98	529
L-ϵ-amino caproic acid	77	568
L-γ-amino butyric acid	90	721
D-arginine	94	973
L-arginine	86	1056
γ-L-ornithine	100	>1250

[a]A mixture-based iterative tetrapeptide library was generated having the N-terminal residue defined with one of 58 amino acids, and the three other positions as equimolar mixtures of 56 amino acids (38). [b]Each of the 58 mixtures making up the library was assayed against *S.aureus* in a microdilution assay. The % inhibition and IC_{50} values were determined following 24h incubation at 37%C (38).

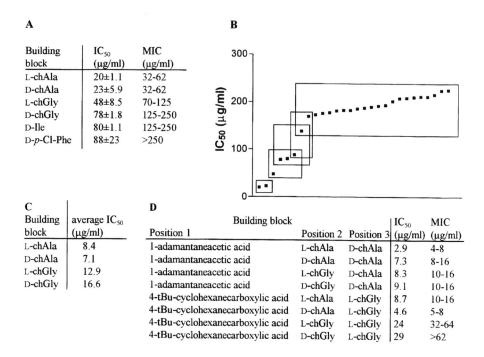

A

Building block	IC$_{50}$ (µg/ml)	MIC (µg/ml)
L-chAla	20±1.1	32-62
D-chAla	23±5.9	32-62
L-chGly	48±8.5	70-125
D-chGly	78±1.8	125-250
D-Ile	80±1.1	125-250
D-*p*-Cl-Phe	88±23	>250

C

Building block	average IC$_{50}$ (µg/ml)
L-chAla	8.4
D-chAla	7.1
L-chGly	12.9
D-chGly	16.6

D

Building block Position 1	Position 2	Position 3	IC$_{50}$ (µg/ml)	MIC (µg/ml)
1-adamantaneacetic acid	L-chAla	D-chAla	2.9	4-8
1-adamantaneacetic acid	D-chAla	D-chAla	7.3	8-16
1-adamantaneacetic acid	L-chGly	D-chAla	8.3	10-16
1-adamantaneacetic acid	D-chGly	D-chAla	9.1	10-16
4-tBu-cyclohexanecarboxylic acid	L-chAla	L-chGly	8.7	10-16
4-tBu-cyclohexanecarboxylic acid	D-chAla	L-chGly	4.6	5-8
4-tBu-cyclohexanecarboxylic acid	L-chGly	L-chGly	24	32-64
4-tBu-cyclohexanecarboxylic acid	D-chGly	L-chGly	29	>62

Figure 5. A bicyclic guanidine positional scanning library was generated using 41 carboxylic acids at position 1, 51 amino acids at position 2, and 49 amino acids at position 3 for a total 102,459 individual compounds. Each mixture was assayed against *Candida albicans* in a microdilution assay (25). The IC$_{50}$s and MICs were determined following 48h incubation at 30°C. **A**. Anticandidal activity of the most active mixtures of the sublibrary having position 2 defined by the listed building block. **B**. Representation of the Tukey test of for the sublibrary having position 2 defined. The IC$_{50}$ values of the 51 mixtures making up the sublibrary are sorted in increasing order. The values with $P \geq 0.05$, i.e., those that are not statistically different from one to another, are grouped in a box. **C**. A set of 32 individual compounds were generated, and the IC$_{50}$ values of those guanidines deriving from the common listed building block at position 2 have been averaged. **D**. Anticandidal activity of 8 representative bicyclic guanidines out of the 32 generated. *t*- Bu: *t*-butyl; L-chAla: L-cyclohexylalanine; D-chAla: D-cyclohexylalanine; L-chGly: L-cyclohexylglycine; D-chGly: D-cyclohexylglycine.

Data analysis of mixture-based or individual compound array libraries is easily handled with a computer spread sheet. To increase the certainty in the selection of active versus non- or less-active mixtures during the screening and deconvolution of libraries, one must first be confident in the screening results by replicating the assays two to three times with replicates within the assays. Screening of replicates enables simple averages, within and between assays, and standard deviations to be determined, which in most cases is sufficient to differentiate active samples, or ultimately to carry out statistical analyses. The differentiation between compounds having similar activities is readily achieved by performing serial-dilution experiments to determine dose-response curves. This is particularly useful when screening mixture-based libraries where activity differentiation drives the deconvolution processes (i.e., the number of iterative mixtures or individual compounds to be generated following the screening of the library). Table 4 illustrates the importance of determining dose-response curves, and consequently IC$_{50}$ values (concentration necessary to inhibit 50% of bacterial growth calculated from the curves), to distinguish active mixtures from the others.

The number of individual compounds or iterative mixtures to be synthesized following the screening of a mixture-based library is based on the statistical activity significance, when required, specificity of the observed activities, and the chemical nature of the defined func-

Table 5. Peptide analogs having specific activities identified from a PS-SCL.

Sequence		MIC (μg/ml)[a]	Tx$_{50}$[b] (μg/ml)
	S.aureus	*C.neoformans*	erythrocytes
YKLLKLLLPKLKPLLFKL-NH$_2$	8	63	22
YKLLKNLLTKLKGLLLKL-NH$_2$	7	55	27
YKLLKNLLDKLKSLLPKL-NH$_2$	15	4	70
YKLLKNLLDKLKGLLPKL-NH$_2$	59	4	92
YKLLKWLLAKLKQLLRKL-NH$_2$	17	68	4
YKLLKWLLAKLKQLLMKL-NH$_2$	33	65	4

[a] MICs were determined following 24h or 72h incubation at 37°C or 26°C for *S.aureus* or *C. neoformans*, respectively. [b]Tx$_{50}$ is the concentration in peptide necessary to lyse 50% of erythrocyte cells following one hour incubation at 37°C

tional groups of the active mixtures. Often mixtures defined at a given position with functional groups of similar chemical character will have similar activities. This may indicate that a number of related analogs of the same compound are responsible for the observed activity. To reduce the number of final compounds needed to be synthesized, similar functional groups are often excluded from selection. One can at a later stage iterate other mixtures or synthesize analogs of the most active individual compounds using the functional groups that were originally excluded.

Lack of high specificity between most active mixtures and the others has often been seen when screening mixture-based libraries for antimicrobial activity in microdilution assays. This is likely due to the number of possible mechanisms of lysis of bacterial cells, and therefore lack of high specificities in active compound sequences. For example, cell lysis through membrane disruption requires overall physico-chemical characters rather than a highly specific compound structure. As mentioned above, statistical analyses have proven to be useful for selection processes in such cases. An example of statistical differentiation is shown in Figure 5 for the identification of bicyclic guanidines having potent anticandidal activity (25). Thus, upon screening a bicyclic guanidine positional scanning library having three diversity positions against *Candida albicans*, the profile of IC$_{50}$ values showed a steady 2-fold decrease in all three sublibraries. For example, for one of the sublibrary (i.e., having position 2 defined), 28 out of 51 mixtures showed IC$_{50}$ values in a 10-fold range with a 2-fold decrease in activity between the two most actives and the third one (Figure 5A). Although, a 2-fold difference in MIC is within the variability of such microdilution assay, this difference was repeatedly seen in separate experiments and confirmed by carrying out statistical analyses of variance (ANOVA) combined with the Tukey post test (Figure 5B). Higher average IC$_{50}$ values were observed for all of the generated individual compounds deriving from L- or D-cyclohexylalanine at this position versus those having L- or D-cyclohexylglycine, which further confirmed the significance of the activity at the mixture level (Figure 5C,D).

Assay Target

The deconvolution of SCLs reported to date were carrying out based on the screening toward one to three microorganisms and were mostly targeted to broad spectrum activity. Furthermore, no toxicity evaluation has been reported at the library screening stage. While toxicity should be a concern when developing novel antibiotics, when using mixture-based libraries, one should be careful in basing the selection process primarily on toxicity level of the mixtures. Based on the screening of over 20 separate SCLs, we have found that only in those cases where nearly all of the mixtures exhibit significant toxicity level, the therapeutical index of the derived antimicrobial compounds is likely to be low (from 1 to 10). On the other

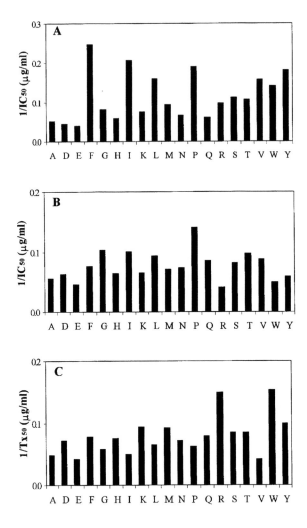

Figure 6. A biased positional scanning library was generated using 19 L-amino acids (L-cysteine was omitted) for each of the four diversity positions (44). The activity of each mixture from the sublibrary having position 16 defined is shown against **A**. *Staphylococcus aureus*, **B**. *Cryptococcus neoformans*, and **C**. human erythrocytes. Each bar represents the inverse IC_{50} or Tx_{50} values of the mixture defined by the amino acid labeled on the x- axis.

hand, a number of individual antimicrobial compounds with low toxicity as determined against human erythrocytes and/or mammalian cell lines were also found to be present in mixtures showing significant level of toxicity at the library screening level. In those cases, the toxicities seen for the mixtures were due to the presence of other cytotoxic compounds within the mixtures. However, if several mixtures show similar high activity, difference in toxicity level may be included as a criteria of selection to minimize the number of iterative mixtures or individual compounds to generate during the deconvolution processes.

We have recently investigated the use of SCLs to generate analogs of a known antimicrobial peptide having specific activity toward one microorganism over others. Thus, a positional scanning SCL was built around a 18-mer peptide scaffold known to adopt an amphipathic α-helix in lipid-like environment by randomizing four positions on the hydrophilic face of the helix (44). This library was assayed against Gram-positive and Gram-negative bacteria, fungi, erythrocytes and mammalian cell lines (total of 6 organisms). As illustrated in Figure 6, different profiles in activity were obtained against the different organisms. The selection process was then based on high activity and high specificity toward one organism over the other five to generate a set of 58 peptides. Most of the 58 peptides exhibit higher activity against a given microorganism type over the others, i.e., either bacteria (*Staphylococcus aureus* and/or *Escherichia coli*), fungi (*Candida albicans* and *Cryptococcus neoformans*), erythrocytes, or mammalian cell lines (McCoy and/or HeLa cell lines) (illustrated in Table 5). These initial studies support the ability of generating individual compounds using combinatorial approaches having distinct activity profiles toward closely related target systems.

Conclusions

Although not thoroughly studied, the reported successes of combinatorial library approaches in novel antibiotic development over the past 10 years show that the ability to synthesize and densely search the molecular space against therapeutically important targets such as drug-resistant microorganisms will permit highly active antimicrobial compounds to be identified. Since its first description, combinatorial chemistry has gained wide acceptance by a large number of pharmaceutical and biotechnology companies as well as academic organizations. The limiting factors in using combinatorial library approaches may include: 1) the availability of standardized chemical protocols although these are rapidly expanding; 2) the capacity (i.e., equipment and cost) to generate and screen libraries having large diversity; 3) the availability of assays formatted for simultaneous screening of multiple samples; and 4) computer capacities for handling and storing large sets of data, which, in turn, may facilitates computer-assisted design capabilities. Nevertheless, the broadly applicable concepts and tools encompassed by combinatorial chemistry have proven to accelerate advances within synthetic organic and medicinal chemical research and discovery, and have been recognized by the journal *Science* as one of nine breakthroughs in scientific research for 1998 (57).

Acknowledgments

I would like to thank Dr. Richard A. Houghten for his helpful discussions in developing novel antimicrobial compounds from combinatorial libraries, and to acknowledge Dr. J. M. Ostresh, A. Nefzi, and C. Pinilla for their extensive contribution in the generation of combinatorial libraries, and Emma Crooks for her technical assistance in performing the described biological assays at Torrey Pines Institute for Molecular Studies.

References

1. Cassell, G.H. 1997. Emergent antibiotic resistance: health risks and economic impact. FEMS Immunol. Med. Microbiol. 18: 271-274.
2. Pinner, R.W., Teusch, S.M., Simonsen, L., Klug, L.A., Graber, J.M., Clarke, M.J. and Berkelman, R.L. 1996. Trends in infectious diseases mortality in the United States. JAMA. 275: 189-193.
3. Silver, L.L. and Bostian, K.A. 1993. Discovery and development of new antibiotics: The problem of antibiotic resistance. Antimicrobial Agents Chemother. 37: 377-383.

4. Maranan, M.C., Moreira, B., Boyle-Vavra, S. and Daum, R.S. 1997. Antimicrobial resistance in staphylococci. Epidiomology, molecular mechanisms, and clinical relevance. Infect. Dis. Clin. North Am. 11: 813-849.

5. Boyce, J.M. and Causey, W.A. 1982. Increasing occurrence of methicillin-resistant *Staphylococcus aureus* in the United States. Infect. Control. 3: 377-383.

6. 1997. Update: *Staphylococcus aureus* with reduced susceptibility to vancomycin. United States, 1997. Morb. Mortal. Wkly Rep. 46: 813-815.

7. Klein, P.A., Greene, W.H., Fuhrer, J. and Clark, R.A. 1997. Prevalence of methicillin-resistant *Staphylococcus aureus* in outpatients with psoriasis, atopic dermatitis, or HIV infection. Arch. Dermatol. 133: 1463-1465.

8. Hiramatsu, K., Hanaki, H., Ino, T., Yabuta, K., Oguri, T. and Tenover, F.C. 1997. Methicillin- resistant *Staphylococci aureus* clinical strains with reduced vancomycin susceptibility. J. Antimicrobial Chemother. 40: 135-136.

9. Grayson, M.L. and Eliopoulos, G.M. 1990. Antimicrobial resistance in the intensive care unit. Semin. Respir. Infect. 5: 204-214.

10. Anonymous. 1991. Nosocomial transmission of multidrug-resistant tuberculosis among HIV-infected persons. Florida and New York, 1988-1991. Morb. Mortal. Wkly Rep. 40: 585-591.

11. Dörner, B., Blondelle, S.E., Pinilla, C., Appel, J., Dooley, C.T., Eichler, J., Ostresh, J.M., Pérez- Payá, E. and Houghten, R.A. 1996. Soluble synthetic combinatorial libraries: The use of molecular diversities for drug discovery. In: Combinatorial Libraries. Synthesis, Screening and Application Potential. R. Cortese, ed. Walter de Gruyter & Co., New York. p. 1-25.

12. Nefzi, A., Ostresh, J.M. and Houghten, R.A. 1997. The current status of heterocyclic combinatorial libraries. Chem. Rev. 97: 449-472.

13. Gordon, E.M., Barrett, R.W., Dower, W.J., Fodor, S.P.A. and Gallop, M.A. 1994. Applications of combinatorial technologies to drug discovery. 2. Combinatorial organic synthesis, library screening strategies, and future directions. J. Med. Chem. 37: 1385-1401.

14. Blondelle, S.E. and Houghten, R.A. 1992. Progress in antimicrobial peptides. Annu. Rep. Med. Chem. 27: 159-168.

15. Bostian, K. and Silver, L. 1990. Screening of natural products for antimicrobial agents. Eur. J. Clin. Microbiol. Infect. Dis. 9: 455-461.

16. Nissen-Meyer, J. and Nes, I.F. 1997. Ribosomally synthesized antimicrobial peptides: Their function, structure, biogenesis, and mechanism of action. Arch. Microbiol. 167: 67-77.

17. Neu, H.C. 1989. Overview of mechanisms of bacterial resistance. Diagn. Microbiol. Infect. Dis. 12: 109S-116S.

18. Chu, D.T.W., Plattner, J.J. and Katz, L. 1996. New directions in antibacterial research. J. Med. Chem. 39: 3853-3874.

19. Blondelle, S.E., Pérez-Payá, E. and Houghten, R.A. 1996. Synthetic combinatorial libraries: Novel discovery strategy for identification of antimicrobial agents. Antimicrobial Agents Chemother. 40: 1067-1071.

20. Reed, J.D., Edwards, D.L. and Gonzalez, C.F. 1997. Synthetic peptide combinatorial libraries: A method for the identification of bioactive peptides against phytopathogenic fungi. Mol. Plant Microbe Interact. 10: 537-549.

21. Hong, S.Y., Oh, J.E., Kwon, M.Y., Choi, M.J., Lee, J.H., Lee, B.L., Moon, H.M. and Lee, K.H. 1998. Identification and characterization of novel antimicrobial decapeptides generated by combinatorial chemistry. Antimicrobial Agents Chemother. 42: 2534-2541.

22. Ostresh, J.M., Husar, G.M., Blondelle, S.E., Dörner, B., Weber, P.A. and Houghten, R.A.

1994. "Libraries from libraries": Chemical transformation of combinatorial libraries to extend the range and repertoire of chemical diversity. Proc. Natl. Acad. Sci. USA. 91: 11138-11142.

23. Houghten, R.A., Blondelle, S.E., Dooley, C.T., Dörner, B., Eichler, J. and Ostresh, J.M. 1996. Libraries from libraries: Generation and comparison of screening profiles. Mol. Div. 2: 41-45.

24. Blondelle, S.E., Nefzi, A., Ostresh, J.M. and Houghten, R.A. 1998. Novel antifungal compounds derived from heterocyclic positional scanning combinatorial libraries. Pure Applied Chem. 70: 2141. Electronic publication at: http://www.iupac.org/symposia/proceedings/phuket97/blondelle.html

25. Blondelle, S.E., Crooks, E., Ostresh, J.M. and Houghten, R.A. 1999. Mixture-based heterocyclic combinatorial positional scanning libraries: Discovery of bicyclic guanidines having potent antifungal activities against *Candida albicans* and *Cryptococcus neoformans*. Antimicrobial Agents Chemother. 43: 106-114.

26. An, H., Haly, B.D. and Cook, D. 1998. Discovery of novel pyridinopolyamines with potent antimicrobial activity: Deconvolution of mixtures synthesized by solution-phase combinatorial chemistry. J. Med. Chem. 41: 706-716.

27. An, H., Cummins, L.L., Griffey, R.H., Bharadwaj, R., Haly, B.D., Fraser, A.S., Wilson-Lingardo, L., Risen, L.M., Wyatt, J.R. and Cook, P.D. 1997. Solution phase combinatorial chemistry. Discovery of novel polyazapyridinophanes with potent antibacterial activity by a solution phase simultaneous addition of functionalities approach. J. Am. Chem. Soc. 119: 3696-3708.

28. Kung, P.-P., Bharadwaj, R., Fraser, A.S., Cook, D.R., Kawasaki, A.M. and Cook, P.D. 1998. Solution-phase synthesis of novel linear oxyamine combinatorial libraries with antibacterial activity. J. Org. Chem. 63: 1846-1852.

29. Silen, J.L., Lu, A.T., Solas, D.W., Gore, M.A., Maclean, D., Shah, N.H., Coffin, J.M., Bhinderwala, N.S., Wang, Y., Tsutsui, K.T., Look, G.C., Campbell, D.A., Hale, R.L., Navre, M. and DeLuca-Flaherty, C.R. 1998. Screening for novel antimicrobials from encoded combinatorial libraries by using a two-dimensional agar format. Antimicrobial Agents Chemother. 42: 1447-1453.

30. Houghten, R.A., Pinilla, C., Blondelle, S.E., Appel, J.R., Dooley, C.T. and Cuervo, J.H. 1991. Generation and use of synthetic peptide combinatorial libraries for basic research and drug discovery. Nature. 354: 84-86.

31. Zwick, M.B., Shen, J.Q. and Scott, J. 1998. Phage-displayed peptide libraries. Curr. Opin. Biotechnol. 9: 427-436.

32. Hermkens, P.H.H., Ottenheijm, H.C.J. and Rees, D. 1996. Solid-phase organic reactions: A review of the recent literature. Tetrahedron. 52: 4527-4554.

33. Terrett, N.K., Gardner, M., Gordon, D.W., Kobylecki, R.J. and Steele, J. 1995. Combinatorial synthesis - The design of compound libraries and their application to drug discovery. Tetrahedron. 51: 8135-8173.

34. Fruchtel, J.S. and Jung, G. 1996. Organic chemistry on solid supports. Angew. Chem. Int. Ed. Engl. 35: 17-42.

35. Thompson, L.A. and Ellman, J.A. 1996. Synthesis and applications of small molecule libraries. Chem. Rev. 96: 555-600.

36. Dooley, C.T., Chung, N.N., Wilkes, B.C., Schiller, P.W., Bidlack, J.M., Pasternak, G.W. and Houghten, R.A. 1994. An all D-amino acid opioid peptide with central analgesic activity from a combinatorial library. Science. 266: 2019-2022.

37. Lam, K.S., Lebl, M., Krchnak, V., Wade, S., Abdul-Latif, F., Ferguson, R., Cuzzocrea, C. and Wertman, K. 1993. Discovery of D-amino-acid-containing ligands with selectide technology. Gene. 137: 13-16.

38. Blondelle, S.E., Takahashi, E., Weber, P.A. and Houghten, R.A. 1994. Identification of antimicrobial peptides using combinatorial libraries made up of unnatural amino acids. Antimicrobial Agents Chemother. 38: 2280-2286.

39. Eichler, J., Lucka, A.W. and Houghten, R.A. 1994. Cyclic peptide template combinatorial libraries: Synthesis and identification of chymotrypsin inhibitors. Pept. Res. 7: 300-307.

40. Lam, K.S., Salmon, S.E., Hersh, E.M., Hruby, V.J., Kazmierski, W.M. and Knapp, R.J. 1991. A new type of synthetic peptide library for identifying ligand-binding activity. Nature. 354: 82-84.

41. Ostresh, J.M., Winkle, J.H., Hamashin, V.T. and Houghten, R.A. 1994. Peptide libraries: Determination of relative reaction rates of protected amino acids in competitive couplings. Biopolymers. 34: 1681-1689.

42. Geysen, H.M., Rodda, S.J. and Mason, T.J. 1986. *A priori* delineation of a peptide which mimics a discontinuous antigenic determinant. Mol. Immunol. 23: 709-715.

43. Pinilla, C., Appel, J.R., Blanc, P. and Houghten, R.A. 1992. Rapid identification of high affinity peptide ligands using positional scanning synthetic peptide combinatorial libraries. Biotechniques. 13: 901-905.

44. Blondelle, S.E., Takahashi, E., Houghten, R.A. and Pérez-Payá, E. 1996. Rapid identification of compounds having enhanced antimicrobial activity using conformationally defined combinatorial libraries. Biochem. J. 313: 141-147.

45. Needels, M.C., Jones, D.G., Tate, E.H., Heinkel, G.L., Kochersperger, L.M., Dower, W.J., Barrett, R.W. and Gallop, M.A. 1993. Generation and screening of an oligonucleotide-encoded synthetic peptide library. Proc. Natl. Acad. Sci. USA. 90: 10700-10704.

46. Kerr, J.M., Banville, S.C. and Zuckermann, R.N. 1993. Encoded combinatorial peptide libraries containing non-natural amino acids. J. Am. Chem. Soc. 115: 2529-2531.

47. Nikolaiev, V., Stierandová, A., Krchnák, V., Seligmann, B., Lam, K.S., Salmon, S.E. and Lebl, M. 1993. Peptide-encoding for structure determination of nonsequenceable polymers within libraries synthesized and tested on solid- phase supports. Peptide Res. 6: 161-170.

48. Ohlmeyer, M.H.J., Swanson, R.N., Dillard, L.W., Reader, J.C., Asouline, G., Kobayashi, R., Wigler, M. and Still, W.C. 1993. Complex synthetic chemical libraries indexed with molecular tags. Proc. Natl. Acad. Sci. USA. 90: 10922-10926.

49. Jayawickreme, C.K., Graminski, G.F., Quillan, J.M. and Lerner, M.R. 1994. Creation and functional screening of a multi-use peptide library. Proc. Natl. Acad. Sci. USA. 91: 1614-1618.

50. Houghten, R.A., Appel, J.R., Blondelle, S.E., Cuervo, J.H., Dooley, C.T. and Pinilla, C. 1992. The use of synthetic peptide combinatorial libraries for the identification of bioactive peptides. Biotechniques. 13: 412-421.

51. Carell, T., Wintner, E.A. and Rebek, J., Jr. 1994. A solution-phase screening procedure for the isolation of active compounds from a library of molecules. Angew. Chem. (Engl). 33: 2061-2064.

52. Blake, J. and Litzidavis, L. 1992. Evaluation of peptide libraries-an iterative stategy to analyse the reactivity the reactivity of peptide mixtures with antibodies. Bioconjugate Chem. 3: 510-513.

53. Pirrung, M.C. and Chen, J. 1995. Preparation and screening against acetylcholinesterase of a non-peptide "indexed" combinatorial library. J. Am. Chem. Soc. 117: 1240-1245.

54. Sepetov, N.F., Krchnák, V., Stanková, M., Wade, S., Lam, K.S. and Lebl, M. 1995. Library of libraries: Approach to synthetic combinatorial library design and screening of

"pharmacophore" motifs. Proc. Natl. Acad. Sci. USA. 92: 5426-5430.

55. Boger, D.L., Chai, W.Y. and Jin, Q. 1998. Multistep convergent solution-phase combinatorial synthesis and deletion synthesis deconvolution. J. Am. Chem. Soc. 120: 7220-7225.

56. Déprez, B., Williard, X., Bourel, L., Coste, H., Hyafil, F. and Tartar, A. 1995. Orthogonal combinatorial chemical libraries. J. Am. Chem. Soc. 117: 5405-5406.

57. Anonymous. 1998. Breakthrough of the year. Science. 282: 2156-2161.

58. Pinilla, C., Appel, J., Blondelle, S.E., Dooley, C.T., Dörner, B., Eichler, J., Ostresh, J.M. and Houghten, R.A. 1995. A review of the utility of peptide combinatorial libraries. Biopolymers (Peptide Science). 37: 221-240.

59. Houghten, R.A., Dinh, K.T., Burcin, D.E. and Blondelle, S.E. 1993. The systematic development of peptides having potent antimicrobial activity against *E. coli* through the use of synthetic peptide combinatorial libraries. In: Techniques in Protein Chemistry IV. R.H. Angeletti, ed. Academic Press, Orlando, Fl. p. 249-256.

60. Blondelle, S.E., Pérez-Payá, E., Dooley, C.T., Pinilla, C. and Houghten, R.A. 1995. Soluble combinatorial libraries of organic, peptidomimetic and peptide diversities. Trends Anal. Chem. 14: 83-92.

From: *Development of Novel Antimicrobial Agents: Emerging Strategies*
ISBN 1-898486-23-9 © 2001 Horizon Scientific Press, Wymondham, UK.

18

The Commercial Development of the Antimicrobial Peptide Pexiganan

Michael Zasloff

Abstract

The development of an antimicrobial peptide from its discovery to its realization as a therapeutic is the subject of this personal account. The story spans at least 12 years and has involved the efforts of hundreds of people, including both scientists and business people, involving disciplines ranging from peptide chemistry to banking, at a cost of about $100,000,000. As yet the antimicrobial peptide remains unavailable for human therapeutic applications.

The Origin of Magainin Pharmaceuticals

In 1987, I reported the discovery of the Magainin peptides (1). At the time I headed the Human Genetics Branch of the National Institute of Child Health and Human Development of the National Institutes of Health (NIH). During the course of surgical removal of oocytes from *Xenopus laevis*, I was struck by the frog's freedom from infection in the setting of a weak inflammatory response. Since the frog was not pouring out white cells or blood borne agents to the site, healing without infection was hypothesized to occur as a consequence of local production of an antimicrobial agent. After several months of failure, in part due to the presence of a metaloenzyme that processed the antimicrobial peptides rapidly (2), the activity was purified. The sequence was determined, the cDNA was cloned, and the peptide was synthesized. The sequence of the peptide, which we called Magainin (Hebrew, for Shield or magain) was novel:

Magainin 1:	GIGKFLHSAGKFGKAFVGEIMKS
Magainin 2:	GIGKFLHSAKKFGKAFVGEIMNS

Indeed a group led by Dr. Dudley Williams had in a prior issue of the JBC published the peptides isolated from the granular glands of Xenopus skin (3). No evidence of Magainin was reported, to some extent confounding my own interpretation of the biological localization of the peptides that I had isolated from Xenopus skin. As the galley proofs from PNAS returned to me, within a day or so, I was advised of the publication in April 1987 by the Williams group in the Biochemical Journal of a "correction" of the data published in the prior

JBC article. The group realized that following collection of Xenopus secretions proteolysis destroyed a considerable fraction of the population of peptides. By adjustment of pH, proteolysis could be controlled and the subsequent profile of peptides deduced by tandem mass spectroscopy, yielded a considerably different population of peptides, now including Magainin 1 and 2 (4). No mention was made of possible function, although previous studies by Günter Kreil had suggested that these amphiphilic molecules could potentially function as antibiotics, based on analogy to cecropin or to several hormones (5).

I bring this story up because at this time in the development of Pexiganan the patent position was of great importance. Without a clear ownership of a substance it is unusual to invest the energy required for development. Had the Williams paper appeared prior to April 1987, it is likely I would not have embarked on the development process of the past 12 years. Indeed, the NIH filed patents by Feb. 1987. No record of disclosure of the Magainin sequence could be documented prior to April 1987 and as such we felt comfortable that the U.S. patent office would grant composition of matter to me, which indeed, it did. Because the composition of matter was "owned" by NIH, no company could develop it without violating our claim to ownership and risking legal action. As a consequence Magainin Pharmaceuticals was formed with venture capital from Wally Steinberg (1934-1997) of Health Care Investment Corporation, a group that had helped form Medimmune and Genetic Therapy based on the work of scientists at the NIH.

At this point in the story a discovery has been made (Magainin), intellectual property protected (patent application), and a company founded to channel money and effort into the development of Magainin as a therapeutic. The date is March 1988.
Within several months a chief executive officer (CEO) was hired. Dr. Barry Berkowitz came to Magainin after a corporate downsizing at SmithKline Beecham (SKB). Some time later, Dr. Len Jacob joined, having been Vice President of worldwide clinical development at SKB. The company began building its quarters and soon its laboratory at 5110 Campus Drive, Plymouth Meeting, Pennsylvania was inaugurated and inhabited.

The Development of Magainin as an Antibiotic

During this period of time the challenge of development of Magainin as an antibiotic became apparent. A limited number of animal experiments were set up in the Company's animal facility to evaluate the efficacy of Magainin as a systemic agent. The usual protocols, in part designed by Dr. Paul Actor (a consultant formerly with SKB, and the developer of the antiparasitic drug, albendazole) involved inoculation of a bacterium into mice, followed by one or two doses of antibiotic. We realized soon into these experiments that high doses of peptide were required, and that the therapeutic index was too narrow for development as a human anti-infective. Mice could be effectively treated with 150 mg/kg of Magainin, but died when treated with 200 mg/kg. Commonly used antibiotics with respect to their therapeutic profiles are truly "magic bullets." Magainin was not in that class, at least when administered systemically to mice. However, we had only begun to explore the combinatorial possibilities offered by a molecule of this type. Indeed over the next few years, we were encouraged by the work of Magainin's senior peptide chemist, Dr. Lee Maloy, formerly of the NIH, that through sequence manipulation (and a lot of trial and error!!) molecules with an improved therapeutic index could be created. In fact, through simple manipulations it was possible to readily increase antimicrobial potency, without increasing in vitro indices of toxicity such as hemolysis. Unfortunately, most of these analogues exhibited little activity in vivo. However, some few, indeed, performed as we had hoped.

In addition to the in vivo performance of Magainin, it also became clear that we faced a major technical challenge in the manufacturing of this molecule. The usual method for synthesis of a 22 amino acid peptide was through solid phase synthesis, a process invented by Bruce Merrifield, for which he received a Nobel Prize in 1963. Dr. Merrified joined our Scientific Advisory Board in 1988. Unfortunately solid phase synthesis was not commercially viable in the case of an antibiotic. Although molecules like calcitonin, produced by Bachem under the leadership of Dr. Rao Makeneini were being used clinically, these entities were hormones, and consumed in doses that amounted to a fraction of a mg/kg. Magainin, on the other hand, functioning as an antibiotic, effectively had to reach μM concentrations in blood, not unlike drugs such as penicillin and erythromycin. As most of us know from our personal experience with antibiotics (as patients) we take upwards of 1 gram of an antibiotic per day. Magainin would be no different. However, because Magainin is a peptide, and absorption of this quantity of intact peptide by the gastrointestinal tract tract is most unlikely, it would have to be given by a non-oral route, either by intramuscular or intravenous injection, the later most likely. Although not ideal for a "blockbuster" drug, successful antibiotics such as gentamycin are delivered this way. The main problem, however, was its cost effective synthesis.

The Merrifield method involved synthesis of a solid phase support employing costly blocked amino acids, removal of the synthetic peptides from the resins, and purification of the desired product. By its nature, the Merrifield method creates a population of contaminants of structure and sequence similar to the product (i.e., molecules one amino acid shorter, etc.) and these impurities can only be purified by costly industrial scale high pressure chromatographic separations, involving expensive matrices, environmentally unpleasant solvents (acetonitrile), and capacity-limiting set-ups (high pressure chromatographic systems). To effectively advance the field, a new method of commercial production of antimicrobial peptides had to be invented.

In late 1990 Magainin Pharmaceuticals became a publicly traded company, offering its stock for sale on the "high tech" US stock trading exchange known as "NASDAQ." At this time a new CEO, Mr. Jay Moorin, had been appointed to replace Barry Berkowitz. Moorin had experience in the financial world having been with Bear Stearns in their health care money management team. His knowledge of the markets was regarded as important for the next growth phase.

A company "goes public," for several reasons. For those who invested initially in the company, the process converts their "paper" into exchangeable securities, and almost always results in a significant amplification of the initial investment. For the Company, however, the process permits the organization to sell shares to the broad public providing a source of the capital needed to advance the development of the product. Whereas in an university, a scientific team depends usually on governmental support, the company requires investment from parties interested ultimately in recovering a significant return on their initial investment. Along with establishing the monetary channels between the company and the public another product of becoming a publicly traded company is the loss of "privacy." Experiments that have "substantial" or material impact on the "value" of the technology (the value of the company) must be reported once they have been deemed "true." The public is given a "real time" view of the ups and down of laboratory progress. Without adequate explanation, this extreme exposure creates a situation that can readily lead to the erosion of credibility between technical personnel and technically naive investors ("I thought you said that such and such was "certainly" going to work"!!).

In 1991 Jacob and Moorin decided that the most reasonable route of development for antimicrobial peptides was as a topical agent. Why? First, the medical community was comfortable with topically applied antibiotic peptides ("Neosporin", Bacitracin). Second, SKB had recently introduced a topical antibiotic, "Bactroban," for the topical treatment of the pediatric infection, impetigo, a contagious disease caused by β-hemolytic *Streptococcus* spp. and *Staphylococcus aureus*. Bactroban had begun to replace oral antibiotics generally in use prior to its introduction, demonstrating that a topical agent could perform as well as a systemic antibiotic, and, more importantly, the medical community could accept a topical therapy in a setting where oral agents were the accepted norm. Because Bactroban had been studied for the treatment of impetigo by SKB, data existed in the Federal Drug Administration (FDA) records regarding the trials and it was possible to develop an entirely new topical antibiotic following in the regulatory footsteps of SKB as documented in their development of Bactroban. It was hoped that Magainin would produce an antibiotic as effective as Bactroban, proving principle, and permitting us to extend its use beyond impetigo (burns, diabetic ulcers, bed sores, etc.). Unlike Bactroban, many of our molecules had a very broad spectrum of action, covering both *Streptococcus* and *Staphylococcus* as well as many other species of Gram positive as well as Gram negative species, including both anaerobes and aerobes.

Which molecule would be developed? A number of Magainin analogs had been created by Dr. Richard Houghton through application of his "tea bag" method of combinatorial synthesis (6). Many of these exhibited potency at least 100 fold greater than Magainin against many bacteria regarded as pathogens. Independently, Dr. Lee Maloy along with Dr. Prasad Kari, a former colleague of Urry at University of Alabama had developed substitution analogs including MSI-78 in their analogue program at Magainin. Maloy and Kari's MSI-78 was an appealing molecule.

MSI-78:   GIGKFALKKAKKFGKAFVKILKK (NH₂)

First, MSI-78 was quite active against known human pathogens. Its antimicrobial spectrum can be appreciated from a recent preclinical study published of its activity against several thousand human clinical isolates (7). As with many antimicrobial peptides the molecule also exhibited little evidence for induction of resistance. Because of its mechanism of action, namely its membrane-disruptive activity, MSI-78 demonstrated no cross-resistance with known antibiotics, and as a consequence, exhibited potency against numerous multidrug resistant bacteria, including methicillin-resistant *Staphylococcus aureus*.

Second, MSI-78 exhibited chemical features that were essential for a commercial product. An antibiotic must be chemically stable under conditions in which it is to be used as a therapeutic. Antibiotics are usually stored at home or in the pharmacy at room temperature. In general they are shipped under conditions in which ambient temperatures can rise above 30-40°C. A commercially viable peptide must exhibit sufficient chemical stability so as not to degrade but a fraction of a percent over a 2 year period under the usual conditions of storage. Unstable chemicals could not be produced, inventoried, distributed and marketed effectively. When a drug is created the FDA requires clear definition of its purity, the nature of all contaminants created over time, and certainty that the product sold will be of a high, predictable purity. The original Magainin peptides contained a single methionine residue highly subject to oxidation. MSI-78, in contrast was composed of chemically inert amino acids. In addition, since MSI-78 was to be developed as a topical antibiotic it would have to be formulated as an aqueous preparation. Thus, the molecule would have to retain its stability in a complex aqueous mixture over this period of time, and not simply as a dry powder.

Third, MSI-78 is composed of an amino acid sequence that could be synthesized by methods other than solid phase. Prasad Kari developed a synthetic strategy employing solution phase chemistry to create MSI-78 by an alternative route.

Step 1. GIG (Segment A) + KFLKKAKKFG (Segment B)
↓
GIGKFLKKAKKFG (Segment AB)

Step 2. Segment AB + KAFVKILKK-Omethyl ester(Segment C)
↓
GIGKFLKKAKKFGKAFVILKK-Omethylester

This synthesis requires coupling of blocked amino acids to form the individual segments, which are then progressively condensed to form the complete molecule. Each of the steps is conducted in standard industrial chemical reactors using the appropriate combination of blocked precursors to insure the correct orientation of successive additions. The sequence of MSI-78 had glycine residues distributed throughout. The synthesis involved condensing fragments bearing terminal functionally activated carboxyl terminal glycine, since racemization, expected to occur via this chemistry would not be an issue with the optically inactive glycine residue. The solution phase approach is considerably more attractive than the classical Merrifield procedure for commercial scale-up. The basic method does not utilize the costly solid phase resins required in the solid phase process. The intermediates can be created independently in different plants and at different times all to be fused at a later date, permitting staging of the synthesis. In contrast, the Merrifield method requires progressive completion of the molecule. The individual segments can be purified prior to the final condensation reducing the complexity of the mixture generated at the end of the synthesis. Indeed the final MSI-78 molecule via this route would be contaminated by impurities that differ dramatically in size and physical properties permitting relatively inexpensive purification by a combination of precipitation and low-pressure column chromatography.

MSI-78, in retrospect, had one chemical feature that should have been engineered out early in its development, namely the presence of a carboxyl terminal amide. Although this modification would seem like a trivial matter in the whole complexity of the synthesis of the peptide, in fact it was to present problems that were not fully appreciated early on. Carboxylterminal amides are present in many naturally occurring peptides. They enhance stability by both interfering with action of processive carboxypeptidases and by stabilization of the alpha helical structure of the molecule in solution (8). Furthermore, we had shown soon after the discovery of Magainin that the terminal amide derivative was several fold more active against a variety of bacterial species. Similarly, MSI-78, lacking its terminal amide, was less potent by 3-10 fold against many human pathogens. In the Merrifield procedure, the terminal amide poses no significant problem, since it is introduced by initiating the synthesis (which begins at the carboxyl terminus) with an amidated amino acid bound to the resin. In the case of the solution phase synthesis created by Kari, the terminal amide had to be introduced after the molecule was completed. To accomplish this, new chemistry had to be devised. In the process eventually commercialized, the terminal amide was added by ammonolysis of a carboxyterminal methyl ester.

Once the process was created and advanced to the gram scale at Magainin, discussions with commercial chemical manufacturers began. Magainin finally chose the Chemical Manufacturing Division of Abbott Pharmaceuticals as its manufacturing facility, in part because of its experience with peptides, its excellent reputation in the pharmaceutical industry, and their certainty that they could produce MSI-78 within the purity specifications required by the FDA at $100/gram. Over the following 6 years we spent in excess of $20,000,000 supporting Abbott's efforts and by late 1998 had brought the cost of MSI-78 (to Magainin) to $200/gram compared to its initial cost by solid phase synthesis in excess of $5000/gram.

Clinical Trials

In June 1993 we began our impetigo trial in a population of children in Puerto Rico. The trial was designed with three arms: a placebo, 1% and 2% MSI-78. Normally clinical development proceeds through 3 phases: In Phase I, safety is evaluated, using the dosing regimen that approximates how you expect the drug to be finally used. In Phase II, a small number of people with the disease to be treated are evaluated with variations on the dose and its regimen to get a sense of efficacy, toxicity in the setting of the actual disease, and to permit you to "fine tune" the regimen for the final or "pivotal" stage, Stage III. Here the drug is administered in a highly regulated protocol, previously agreed upon with the FDA, with expectations of results fully laid out. The protocol duplicates the manner a drug will be administered to the public if the agent is approved. In the case of MSI-78 the FDA permitted us to "skip" much of the Phase II process, in large part because MSI-78 proved itself to be quite safe when applied to the skin of normal people or to abraded and infected wounds in animals. In other words, the drug would be evaluated for efficacy after its first human trials against impetigo.

About 300 children were enrolled in the impetigo study over the course of 1.5 years. Since the studies were "blinded", in the sense that physicians were not informed if a patient was receiving placebo (cream base alone), we had no sense of the progress of the trial. However, it was clear who recovered and who did not. Treatment failures approached about 20%, roughly the fraction of the population associated with the placebo arm. As the trial proceeded we anticipated that when all data was revealed, the placebo group would represent predominantly failures while the two treatment arms predominantly success.

An independent data-monitoring group made the data known to us in April 1994. We learned that each arm of the protocol, whether we provided antibiotic or not demonstrated an 80% recovery. In fact when we reviewed the history of the children who had "failed" therapy, and were subsequently treated with oral antibiotics, about 20% of these children failed to respond to therapy. Microbiological analysis of children treated with MSI-78, placebo, or oral antibiotics provided very little distinction between groups.

It appeared that during the years between the time when SKB had evaluated Bactroban in Puerto Rico and the time we ran our trial, public health had dramatically improved. Impetigo, always regarded as a self-limiting infection, resolved effectively in our population of children with no need for medicine if the infected lesions were cared for regularly with soap and water.

The business community responded very negatively and very abruptly. Within minutes of our announcement of the trial results our stock price fell from $18.00 to $3.00, and continued progressively down over the following months. This response was a clear message of no confidence from those in the public who had supported us. In the scientific community the impetigo trial could not have been published; it was simply inconclusive and neither proved nor disproved the efficacy of MSI-78 as an antibiotic. However, in the business community the first antimicrobial peptide to be developed as a drug, MSI-78 was regarded as a "failure." Despite the profound setback Magainin maintained its confidence in MSI-78. The impetigo trial had demonstrated rather clearly that topically administered MSI-78 was safe, free of side reactions either locally or systemically. In addition, the public was becoming increasingly more aware of the growing problem of antibiotic resistance. We decided to redirect our efforts from the treatment of the relatively "mild" problem of impetigo, to the very difficult and medically substantial problem of infected diabetic leg ulcers.

As people with diabetes age they commonly develop ulcers on their feet. It is thought that skin breaks down because of a loss of sensation that accompanies diabetes, which, in turn, interferes with an individual's ability to sense local pain. Subconscious repositioning of the foot is reduced, pressure is repeatedly applied to weight bearing points on the foot, and the skin breaks down due to locally impaired blood flow. Once the skin has broken these lesions become extremely difficult to heal and become infected. Once infected the lesions grow deeper, invading the soft tissue, muscle, and finally bone. All too often the limb must be amputated. Amputation leads to progressive disability, poor balance, and ultimately to progressive loss of all or portions of the same or opposite limb. In truth, the actual pathophysiology of infection in diabetes is not understood, and I for one, believe that the striking proclivity is a reflection of the malfunctioning in the diabetic of a normally robust process of host defense. My suspicions point to the role of sensory nerves in the regulation of the expression of epidermal antimicrobial peptides, like defensins.

Although therapy currently exists for the treatment of infected diabetic foot ulcers, it is less than optimal. Broad-spectrum antibiotics are required and must be taken, usually, for at least a month. In general the side effects associated with these agents leads to patients stopping therapy before effective control has been achieved. In addition, surveys of the organisms cultured from the infected ulcers of these patients revealed a growing percentage of isolates resistant to the usual antibiotics. Furthermore, physicians express an interest in a therapy that focuses attention on the ulcer, as required in the case of an agent that had to be applied directly on the wound, versus one that need be taken orally.

The clinical program was organized under the close guidance of the FDA. It was decided that the most appropriate way to evaluate MSI-78 for topical use in this disease was to compare it, "head-to-head", with an oral antibiotic. In addition, normal supportive care appropriate for infected diabetic ulcers, namely surgical curetting of the infected wound, prior to antibiotic therapy, would be included in both arms of the study. In this condition, as in many soft tissue infections, dead tissue must be removed from the body in concert with antibiotic therapy. The FDA argued that a placebo trail, comparing MSI-78 with debridement to debridement alone was not an ethically acceptable alternative, given the possibility that an untreated infection could, if not controlled, result in the loss of a limb.

In August 1994 our clinical trials began. Patients were treated with either 1% MSI-78 or 2% MSI-78 (twice a day) for up to 28 days. A third cohort received oral ofloxacin. Each group was also given a corresponding "dummy" drug, either placebo pills in the case of those receiving MSI-78 cream, or placebo cream in the case of those receiving ofloxacin. Patients were evaluated clinically on the first few days of therapy, Day 10, at the end of therapy, and two weeks after antibiotic therapy had ended. Parameters such as clinical appearance, wound size and depth, signs of infection, along with microbiological cultures were collected. The trial was headed and designed by the eminent expert in infected diabetic foot ulcers in the United States, Dr. Ben Lipsky, of the University of Washington Medical Center. At Magainin Dr. Paul Litka directed its execution.

In 1995 Magainin was given a planned interim look at the data. This involved a comparison of the three arms of the study, without the breaking of the "blind". An independent data monitoring group accessed the code, and when they compared those treated with oral antibiotic to those receiving either 1 or 2% MSI-78, no differences could be appreciated. In other words, topical therapy was working as well as oral therapy, and both were achieving results in the 80-90% range. Since no difference could be appreciated between the 1% and 2% groups, it was decided to suspend studies with the more concentrated preparation and proceed with the 1% formulation, representing a less costly (in terms of drug expense) regimen. It is fair to say that Magainin was encouraged by this turn of events and efforts were redoubled to complete the planned clinical trials by midyear 1997. Ultimately we enrolled close to 1000 patients in two independent clinical trials involving over 80 different clinical sites in the US.

On March 18, 1997 we announced the completion of the final clinical trial. Both studies had demonstrated that MSI-78 was a reasonable alternative to oral agents. In one of the trials, MSI-78 was not quite as good as the oral agent, but in the other, it was. It was clear that MSI-78 had a broad and powerful spectrum, did not cause emergence of resistance, and was associated with no directly related side effects. At this time SKB decided to join Magainin as a development partner. It was anticipated that SKB would help market and promote the use of this new topical antibiotic as effectively as they had done for Bactroban, the only other FDA approved topical for use in the treatment of infection. MSI-78 was given the trade name **Locilex** in agreement with SKB. Its generic name was **Pexiganan.**

The FDA Panel

On July 28, 1998 we submitted documentation of the entire development program to the FDA, a massive document comprising **304,000 (!)** pages. This document, called an "NDA" (New Drug Application) had taken much longer than anticipated to assemble. In August 1998 Michael Dougherty, our CFO replaced Jay Moorin as CEO; Len Jacob had been replaced by Dr. Roy Levitt; and Paul Litka had been replaced by Dr. Ken Holroyd. A new team was now at the helm.

On March 4, 1999 Magainin appeared before an advisory panel convened by the anti-infective branch of the FDA. This group of academic experts was given the charge of evaluating the data presented in the NDA and arriving at an opinion regarding the efficacy and safety of MSI-78. The session was convened in Silver Spring, Maryland and was open to the public. In the morning an expert appointed by the FDA presented "background" information outlining the natural history and treatment of **uninfected** diabetic ulcers. I wish to emphasize the fact that this individual **did not** discuss the natural history and treatment of the disease **we** had developed Pexiganan to treat, namely, **infected** diabetic ulcers. Rather he talked to the panel about a condition he had studied. He concluded that **uninfected** diabetic ulcers **need not be treated with antibiotics.** With respect to the treatment of infected diabetic ulcers this expert had not presented any substantive guidelines, but left the panel with the impression that a diabetic foot ulcer, whether associated with infection or not, likely need not be treated with antibiotics except in the most advanced cases; and in these severe infections a topical antibiotic would most certainly not be appropriate. As it turned out virtually no one on this panel had actual clinical experience treating infected diabetic ulcers. One pediatrician in fact acknowledged as such but presented her opinion regardless.

Ken Holroyd next presented the MSI-78 story. Ben Lipsky presented a brief overview of infected diabetic ulcers and the rationale for topical treatment, but the impression left by the FDA expert was palpable and could not be erased. Ken Holroyd did a masterful job of presenting the body of the NDA. He was followed by a spokesperson from the FDA who basically corroborated Magainin's data and concluded that this topical therapy was essentially as good as the oral agent. By lunch we were confident that MSI-78 (Pexiganan)(Locilex) would become a drug.

In the afternoon the tide suddenly turned against us. The FDA asked the panel if they could conclude from all they heard in the morning that (1) MSI-78 was effective in the treatment of infected diabetic ulcers and (2) was safe. A physician suggested that in her opinion (not founded on clinical experience), it seemed that removing dead, infected tissue from an ulcer was all that one needed to do to treat the condition. She argued that as far as she was concerned, neither the oral antibiotic nor the topical were of any value whatsoever, so it was not, therefore, a surprise that MSI-78 and the "wonder drug" ofloxacin behave clinically equivalent. It was the curettage, the surgery that really mattered, she argued. One by one, members of the advisory panel stood up and concurred. Those who disagreed with this position included the Panel's Chairman, Dr. William Craig. Several of the panelists who voted against approval of MSI-78, summarily left the room hours before the scheduled end of the meeting, eliminating any opportunity for further debate or discussion. The final vote was 7-4, **against** approving MSI-78 as a human therapeutic (actually 7-5 counting one member of the panel who had failed to submit his "paperwork" in time to be officially included in the vote). The panel suggested that we evaluate this antibiotic on infected diabetic ulcers comparing it to surgical debridement. Should patients deteriorate, and infection spread, they suggested we stop the study (in the case of that individual and introduce intravenous antibiotics). Subsequently other divisions of the FDA visited Magainin, Abbott and the company involved in the formulation of the actual cream base. During these inspections concerns arose regarding highly technical regulatory details surrounding the manufacturing of the cream and the quality controls in place over the past 5-6 years involved in its manufacture.

On July 28, 1999 the FDA advised Magainin that it could not approve MSI-78 for the treatment of infected diabetic ulcers. SKB, upon learning of the FDA's decision, informed Magainin that it needed time to evaluate whether MSI-78 should be developed as a drug, despite their having invested millions of dollars in its anticipated marketing. Ironically, at about the same time an oral antibiotic, trovafloxacin, the only antibiotic approved by the FDA for the treatment of infected diabetic ulcers, was withdrawn by Pfizer, under pressure by the FDA, because of numerous cases of fatal liver disease associated with its use. Without adequate funds to proceed, Magainin was forced to halt its development of MSI-78. The vast inventory of MSI-78 powder now sits in a temperature and humidity-controlled room awaiting a decision regarding its fate.

Acknowledgements

I wish to thank everyone who has participated in this complex project. This includes my many current and former colleagues at Magainin; the scientists at Abbott Laboratories; the enthusiastic team at SmithKline Beecham; Ken Holroyd, Prasad Kari and Lee Maloy for reading the manuscript; and Barbara Zasloff for her continued and unflagging support at all times.

References

1. Zasloff, M. 1987. Magainins, a class of antimicrobial peptides from Xenopus skin: isolation, characterization of two active forms, and partial cDNA sequence of a precursor. Proc. Natl. Acad. Sci. USA. 84:5449-53.
2. Resnick, N.M., Maloy, W.L., Guy, H.R. and Zasloff, M. 1991. A novel endopeptidase from Xenopus that recognizes alpha-helical secondary structure. Cell 66:541-554.
3. Gibson, B.W., Poulter, L., Williams, D.H. and Maggio, J.E. 1986. Novel peptide fragments originating from PGLa and the caerulein and xenopsin precursors from *Xenopus laevis*. J. Biol. Chem. 261:5341-5349.
4. Giovannini, M.G., Poulter, L., Gibson, B.W. and Williams, D.H. 1987. Biosynthesis and degradation of peptides derived from *Xenopus laevis* prohormones. Biochem. J. 243:113-120.

5. Hoffmann, W., Richter, K. and Kreil, G. 1983. A novel peptide designated PYLa and its precursor as predicted from cloned mRNA of *Xenopus laevis* skin. EMBO J. 2:711-714.

6. Cuervo, J.H., Rodriquez, B. and Houghton, R.A. 1988. The Magainins: sequence factors relevant to increased antimicrobial activity and decreased hemolytic activity. Peptide Research 1:81-86.

7. Ge, Y., MacDonald, D.L., Holroyd, K.J., Thornsberry, C., Wexler, H. and Zasloff, M. 1999. *In vitro* antibacterial properties of pexiganan, an analog of magainin. Antimicrobial Agents Chemother. 43:782-788.

8. Chen, H-C., Brown, J.H., Morell, J.L. and Huang C.M. 1988. Synthetic magainin analogues with improved antimicrobial activity. FEBS Letts. 236:462-466.

Appendix

Selection of important organizations and networks (in alphabetical order) dealing with antimicrobial resistance, and their world-wide-web Internet links.

Organization	Internet address / Aims
APUA Alliance for the Prudent Use of Antibiotics	http://www.healthsci.tufts.edu/apua/index.html APUA is a non-profit organization dedicated to preserving the power of antibiotics, and has been monitoring the worldwide emergence of such strains since 1981. The group shares information with members in more than 90 countries. It also produces educational brochures for the public and for health professionals.
CDC Centers for Disease Control and Prevention	http://www.cdc.gov/ The CDC is an agency of the Department of Health and Human Services of the U.S.A. Epidemiology and disease surveillance directed towards disease prevention and control of disease outbreak.
CISET Committee on International Science, Engineering and Technology	http://www.state.gov/www/global/oes/health/task_force index.html Task Force on Emerging Infectious Diseases; main function of CISET is to develop, on an interagency basis, policies for furthering international science and technology cooperation in the national interest.
CSR Communicable Disease Surveillance and Response	http://www.who.int/emc/amr.html International surveillance system of the World Health Organization on anti-infective drug resistance and containment. The goal of the WHO team, Anti-infective Drug Resistance Surveillance and Containment (DRS), is to contribute to reducing the morbidity, mortality and costs associated with infections caused by drug resistant organisms.
EARSS European Antimicrobial Resistance Surveillance System	http://www.earss.rivm.nl/ EARSS is an international network of national surveillance systems, which aims to aggregate comparable and reliable antimicrobial resistance data for public health purposes; collects data since 1998 on *Staphylococcus aureus* and *Streptococcus pneumoniae* infections.

ENARE

European
Network for
Antimicrobial
Resistance and
Epidemiology

http://www.accu.nl/ewi-enare/enare/index.html

A research group developing a program for the understanding, control and prevention of the appearance and spread of multiresistant microbes, with funding from the European Union. ENARE is solely devoted to approaching these problems on a European level, and to develop strategies to fight this major threat to public health. Since beginning of 1997 ENARE became part of SENTRY, a global surveillance network.

NIAID

The National
Institute of
Allergy
and Infectious
Diseases

http://www.niaid.nih.gov/

NIAID, one of 25 separate Institutes and Centers of the National Institute of Health (http://www.nih.gov/) provides major support for scientists conducting research aimed at developing better ways to diagnose, treat and prevent the many infectious, immunologic and allergic diseases that afflict people worldwide.

ROAR

Reserviors of
Antibiotic
Resistance
Network

http://www.healthsci.tufts.edu/apua/roarhome.htm

ROAR is a unique network dedicated to generating a new impetus wordwide for research on commensal bacteria as reservoirs of resistance that can be transferred to human pathogens. A joint project of the University of Illinois, Urbana, and APUA.

SENTRY

http://www.accu.nl/ewi-enare
/enare/sentry_antimicrobial_surveillance.html

The SENTRY Antimicrobial Surveillance study was launched in early 1997. The Sentry program is global, currently operative in North and South America and Europe but expanding; longitudinal, able to detect trends in antimicrobial resistance; and modular, allowing implementation from key reference centres worldwide.